A MONOGRAPH ON WHITEFLIES

N.S. Butter
Former Professor and Head
Department of Entomology
Punjab Agricultural University
Ludhiana, India

A.K. Dhawan
Former Additional Director of Research
& Head Department of Entomology
Punjab Agricultural University
Ludhiana, India

CRC Press
Taylor & Francis Group
Boca Raton   London   New York

CRC Press is an imprint of the
Taylor & Francis Group, an **informa** business

A SCIENCE PUBLISHERS BOOK

First edition published 2021
by CRC Press
6000 Broken Sound Parkway NW, Suite 300, Boca Raton, FL 33487-2742

and by CRC Press
2 Park Square, Milton Park, Abingdon, Oxon, OX14 4RN

ISBN: 978-0-367-55903-8 (hbk)
ISBN: 978-0-367-55907-6 (pbk)
ISBN: 978-1-003-09566-8 (ebk)

Typeset in Times New Roman
by Radiant Productions

DEDICATED
to
Late Sardar Naginder Singh Butter
(Village: Hakam Singh Wala)

A great visionary and farsighted personality who has contributed immensely towards promoting education in remote areas.

Foreword

Whiteflies are economically the most important pests occurring in the open as well as affecting the protected cultivation of crops. Their incidence has become widespread, acquiring serious proportions in a range of crops during the last two decades, due to a shift in global climate favoring its build-up. Hitherto, considered a minor pest in the greenhouse environment, whiteflies have now turned into a serious threat to the agroecosystem in tropical, subtropical, and temperate regions, causing great concern globally. If left unmanaged, they may lead to the complete failure of the crop.

The ecology and epidemiology of the pest studied by researchers in different parts of the world has helped in the identification of weak links in the pest cycle that could be strategically targeted to avoid economical losses.

Additionally, any effective whitefly management strategy must rely on sound knowledge about the factors leading to the outbreak, the availability of tolerant genotypes, pest-suppressive cultural practices, and other control measures available.

This compendium attempts to compile a comprehensive review of scientific literature on different aspects of the whitefly, which can be a meaningful resource for all those concerned with the understanding and management of this pest, and be of immense academic and research value for students and researchers of entomology and crop husbandry.

I trust this monograph, painstakingly prepared by Dr. N.S. Butter and Dr. A.K. Dhawan, will be of tremendous value since it introduces the reader to diverse whitefly (*Bemisia tabaci*)–*Geminivirus* systems, along with recent findings and research endeavors focused on this pest. It is a source of updated information on the whitefly's taxonomy, biotypes, identification, feeding mechanism, biology, ecology, pesticide resistance/resurgence, host resistance, natural enemy complex, physical and bio-intensive measures, and IPM modules.

Based on their rich research experience, the authors have suggested ways and means to effectively tackle this pest from various possible dimensions. As the eventual goal is to utilize this research and experience-based understanding into sustainable management practices of whiteflies and the viruses they transmit, this monograph is anticipated to command a wide readership among academicians, researchers, and professionals in the agricultural and life sciences. Thus, I am confident the monograph will prove to be an excellent reference source for farm scientists, policymakers, extension functionaries, and entomology students at large.

I wish the authors great success in their academic and agrarian pursuit and look forward to many more of their publications intended to benefit the farmers and the farming practices of our country.

Dr. N.S. Malhi

Ex Vice-Chancellor, Guru Kashi University

Bathinda, Punjab, India

Preface

The whitefly is one of the most notorious and serious pests of this century and can be found in all tropical, subtropical, and temperate regions. The insect used to be considered a pest of minor importance since time immemorial. However, it has now become a serious threat and can be classified as a pest of regular (and worrying) occurrence. Efforts on war footing are being made to understand this pest and identify the exact reasons for its rapid spread.

Thus, as part of the authors' contribution towards countering this global threat, they have put forward this monograph that provides a thorough overview of the literature on this pest to assess its build-up. Their critical analysis of this literature has taken into account the role of the changing climate, the polyphagous nature of the pest, and the haplodiploid mode of its reproduction along with numerous regional factors.

Thus, information on different aspects of the whitefly has been compiled starting with an introductory chapter exploring its origin, rapid spread, taxonomic status, and development of biotypes (as this aspect remained largely unexplored). Concentrated efforts have been made to bring about clarity regarding its taxonomic position and the existence of its biotypes. In all, the monograph contains fourteen topics covering various aspects of the whitefly. The important parameters of whitefly research including feeding apparatus, biology, bioecology, speciation, resurgence, development of resistance to pesticides, determination of economic thresholds, nature and extent of damage to crops, vector transmission of viruses, and different tactics of integrated pest management have been included in this write-up.

Additionally, relevant information on various tactics tried against the whitefly at various locations has also been included. The IPM based on the latest development of literature was analyzed for field crops and the greenhouse cultivation of crops. The tactics utilized for the management of whiteflies along with protection of the fragile environment which had hitherto remained neglected have also been given due importance.

The major focus in the write-up is on pesticide resistance, factors of pest resurgence, host plant resistance, transgenesis, and various control tactics (chemical, biological, physical, cultural control). The role of endosymbionts harbored by whiteflies since time immemorial has been made crystal clear as well. This is crucial as the identification of endosymbionts in whiteflies has paved the way towards understanding the existence of its biotypes. Additionally, small measures involving non-chemical methods contributing towards the management strategies of whiteflies have been re-evaluated for the development of an effective, strong strategy causing the least disturbance to the ecosystem.

Thus, the information on the new paradigms of different tactics that have been compiled and judiciously presented will be highly beneficial to scientists engaged in whitefly research to restructure research direction and improve management strategy required for extension functionaries. This monograph is therefore a rich source of key information for scientists, extension functionaries, and students, and will go a long way towards providing effective solutions to mitigate whitefly-related problems in cropping systems the world over.

The authors believe this selection of topics of international importance to be highly comprehensive and that the information contained in this monograph will be utilized to develop a strong strategy to check complex whitefly-related problems. It is the authors' hope that this monograph will be a rich source of scientific information for all categories of readers, and become a powerful weapon for researchers, extension functionaries, and policymakers to utilize to mitigate problems related to this notorious pest.

N.S. Butter

A.K. Dhawan

Acknowledgments

The numerous whitefly outbreaks that have occurred throughout the globe have forced scientists to address this alarming problem. Given this situation, we believe it is vital to have a compendium that will contain information addressing all aspects of whitefly research. We are thankful to the scientific community for selecting this ticklish insect, foreseeing its importance in the crop ecosystem and for gathering in-depth information of immense value. On the valuable advice and encouragement from Dr. B.S. Dhillon, the Hon'ble Vice Chancellor, Punjab Agricultural University, Ludhiana, we have undertaken the endeavor to compile this important work which seeks to fill in any gaps in the readers's understanding of this pest. We are also grateful to our faculty of entomology for expressing their desire for such a publication on whiteflies. We sincerely thank Dr. Vijay Kumar, Sr. Entomologist and Dr. Chander Mohan, Ex. Sr. Plant Pathologist, Punjab Agricultural University, Ludhiana for their help in arranging for the colored photographs. The authors are also indebted to Dr. Vikas Jaiswal for extending a helping hand in the preparation of the line drawings. The authors are profusely thankful to Mrs. Shawinder Pal Kaur for her help throughout the preparation of this manuscript. Lastly, the support provided by Mr. Manjit Singh for the computer work is also gratefully acknowledged.

N.S. Butter

A.K. Dhawan

Contents

Foreword   v

Preface   vii

Acknowledgments   ix

List of Plates   xv

List of Figures   xvii

List of Tables   xix

Glossary   xxi

Abbreviations   xxiii

1. Introduction, Identification, and Feeding Mechanism   1

    1.1   Introduction   1
    1.2   Identification   4
    1.3   Feeding Mechanism   7

2. Bioecology and the Extent of Damage   10

    2.1   Bioecology   10
        2.1.1 *Bemisia tabaci*   12
        2.1.2 *Trialeurodes vaporariorum*   16
        2.1.3 *Trialeurodes abutilonea*   18
        2.1.4 *Tetraleurodes perseae*   18
    2.2   Extent of Damage   20

3. Taxonomic Status, Biotypes, and Endosymbionts   25

    3.1   Taxonomic Status   25
    3.2   Biotypes   34
    3.3   Endosymbionts   39

4. Population Sampling and Economic Thresholds   47

    4.1   Population Sampling   47
    4.2   Sticky Traps/Application of Remote Sensing   48
    4.3   Economic Thresholds   50

5. Pesticides—Resistance and Resurgence   **54**

    5.1   Pesticides   54
    5.2   Pest Build-up   59
    5.3   Pest Resurgence   62
    5.4   Pest Resistance   63
    5.5   Selective Use of Insecticides   71
        5.5.1   Neonicotinoids and New Molecules   72

6. Host Plant Resistance   **75**

    6.1   Screening Techniques   76
    6.2   Bases of Resistance   80
    6.3   Resistant Varieties   85
    6.4   Biotechnology   90

7. Natural Enemy Complex, Entomopathogenic Organisms,   **93**
and Botanicals

    7.1   Predators   94
    7.2   Parasitoids   97
    7.3   Entomopathogenic Organisms   101
        7.3.1   Biopesticides   103
    7.4   Botanicals   104

8. Insect Behavior Alteration Approaches   **109**

    8.1   Mulches   109
    8.2   Semiochemicals   111
    8.3   Tissue Culture   116

9. Mechanical and Physical Measures   **117**

    9.1   Mechanical Measures   117
    9.2   Physical Measures   119
        9.2.1   Release of Natural Enemies   119
        9.2.2   Pest Monitoring/Surveillance   120
        9.2.3   Blasting with Water   120
    9.3   Use of Botanicals   120

10. Cultural Measures   **122**

    10.1   Altering Sowing/Harvesting Time   122
    10.2   Crop Geometry   122
    10.3   Crop Thinning and Roguing   123
    10.4   Crop Sanitation/Crop Free Period   123
    10.5   Isolation of Crop   123
    10.6   Use of Fertilizers and Irrigations   124
    10.7   Barrier Crop/Eradication of Overwintering Hosts   125
    10.8   Elimination of Weeds   125
    10.9   Use of Mixed Crops/Inter-Crops/Trap Crops   126
    10.10 Border Plantation/Wind Breaks/Modification of Habitat   126

11. Non-Chemical Measures   **128**

12. Weather Parameters and Forecasting Models   **132**

12.1   Weather Parameters and Whitefly   132

12.2   Forecasting Models   135

13. Vector of Plant Viruses   **138**

13.1   Plant Virus Transmission   138

13.2   Coat Protein (CP)   140

13.3   Virus Diseases   140

14. Management Strategies   **145**

14.1   Pest Monitoring and Surveillance   146

14.2   Breeding/Miscellaneous Strategies   147

14.3   Selection of Insecticides   151

     14.3.1   Neonicotinoids   151

     14.3.2   Entomopathogenic Fungi   151

     14.3.3   Botanicals   151

14.4   Suggested Integrated Whitefly Management Model (SIWMM)   152

References   **160**

Subject Index   **207**

About the Authors   **211**

List of Plates

1.1   Whitefly adults feeding on cotton leaves   3
2.1   Whitefly adult   15
2.2   Red eye nymphs   15
2.3   Damage of whitefly   21
13.1   Cotton Leaf Curl Virus disease   142

List of Figures

3.1   Endosymbionts of whitefly   40
5.1   Changing scenario of pests & pesticide use in punjab, India   67
13.1   Whitefly vector species with virus genera and viruses vectored   138
14.1   Suggested integrated whitefly management model   150

List of Tables

2.1	Important host plants of *Bemisia tabaci*	13
2.2	The effect of whitefly feeding on seed germination and fiber qualities in cotton	22
4.1	Economic thresholds of whiteflies in different crops	51
5.1	Pesticides evaluated/recommended against whiteflies	57
5.2	Pesticide-induced resurgence in whiteflies	58
5.3	Pesticide resistance in whiteflies/biotypes	65
6.1	Characteristic features/factors imparting resistance against whiteflies in crop plants	82
6.2	Host plant resistance in crop plants to *Bemisia tabaci*	86
7.1	Important parasites (Parasitoids and predators) of whitefly	96
7.2	Use of biorational remedial measures against the whitefly	105
12.1	Impact of weather parameters on the whitefly population	134
13.1	Virus genera transmissible through whiteflies	141
14.1	Suggested Integrated Whitefly management model	153

Glossary

Allelochemicals: These are the chemicals produced by an organism which elicit a response in the receiving individuals of the same or of a different species.

Allomones: These are the allelochemicals produced by an organism which trigger a response in the receiving individuals of a different species and the benefit goes to the emitter.

Antibiosis: The feeding of insect pest on resistant/immune plant species adversely affects its biology.

Antixenosis: It is a synonym of non-preference.

Arrhenotokous: It is a type of reproduction in insects in which males are produced from unfertilized eggs and females from fertilized eggs.

Biotypes: A population of insects with a similar genetic make-up.

Clades: The populations of insect species which have evolved from a common ancestor.

Cryptic species: It is one of the two or more morphologically indistinguishable groups of population lack inter-breeding. Or It is a group of different populations of species but lack inter-breeding.

Economic damage: It is the amount of damage done by the insect pest that justifies the cost of control operations.

Economic injury level: It is the lowest pest population that inflicts economic loss.

Economic threshold: The population density at which the pest control is undertaken to prevent the increasing population from reaching the economic injury level.

Entomopathogenic organisms: It is the application of products containing microorganisms which are available in the market for use against insect pests to prevent damage. It is highly safe for the environment.

Haplodiploid reproduction: It is a type of reproduction in insects in which females are produced from the fertilized diploid eggs while the males are emerge from unfertilized haploid eggs.

Invasive species: These are the introduced species in a specific location and possess the ability to spread fast and inflict economic loss.

Kairomones: These are the allelochemicals released by an organism in the environment and perceived by the organism of different species and the benefit goes to the receiver.

Lingula: It is a tongue-like structure in vasiform orifice that covers the operculum in members of Aleyrodidae.

Non-preference: It is the quality of plant due to which it becomes unattractive to the insect species for feeding, oviposition, and shelter.

Operculum: It is an opening in the vasiform orifice through which the whitefly excretes honeydew and it is covered with a flap-like structure called lingula.

Parasitoids: It is the word that covers both the parasites and the predators.

Parthenogenetic means: It is a type of reproduction in which the males are produced without fertilization of eggs as in whiteflies.

Plant resistance: It is the relative amount of heritable qualities possessed by the plant that ultimately decide the degree of damage caused by the insect pest.

Poikilothermic: The organisms mainly insects in which the internal body temperature does not remain constant, thus to overcome the kind of adverse condition, these creatures go in to diapause.

Predator: It is a free-living insect which normally kills the host and consumes more host individuals to complete its life cycle.

Putative species: The species with a nucleotide gene sequence of open reading frames and its protein coding has not been completely understood. Or It is widely accepted supposition without any proof.

Pest resurgence: The rapid rebound in the population of insect pests after the drastic reduction in numbers with management action.

Pest resistance: It is the end result of the process of genetic selection with which the insect population survives the deleterious effect of pesticide or mixture of pesticides.

Semiochemical: It is chemical produced by an organism eliciting a behavioral response in the receiving individuals of a species.

Sieve tubes: The elongated cells of phloem which act as conduits of sugar produced during photosynthesis.

Tolerance: It is the plant's level of resistance in which it is able to withstand or recover from damage caused by insect pests through growth and other physiological processes.

Thelytokous: It is the process of parthenogenesis reproduction in which the mother produces only daughters from unfertilized eggs in a parasitoid wasp.

Vasiform orifice: It is a sort of opening on the dorsal surface of 8th and 9th abdominal segments meant for exuding honeydew.

Abbreviations

AFLP: Amplified Fragment-length Polymorphism
CABI: Commonwealth Agricultural Bureaux International
CMRAB: California Melon Research Advisory Board
CP: Protein Coat
CVYV: Cucumber Vein Yellowing Virus
CYSDV: Cucumber Yellow Stunting Disorder Virus
DDT: Dichloro Diphenyl Trichloroethane
ED: Economic Damage
EPPO: European Plant Protection Organization
ETH: Economic Threshold
ETL: Economic Injury Level
ICAC: International Cotton Advisory Committee
IGR: Insect Growth Regulators
IWMM: Integrated Whitefly Management Model
IPM: Integrated Pest Management
JA: Jasmonic Acid
LPM: Lipophyllic Molecules
MEAM I: Middle East Asia Minor I
MED: Mediterranean
mt COI: mitochondrial Cytochrome Oxidase I
OLO: Orentia-like Organisms
PAGE: Polyacrylamide Gel Electrophoresis
PCR: Polymerase Chain Reaction
PB: Peta Byte
RAPD: Random Amplification of Polymorphic DNA
SA: Salicylic Acid
SARAH: Software for Assessment of Antibiotic Resistance to Aleyrodidae in Host Pants
SMW: Standard Meteorological Week
SGAG: Sucrose Gene Alpha Glucosidase
SMLs: Small Lipophylic Molecules
SSA: Sub-Saharan Africa
TTV: Tomato Torrado Virus

CHAPTER 1

Introduction, Identification, and Feeding Mechanism

◇◇

1.1 Introduction

The whitefly, *Aleyrodes proletella*, was first described as a moth in 1734 by Reaumur, a French scientist (Reaumer 1734), placing the insect (wrongly) in the order Lepidoptera. In 1796, Latreille, a French zoologist, corrected the mistake and placed it under the right order, Hemiptera, in the class Insecta in the phylum Arthropoda (Latreille 1796). Up till now, a total of 1556 species covering 161 genera of whiteflies have been identified (Forero 2008). Of these whiteflies, *Bemisia tabaci*, *Trialeurodes vaporariorum*, and *Trialeurodes abutilonea* were important from the point of view of agricultural pests. In particular, *Bemisia tabaci* (Gennadius) (Hemiptera: Aleyrodidae) is considered a globally important, polyphagous, highly destructive, notorious and invasive pest. Above all, it is a highly significant vector of plant viruses belonging to the *Begomovirus, Crinivirus, Carlavirus, Torradovirus* and *Ipomovirus* genera (Navas-Castillo et al. 2011, Jones 2003). It is also known to be a rapidly spreading insect. The *Bemisia tabaci* is considered as a complex of many species or biotypes in addition to the already named New World (Biotype-A), Middle East Asia Minor 1 (MEAM1) (Biotype-B/*Bemisia argentifolii*), and the Mediterranean (MED) (Biotype-Q) depending on the geographical location (Cuthbertson and Vanninen 2015, Boykin 2014, Dinsdale et al. 2010). Being polyphagous, it devours a variety of host plants and is responsible for large scale devastation. Rex Dalton, an American scientist, while observing the enormity of whitefly infestation on poinsettia, termed it "The Christmas Invasion" Larry Antilla, an entomologist with the Arizona Cotton Research and Protection Council in Phoenix once said, "It scares me to death. It is one pest that could completely bury us". Expressing deep concern over the uncontrollable and unstoppable outbreak of *Bemisia tabaci* on many commercial crops, Dennehy remarked, that the Q-biotype "is resistant to every pesticide we've tested." Robert Gilbertson, a plant pathologist from the University of California, went on to name it "the worst (agricultural pest) problem in regions of Africa, Asia, and South America" (Dalton 2006).

A tiny adult measuring about 2–3 mm in size is mainly confined to the tropical and subtropical regions in the whole world. However, its prevalence in the temperate region cannot be ignored as these insects thrive better on protected crops raised inside greenhouses. This is especially significant as nowadays, the area under greenhouses is on the increase and the chances of further build-up in a population of whiteflies are in the offing. However, the most important and peculiar feature of this pest is the difficulty in detecting the damage it inflicts on the crops as this pest continues to thrive and hide underneath the leaves throughout its life. It remains on the underside of the surface of leaf, starting with egg-laying, until it emerges as an adult. Whiteflies are hemimetabolous insects and go through the egg, nymph, and adult stages. The pseudo nymphal stage is also present in its development. The adult female (after emergence) lays ovoid shaped hundred eggs, each tapers from one side. An egg is attached to the plant through a stalk called a pedicel. To avoid desiccation, the eggs draw water through this pedicel until they hatch. The eggs are white at the time of laying and turn brown near hatching. They are either spindle-shaped or sub-elliptical and laid in a circular or semi-circular pattern. The egg pedicel of the whitefly species is inserted into the host plant stomata during oviposition in several species, including *Orchamoplatus mammaeferus* (Quaintance and Baker) (*Aleuroplatus samoans/Orchamus samoanus*), *Aleurothrixus floccosus* (Maskell), *Singhius hibisci* (Kotinsky), *Aleurodicus dispersus* (Russell), *Crenidorsum aroidephagus*, *Dialeurodes citrifolii* (Morgan), *Dialeurodes kirkaldyi* (Kotinsky), *Aleurocanthus spiniferus* (Quaintance), *Odontaleyrodes rhododendri* (Takahashi), *Paraleyrodes naranjae* (Dozier), *Paraleyrodes perseae* (Quaintance), and *Aleyrodes shizuokensis* Kuwana. However, the egg pedicel of *Bemisia tabaci* (Gennadius) and *Trialeurodes vaporariorum* (Westwood) is inserted through the leaf surface directly into host plant tissues but not into the stomata (Paulson and Beardsley 1985). After fertilization, the pedicel is converted into the stalk. The eggs hang with pedicel from the lower surface of the leaf. The whitefly's process of reproduction is unusual (parthenogenetic mode) and they are able to reproduce without males, which, on a side note, are often smaller than the females.

The generalized biology of whiteflies shows that it remains confined to a point on the lower leaf surface during feeding and oviposition. The detailed biological parameters of whitefly were worked out and discussed by several researchers, namely Gangwar and Gangwar (2018), Butter and Vir (1991), Reddy et al. (1986), Pollard (1955), and Husain and Trehan (1933). These studies traced the reaction of whiteflies to different germplasms of cotton. Of these, study carried out in Punjab was taken as a base to explain the biology of whitefly (Butter and Vir 1991). After laying eggs, the hatching takes place in about 3–5 days in north India. On hatching, transparent light green crawlers move for a short period and settle down on suitable spots on the lower surface of the leaves (Husain and Trehan 1933). The nymphal stage is completed in about 7–14 days while the pupal stage lasts for 2–8 days (Husain and Trehan 1933, Butter and Vir 1991). The crawlers remain sedentary during the second and third instars. The oval-shaped nymphs are formed during the pseudo pupal stage (Reddy et al. 1986). This stage is characterized by the presence of two conspicuous red eyes. The white adults with pale yellow wings emerge in the morning from the pseudo pupae. The wings cover the body, in an inverted 'V' position, and the sitting whitefly looks like a tent.

The pest in the egg stage is invisible to the naked eye and thus, the crop manager is unable to detect the infestation. The pest passes through four instars. During all the four instars, the insect is almost sessile and remains attached to the lower leaf surface. While the pest is sessile, it feeds on the lower leaf surface and yellow spots become conspicuous on the upper surface of the leaf as a result of feeding at one spot. Therefore, critical observation can be used to detect the yellow spots on the upper surface of the leaf, which is the only indication of the whitefly's attack. The adult is a poor flier, flying away at the slight disturbance, and settles back down on the same plant after making whirling movements. The adults remain congregated on the lower leaf surface (Plate 1.1). The whiteflies are sap-sucking insects and draw a lot of plant sap. The filter chamber is an adaptation in the digestive system responsible for the elimination of excessive fluid from the system. Its presence is felt only when these tiny creatures exude the honeydew on the upper surface of a leaf while sucking the sap; subsequently, the leaf gets covered with sooty mold. That is, the insect feeds and secretes honeydew from the vasiform orifice located on the dorsal surface of its last abdominal segment, which then falls on the upper surface of leaves on which black sooty mold develops. This damage can be detected from a distance. The presence of ants is another indication of the attack of sucking pests including the whiteflies. The ants visit these plants for feeding on sugary substance (honeydew) and can be seen on infested plants. Additionally, the presence of red-eye nymphs or pseudo-empty pupal cases on the lower leaf surface helps to differentiate the damage of whiteflies from other sucking pests like aphids, psyllids, mealybugs, etc.

Plate 1.1. Whitefly adults feeding on cotton leaves.

1.2 Identification

Earlier taxonomic studies contained scanty information with respect to this peculiar insect. However, this situation changed when an eminent entomologist, Dr. A.L. Quaintance, studied this ticklish insect for taxonomic studies in 1900 and contributed significantly towards its taxonomic aspect (Quaintance 1900). The whiteflies were included in the family Aleyrodidae in 1907. This category of insects has a great resemblance to psyllids as the whiteflies have fewer segments in antennae and poor wing venation. However, the Aleurodes differ from psyllids due to the presence of the vasiform orifice on the dorsal side (in the ninth abdominal segment in males and ninth and eighth segments in females) (Cahill et al. 1995). Currently, these insects are placed in the super-family Aleyrodoidea in Phylum: Arthropoda (order Hemiptera; suborder Homoptera) and account for 30% of the total population (Martin and Mound 2019, Sundararaj 2014, Martin 1987). The Aleyrodoidea family is further divided into three sub-families (Aleyrodinae, Aleurodicinae, and Udamoselinae). Of the three subfamilies, two families, namely Aleyrodinae and Aleurodicinae, are important ones (Dooley 2006). The sub-family Aleyrodinae is more common and contains a large number of genera and families. Before 1987, all whiteflies discovered earlier in India were reported to be belonging to Aleyrodinae. It was only in 1987 that Vasantharaj David discovered the presence of whiteflies of sub-family Aleurodicinae in India (David 2020). The two important sub-families can be separated from each other by the presence of wax-producing pores on the pupal cases. That is, while the sub-family Aleyrodinae is more common and contains a large number of genera and families, the wax pores generally present on pupal cases are absent in this sub-family. However, the insects in the sub-family Aleurodicinae develop one or more wax pores on the pupal case. Waxy material is exuded from these pores, which are present on both males and females (two pairs in females and four pairs in males). It appears as a curly thread, which on drying, gets broken. The body gets completely soiled because the wax spreads and covers everything except the eyes. The subfamily Udamoselinae has been described considering characteristics noted from one adult speciman collected from Equador (Martin 2007). The species was named as *Udamoselis estrallamarinae* based on the characteristic features of wing venation, paronychium structures, and distribution of wax glands in the abdomen. These specimens were never recorded subsequently. As their features greatly resemble the characteristics of the sub-family Aleurodicinae, this family was generally regarded as a synonym of it.

The name whitefly given to these insects (regardless of family) is sometimes misleading as all the whiteflies are not white in color. The whitefly species possess wings of different colors like black (citrus whitefly/*Aleurocanthus woglumi*) (Dietz and Zetak 1920), red (coffee whitefly/*Dialeurodes kirkaldyi*), cloudy (bayberry whitefly/*Parabemisia myricae*), and pale yellow (*Bemisia giffardi*), in addition to whiteflies with white wings. More than 200 species are identified out of which a few are important in latin America, less than 12 species are agriculturally viable as these inflict losses. The commonly available and economically important whiteflies are the cotton whitefly (*Bemisia tabaci*), ash whitefly (*Siphoninus phillyreae*), banded whitefly (*Trialeurodes abutilonea*), greenhouse whitefly (*Trialeurodes vaporariorum*), spiralling whitefly (*Aleurodicus dispersus*), sugarcane whitefly (*Aleurolobus barodensis*), another species of sugarcane whitefly (*Neomaskellia bergii*), cardamom whitefly (*Aleuroclava cardamom*), betelvine whiteflies (*Singhiella pallida, Aleurocanthus rugosa, Acaudaleyrodes rachiposa*), arecanut whitefly (*Aleurocanthus arecae*), cabbage whitefly (*Aleyrodes proletella*), greenhouse whitefly (*Bemisia tabaci* Q-biotype), citrus black whitefly (*Aleurocanthus woglumi*), citrus whitefly (*Dialeurodes citri*), cardin's whitefly (*Metaleurodicus cardini*), bayberry whitefly (*Parabemisia myricae*), and cloudy-winged whitefly (*Singhiella citrifolii*). More and more species are discovered and identified every passing day and added to the list. In India commendable work has been done on the identification of genera and species of whitefly. In all 60 species belonging to 24 genera were examined using the identification keys (David and Subramaniam 1976). Of the assessed species 30 were discovered as new to science and 4 were new to India.

Six more new whiteflies species namely, *Aleuroclava ficicola, Aleurotrachelus camelliae, Bemisia afer, Bemisia tabaci,* and *Pealius machili* have been discovered and identified from mulberry. All species are new to the region in China (Wang et al. 2014). The new species, *Acaudaleyrodes rachipora, Aleurolobus niloticus, Aleurolobus orientalis, Dialeurodes kirkaldyi, Paraleyrodes indicus, Pellius simplex, Singhiella bassiae, Trialeurodes ricini,* and *Trialeurodes vaporariorum* have been added to the list.

Geographically speaking, of the 200 whitefly species devouring a wide variety of crops (agricultural, horticultural, ornamental) in China, the significant ones are *Aleurocanthus spiniferus* (Quaintance), *Trialeurodes vaporariorum* (Westwood), *Dialeurodes citri* (Ashmead), and *Bemisia tabaci* (Gennadius) (Fengming 2008). Out of these, *Aleurocanthus spiniferus* and *Dialeurodes citri* are serious pests of orange, rice, and other crops. *Bemisia tabaci* was first recorded in China in 1940. Now with the introduction of Bt kinds of cotton, the amount of non-target insects including whiteflies surged and are a cause of concern. Of the total known species, 150 (10%) are in the Southwestern USA and constitute 33% of the economically important species (Hodges and Evans 2005). *Trialeurodes vaporariorum* is present in the whole of USA in crop plants in screen houses. *Bemisia tabaci* B-biotype (*Bemisia argentifolii*/silver leaf whitefly) in the eastern region is known to thrive on crops and ornamentals. The other species of whitefly identified in the US are the citrus whitefly (*Dialeurodes citri*) (also known as *Dialeurodes citrifolii*), banded whitefly (*Trialeurodes abutilonea*), and giant whitefly (*Aleurodicus dugesii*) and these species are from two sub-families, namely Aleyrodinae and Aleurodicinae. The most destructive species of whiteflies are *Trialeurodes vaporariorum* and *Bemisia tabaci*. Of these two species the former infests 859 plant species of 121 families while latter is known to devour mainly 6 families such as Fabaceae, Asteraceae, Malvaceae, Solanaceae, Euphorbiaceae, and Cucurbitaceae (CABI 2020a).

Generally, the identification of whitefly is based on the characteristic features of the fourth instar nymph/pupa (red-eye nymphs) (The immature stages being sessile, it is, therefore, easy to handle for identification purposes in India) (Sundararaj 2014, Forero 2008). The emergence of male adults is from unfertilized haploid eggs while fertilized diploid eggs give rise to females. Other important characteristics (such as shape, size and position of setae on puparium, size of vasiform orifice, caudal furrows, length of submarginal setae, number of setae, number of tubercles of male puparium, row of papillae, etc.) of the pseudo pupa/adult are taken. Entomologists working on the whitefly's taxonomic aspect presented an excellent pictorial key for its identification (Martin and Mound 2019, Sundararaj 2014, Dooley 2006, Martin 1987, Hodges and Evans 2005, Martin 2004, Miller et al. 2000). The taxonomist of whiteflies should have prepupal stage specimens, pupal cases, and have knowledge of characters and keys for identification of genus and species of whitefly (Hodges and Evans 2005). The pictorial keys mainly highlighted the characteristics of nymphal instars. For identification 4th instar nymph (red eye nymph) was taken into account. The important characteristics include the presence or absence of compound pores on pupa, thoracic claws (presence/absence), lingula length, and shape, setae on the apex, pores on a dorsal disc, presence of papillae, the shape of vasiform orifice, etc.

The molecular techniques for identification are taken into account now a days as these are more precise than any other method. These are polymerase chain reaction (PCR) amplification and polymorphic analysis of restrictive enzyme, digestion of mitochondrial cytochrome oxidase 1 (COI) gene and are used for differentiation of the species in the Mediterranean basin (Ovalle et al. 2014, Domenico et al. 2006). The mitochondrial (mt COI) DNA marker, the ribosomal RNAs (Caterino et al. 2000) and a ribosomal nuclear marker of the internal transcribed spacer I (ITSI) region sequences (De Barro 2005) are of common use. Of these, mitochondrial cytochrome oxidase I (mt COI) is in use (Boykin et al. 2007, Sseruwagi et al. 2006). To study genetic variability and evolution, the mitochondrial cytochrome oxidase I (mt COI) marker (Frohlich et al. 1999) and ITSI region sequences (De Barro 2005) are preferred in various locations (Lee et al. 2013). Molecular-based assays have also been used to distinguish various species of whitefly through RAPD-PCR via the amplification of fragments in the insect genome. These fragments can be seen easily as variable sizes bands in gel electrophoresis. These variable size bands are considered as biotypes. Additionally, of all the species of whiteflies, the banded whitefly (*Trialeurodes abutilonea*), greenhouse whitefly (*Bemisia tabaci*), and sweet potato whitefly (*Trialeurodes vaporariorum*) are primarily greenhouse insects. *Bemisia tabaci* is more complex and has biotypes such as B-Biotype and Q-Biotype—the B-biotype is now named as *Trialeurodes abutilonea*. The adults of this whitefly are conspicuous due to band of a zig-zag pattern on forewings, while these bands are missing in other species. While resting, the adults of *Bemisia tabaci* sit with folded wings in triangular fashion while the linear pattern is apparent in sweet potato whitefly. The identification is based on the presence of long, small, and small/curly waxy filaments in *Bemisia tabaci*, *Trialeurodes vaporariorum*, and *Trialeurodes abutilonea*, respectively.

At present, the taxonomic hierarchy of whitefly of *Bemisia tabaci* is Animalia (Kingdom) > Bilateria (Sub-kingdom) > Protostomia (Infra-kingdom) > Ecdysozoa (Super-phylum) > Arthropoda (Phylum) > Hexapoda (Sub-phylum) > Insecta (Class) > Pterygota (Sub-class) > Neoptera (Infra-Class) > Paraneoptera (Super-order) > Hemiptera (Order) > Sternorrhyncha (Sub-order) > Aleyrodoidea (Super-family) > Aleyrodidae (Family) > *Bemisia* (Genus) > tabaci (Species). The sub-order includes aphids, mealybugs, and scale insects (including whiteflies). The whiteflies are separated from the other members and classified in a separate super-family. For example, the members of the super-family Aleyrodoidea (Homoptera; Hemiptera) are separated from other hemipterous insects by the segmentation of the antennae, structure of pre-tarsus and tarsus, wing venation, ocelli, and compound eyes. The presence of seven-segmented well-developed antennae in most members is peculiar to the identification of whitefly. The antennae comprise five zonotrichia and five sensilla (chaetica, campaniform, coeloconic, pegs, and basiconic) and perform the olfactory and thermos/hygroreceptory functions (Mellor and Anderson 1995). All the basiconic sensilla are ultrastructural and perform the olfactory function. Likewise, small digitate-tipped sensory pegs are there to perform the thermo/hygroreceptory functions. The tarsus of the leg is divided into two segments of equal length while the pre-tarsus is clawed and encloses a flap-like empodium (which is bristled in some species of whiteflies). The whitefly adults have two pairs of membranous wings of almost equal size. Although the adult flies have poor wing venation, the presence of a long conspicuous single radial vein in the wings is a characteristic feature of whiteflies in Sternorrhyncha. Another identification mark is the location of two ocelli at the anterior margins of the compound eyes. The unique constriction in the compound eye divides the eye into two halves. The two halves of the compound eye are separated by an ommatidium in *Bemisia* but Aleurodes don't have the ommatidium. The presence of the anus on the dorsal side on the last abdominal segment is another feature of the Aleyrodoidea super-family. These whiteflies have a triangular vasiform orifice through which these insects exude honeydew. The vasiform orifice is covered with a flap-like structure called the operculum. At the base of the orifice, another tongue-like structure called the lingula is present. These insects feed on a liquid diet containing excess water. To get rid of excess water, these insects have a modified digestive tract, a structure, called the filter chamber. It is important to note that these characteristic features are typically of the whiteflies belonging to the family Aleyrodidae; the whiteflies belong to this order and can thus be distinguished from other insect species in the sub-order Homoptera.

1.3 Feeding Mechanism

The whiteflies are sap-sucking (sheath-forming) insects with mouthparts that pierce and suck. They puncture the cuticle, penetrate the mesophyll, after which the stylet finally touches the phloem. In Hemiptera, the insect feeding is both without and with sheath formation, and the whitefly falls in the category of sheath-forming insects. The whitefly selects those plant parts which are more suitable for feeding and oviposition. The selection process consists of various phases (Hassel and Southwood 1978) mediated by visual (Prokopy and Owens 1983), olfactory (Visser 1988, Visser and de Jong 1988), and gustatory cues (Stadler 1986). The selection of plant species/ host is mainly through the color as the whiteflies are attracted to yellow/green color followed by yellow, orange/red, dark green, and purple (Husain and Trehan 1940). Before feeding, the adults recognize the host using photoreceptors located on both the dorsal and ventral sides of the compound eyes (Doukas and Payne 2007). During a forage flight, green visual stimuli (light reflected at 550 ± 10 nm) are responsible for directing the whiteflies. The attraction of whiteflies to the yellowish leaves is on account of the ventral half of the eye. *Bemisia tabaci* is more sensitive to long-green-yellow wavelengths between 500–580 nm, while the upper part is sensitive to the short wavelength of UV range between 340–380 nm. The whiteflies thus locate hosts with visual stimuli from a distance (Issacs et al. 1999). *Bemisia tabaci* favors green light; this attraction is highly enhanced by the acquisition of the plant virus, Tomato Yellow Leaf Curl Virus (Jahan et al. 2014).

The feeding apparatus of *Bemisia tabaci* comprises two insect mandibles and two maxillae in which the maxillae interlock to form salivary and food canals. In whiteflies hair covered triangular labrum and 4 segmented labium are present. When the whitefly is not feeding, the stylet is contained in labium. The stylet bundle in the labial grove consists of two external mandibular stylets and two interlocking maxillary stylets responsible for construction of food and salivary canals. The protractor muscles are meant for movement of stylets. The salivary sheath is formed during penetration. The stylets are withdrawn prior to moulting and new are formed and inserted following the molt (Thomas et al. 2008).

The head contains a pair of primary and accessory salivary glands. The salivary glands are present in the head and responsible for the production of saliva which is taken to the cibarial pump. There are two maxillary canals (food and salivary) which join at a point to form a single canal (acrostyle), and concentrate on feeding on the phloem. The food canal formed with maxillae receives saliva through the efferent duct via the cibarial pump to the esophagus, midgut, and hindgut (Harris et al. 1996). The prolonged penetration is similar both in aphids and whiteflies. During penetration, the saliva is simultaneously discharged (Moreno et al. 2011). The anterior and posterior regions of the midgut join at a point to form the filter chamber. Just before the filter chamber, two finger-like structures called caeca are present. The malpighian tubules are absent in the whitefly; instead, specialized cells are present in the filter chamber to perform the function of malpighian tubules (Ghanim et al. 2001). A pair of mandibles is also present in whiteflies.

The feeding of whiteflies is concentrated on vascular tissues. On landing on a suitable plant species, the whitefly releases watery saliva to dissolve the cuticular wax for easy penetration of the stylet (Miles 1999). The feeding is on sieve tissues and stylet follows an intercellular pattern: Pollard (1955) conducted a morphological study on the penetration of stylet behavior of *Bemisia tabaci* and observed that the stylet usually follows intracellular/intercellular path through the epidermis or mesophyll cells. The stylet penetration normally becomes intercellular once the whitefly starts feeding on sieve tissues of the phloem. The feeding initiation thus requires 15–30 minutes due to this process. The whiteflies make fewer shallow probes lasting for a duration of some seconds in the epidermis or mesophyll to judge the suitability of sap. Additionally, The whiteflies usually land on the upper side of the leaf, move to the edge of the leaf, and settle on the underside for feeding in response to light intensity. On exposure to high intensity, the adults move to the other side and resume feeding. In certain hosts, the whitefly does not move on the plants (eggplant and cucumber) after probing while on some other plants (tomato and melon), the whitefly frequently changes its feeding position. The age of the leaf is another factor for the selection of feeding and oviposition. For example, *Bemisia tabaci* prefers young leaves and the selection occurs with small probes in the mesophyll. Many physical factors like trichome on the leaf surface act as barriers for penetration. *Bemisia tabaci* does not use apical young leaves for feeding due to the high leaf density of trichomes. Similarly, the thickness of the cuticle determines stylet penetration. The distance between the leaf cuticle and vascular tissues is instrumental too in affecting the feeding behavior of the whitefly. Also, the defensive behavior of host plants in response to adverse climatic conditions (like extreme temperature and variable rainfall) determines the whitefly's feeding behavior. The whiteflies do not have overwintering stage in their life history,therefore these insects cannot thrive in the freezing temperatures on host plants in the open fields. However these insects can build up in the screenhouses at freezing temperatures. The chemical factor is another parameter that affects probing by whiteflies in old citrus leaves (Walker 1988). The nutritional quality of the plant also affects the insect's feeding preference. Feeding on multiple hosts is an adaptation by the insect to survive in nature. The feeding behavior of such insects is crucial to defining host-plant interactions accurately (Miles 1972). More information pertaining to insects on various subdisciplines taxonomy systematics, physiology, toxicology, morphology, anatomy, biological control, techniques of integrated pest management, toxicology, vector entomology, medical entomology, storage pests, etc. can be had from this compilation (John 2008).

To recap, this chapter highlights the introductory information, taxonomic position, and feeding apparatus of whiteflies. The whiteflies are tiny sucking insects with a polyphagous nature belonging to the order Hemiptera. They have the ability to rapidly spread (due to air currents and human activity) and occupy an important place in tropical, subtropical, and temperate regions. Introductory information on identification of the insect and its feeding mechanism have been presented. Despite the difficulty in the study of taxonomy, more than 1,556 species of whiteflies have been identified; pictorial keys have been developed to facilitate identification. This identification is based on the fourth instar nymph (Red-eye nymph); however, the adult has two pairs of wings that are equal in size but devoid of wing venation and seven-segmented antennae. The presence of zonotrichia and sensilla on the antennae perform the olfactory and hygroreceptory functions. With regards to feeding, whiteflies are attracted to yellow color and make use of visual, olfactory, and gustatory cues to select a host plant, upon which these insects feed on the phloem. The stylet takes 15 to 30 minutes to reach the phloem and while initiating feeding, the insect releases saliva to facilitate the penetration of stylets. The whiteflies, suck the sap, secrete honeydew through a vasiform orifice located on the dorsum of the insects' eighth and ninth abdominal segments and make the substratum suitable for the development of black sooty mold—an important clue in identifying a whitefly infestation. Lastly, factors affecting the feeding habits of the whitefly have also been mentioned.

CHAPTER 2

Bioecology and the Extent of Damage

◇◇

The bioecology of whiteflies and the damage inflicted by them are almost similar for most species of whiteflies. The important species are greenhouse whitefly, *Trialeurodes vaporariorum*, tobacco whitefly, *Bemisia tabaci*, spirally whitefly, *Aleurothrixus dispersus*, citrus whitefly, *Aleurothrixus floccosus* and cabbage whitefly, *Aleyrodes proletella*. For the purpose of discussion, the most common and agriculturally important species have been taken for exploring various aspects.

2.1 Bioecology

The whiteflies are poikilothermic insect pests affecting field and greenhouse crops throughout the globe. The selection of the host plant is done taking into account the color of the substratum as these insects have a strong attraction to yellow color and a lesser attraction towards other colors (green > dark green > red > purple ones). There are of course some variations within different species but broadly, they follow the same pattern. To cite an instance, the *Trialeurodes* species and *Bemisia tabaci* have an attraction to blue ultraviolet rays (shorter wavelength) and yellow color rays (longer wavelength) of light. Similarly, *Aleyrodes proletella* has altogether different preferences based on odor (Trehan 1944). After landing, further changes in the insects' behavior are apparent. The feeding is initiated and the penetration of stylet inside the plant is generally through stomatal openings. Initially, the stylets follow the intercellular path in non-vascular tissues like the cuticle/mesophyll/parenchyma. At this stage, the stylet is withdrawn in case the host is incompatible, otherwise it continues to withdraw and penetrate the plant. Once the stylet touches the phloem tissues, it follows the intercellular path and commences the feeding. The penetration of the stylet continues until the adult's hunger is fully satisfied. The penetration time taken by stylet to reach the phloem is between 15–30 minutes. However, the stylet penetration continues to be generally intercellular once the entry into the phloem has taken place. After the feeding, the development of the pest is an important parameter to be looked into. Both *Bemisia tabaci* and *Trialeurodes vaporariorum* prefer young glabrous leaves for egg-laying (Mound 1984). The development of two whiteflies (*Bemisia tabaci* and *Trialeurodes vaporariorum*) was studied and the ideal temperature range determined was 30–33°C. With the increase in temperature, the increase in fecundity and decrease in generation period was recorded. This resulted in the production of more generations (Hilje and Morales 2008). The changes in the behavior of whitefly adults were conspicuous with respect to different host plants under various conditions. In this context, eggplant, cucumber, tomato, and pepper were selected as hosts to study the biology of whiteflies. Out of these four host plants; cucumber was the most preferred host plant for oviposition and feeding while pepper was the least preference for *Bemisia tabaci* (Sharma and Budha 2015). The influence of temperature on the biology of two important species was also examined. The biological parameters include adult longevity, fecundity and mortality of immatures. The females of both the species were more tolerant to high temperature (39°C). *Bemisia tabaci* was more tolerant to high temperature than *Trialeurodes vaporariorum* as the former was able to withstand high temperature of 41°C or beyond while the latter perished at this temperature (Cui et al. 2008). The fecundity of *Trialeurodes vaporariorum* declined with the increase in temperature with the result that it was nil at 43°C. Similarly, the development of the offspring towards adult formation of both the species declined as the temperature rose (Cui et al. 2008). The influence of temperature on immature stages is more pronounced and the immature development is completed in about 18 days at 21°C. The generation from egg to adult takes about 27days (CABI 2020). While analysing the development of these two species of whiteflies on different hosts, the short development period of these insects on brinjal (eggplant) versus the much longer period on sweet pepper were recorded. The generation is completed in much shorter time as the eggplant is much preferred host of this species (CABI 2020). It was further demonstrated that stylet of *Bemisia tabaci* while feeding on a lower surface of the leaf remained within the leaf, whereas it could reach up to the upper surface of the leaf in *Trialeurodes vaporariorum* (Pollard 1955). Due to the proximity to the phloem on the lower surface, the *Bemisia tabaci* concentrates on the lower side of the leaf. However, as the gap from the phloem to upper cuticle is more, the length of the stylet can manage to reach the phloem from the upper surface, thus the species of the whitefly, *Trialeurodes vaporariorum* concentrates on the upper surface of the leaf. However, both species avoid rain and thus tend to localize themselves on the lower leaf surface in such weather.

The response of both the species of whiteflies to low temperatures was evaluated to clarify the interspecific differentiation in adaptation. It was found that the cold-tolerant ability of all developmental stages of *Trialeurodes vaporariorum* was much higher than that of *Bemisia tabaci* (B-Biotype) on exposure to a temperature of 2°C for 1–12 days. When the species were kept at a cold temperature (5°C for 48 h), the survival of *Trialeurodes vaporariorum* was 78% (at −8°C for 10 h), that is, greater than that of the *Bemisia tabaci* B-biotype. It was further indicated that the optimal temperature required for the development of *Trialeurodes vaporariorum* was lower than that for the B-Biotype. On account of this, the development period from egg to adult stage of the B-biotype was demonstrated to be 1.5 times longer than that of the *Trialeurodes vaporariorum* (at 18° and 15°C, respectively). These findings further indicated the faster development of B-Biotype over *Trialeurodes vaporariorum* at a constant temperature of 24°C. It was found that the ability of the latter species of whitefly to tolerate lower temperature was also greater than B-Biotype in a detailed study that was carried out (Cui et al. 2010).

After emergence, the adults soon feed and mate within one to two days through arrhenotokous mode of reproduction in which the unmated females produce male progeny. After 1 to 2 days (of the pre-oviposition period), eggs are laid one by one on the underside of the leaf in summer. The developmental rates vary, depending upon the temperature (Butler and Henneberry 1986) and host plants (Coudriet et al. 1985). In a study by Gameel (1978), the hatching of eggs took 21 and 5 days at 15°C and 40°C, respectively. The total time of development from egg to adult at 27°C was 16 days when the insect was attached to a sweet potato and 38 days on carrots (Coudriet et al. 1985). Depending upon the temperature, the longevity was between 2 to 34 days for males and 8 to 60 days for females (Butler and Henneberry 1986). The population build-up and development of *Trialeurodes* was lower than that of *Bemisia tabaci* (B-biotype) at 2°C for 1–12 days/5°C for 48 h. The survival of the former was greater (78%) at 8°C for 10 h than B-biotype. The species *Bemisia tabaci* possessed wider adaptability than the other species.

2.1.1 *Bemisia tabaci*

The genus *Bemisia* contains thirty-seven species and is thought to have originated from Asia (Mound and Halsey 1978). *Bemisia tabaci*, being possibly of Indian origin (Fishpool and Burban 1994), was described by numerous names before its morphological variability was recognized. These names include *Aleurodes tabaci* (Gennadius), *Bemisia achyranthes* (Singh), *Bemisia bahiana* (Bondar), *Bemisia costa-limai* (Bondar), *Bemisia emilie* (Corbett), *Bemisia goldingi* (Corbett), *Bemisia gossypiperda* (Misra & Lamba), *Bemisia gossypiperda* var. *mosaicivectura* (Ghesquiere), *Bemisia hibisci* (Takahashi), *Bemisia inconspicua* (Quaintance), *Bemisia longispina* (Priesner & Hosny), *Bemisia lonicerae* (Takahashi), *Bemisia manihotis* (Frappa), *Bemisia minima* (Danzig), *Bemisia minuscule* (Danzig), *Bemisia nigeriensis* (Corbett), *Bemisia rhodesiaensis* (Corbett), *Bemisia signata* (Bondar), *Bemisia vayssieri* (Frappa). These species of *Bemisia tabaci* were also endorsed as synonyms (Russell 1957). The species almost has worldwide prevalence (Zhang et al. 2019a, CABI 2020, 2017, Ghahari et al. 2013) except for North Europe (Sweden, Finland, Republic of Ireland, United Kingdom) and has a wide variety of 900 host plants (Mckenzie et al. 2014, 2004). The details about important hosts from all categories along with their families is presented (Table 2.1).

The whitefly (*Bemisia tabaci* Gennadius) (Aleyrodidae, Homoptera) originated from Asia (Mound and Halsey 1978) as mentioned earlier and was first recorded by Gennadius (1989) when it was found to attack tobacco in Greece as *Aleyrodes tabaci* (now *Bemisia tabaci*) and cotton in India/Asia (1905) (Cock 1993, Anonymous 1989, Cock 1986). Now it affects southern Europe, Africa, India, and Australia. In India, right from Jammu and Kashmir to Punjab, Haryana, Uttarakhand, Uttar Pradesh, Rajasthan, Gujarat, Tamil Nadu, West Bengal, Orissa, Chhattisgarh, Bihar, Andhra Pradesh, Assam, Madhya Pradesh, Karnataka, Kerala, Meghalaya, Lakshadweep, Delhi, and even the Andaman Nicobar Islands, the *Bemisia tabaci* is

Table 2.1. Important host plants of *Bemisia tabaci* (Dubey et al. 2009, Dubey and Ko 2008, Dhawan and Simwat 1997, Anonymous 1989).

S No.	Plant Species	Family	Kind
1	*Achyranthes aspera*	Amaranthaceae	Flower
2	*Amaranthus viridis* (pigweed)	Amaranthaceae	Weed wasteland
3	*Beet*	Amaranthaceae	Weed
4	*Fernaldia pandurata*	Apocynaceae	Vine
5	*Ageratum conyzoides*	Asteraceae	Wild host
6	*Gerbera jamesonii* (African daisy)	Asteraceae	Flower
7	*Lactuca sativa* (lettuce)	Asteraceae	Leafy vegetables
8	*Parthenium hysterophorus* (Carrot grass)	Asteraceae	Weed wasteland
9	*Sonchus* (Sowthistle)	Asteraceae	Weed
10	*Sunflower*	Asteraceae	Crop
11	*Brassica oleracea* var. *botrytis* (Cauliflower)	Brassicaceae	Main
12	*Cleome viscosa* (Hulhul)	Cleomaceae	Weed
13	*Ipomoea*	Convolvulaceae	Flower
14	*Ipomoea batatas* (Sweet potato)	Convolvulaceae	Crop main
15	*Brassica oleracea* var. *gemmifera* (Brussels sprouts)	Cruciferae	Main
16	*Brassica oleracea* var. *italica* (Broccoli)	Cruciferae	Main
17	*Brassica oleracea* var. *viridis* (Collards)	Cruciferac	Other
18	*Cole crops*	Cruciferae	Root crops
19	*Mustard (Brassica campestris)*	Cruciferae	Oilseed crop
20	*Cucumis sativus* (Cucumber); Cucurbits	Cucurbitaceae	Main
21	*Acalypha indica*	Euphorbiaceae	Herb
22	*Euphorbia pulcherrima* (Poinsettia)	Euphorbiaceae	Weed
23	*Manihot esculenta* (Cassava) *Euphorbia characias*	Euphorbiaceae	Main
24	*Chrozophora verrucosa*	Euphorbiaceae	Weed
25	*Cyamopsis tetragonoloba* (Guar)	Fabaceae	Gum Crop
26	*Arachis hypogaea* (Groundnut)	Fabaceae	Main
27	*Glycine max* (soybean)	Fabaceae	Main crops
28	*Vigna unguiculata* (Cowpea)	Fabaceae	Crop (fodder)
29	*Leucaena leucocephala* (Leucaena)	Fabaceae	Small tree
30	*Crotalaria juncea*	Fabaceae	Sunn hemp
31	*Sinningia speciosa* (Gloxinia)	Gesneriaceae	Flower
32	*Origanum majorana* (Sweet marjoram)	Lamiaceae	Herb
33	*Rhynchosia minima*	Leguminosae	Weed
34	*Phaseolus* (beans)	Leguminosae	Pulse crop
35	*Abutilon indicum* (China jute)	Malvaceae	Flower
36	*Abelmoschus esculentus* (Okra)	Malvaceae	Main

Table 2.1 Contd. ...

...Table 2.1 Contd.

S No.	Plant Species	Family	Kind
37	*Gossypium* sp. (Cotton)	Malvaceae	Fiber crop
38	*Hibiscus impatiens* (Balsum)	Malvaceae	Flower
39	*Malva parviflora*	Malvaceae	Weed
40	*Sida species*	Malvaceae	Weed
41	*Morus alba* (Mora)	Moraceae	Other
42	*Cajanus cajan* (Pigeon pea)	Papilionaceae	Main
43	*Piper nigrum* (Black pepper)	Piperaceae	Main
44	*Stemodia viscosa*	Plantaginaceae	Weed
45	*Digitaria arvensis*	Poaceae	Weed
46	*Nicotiana tabacum* (Tobacco)	Solanaceae	Main
47	*Nicotiana debneyi*	Solanaceae	Other
48	*Datura stramonium*	Solanaceae	Wasteland weed
49	*Solanum* (Nightshade)	Solanaceae	Wild host
50	*Solanum aethiopicum* (African scarlet eggplant)	Solanaceae	Wild eggplant
51	*Solanum lycopersicum* (Tomato)	Solanaceae	Main vegetable
52	*Solanum melongena* (Aubergine)	Solanaceae	Wild
53	*Solanum tuberosum* (Potato)	Solanaceae	Main crop
54	*Solanum nigrum* (ground cherry)	Solanaceae	Weed
55	*Physalis minima*	Solanaceae	Weed
56	*Daucus carota sativus* (Carrot)	Umbelliferae	Root vegetable
57	*Lantana camera*	Verbenaceae	Shrub

found throughout the country. The outbreaks of this pest in South India (1985–87) and North India (1987–95) on various crops, especially cotton and tobacco, have been recorded (Sharma and Batra 1995).

It is a polyphagous insect with global occurrence, thriving on 500 host plants. It colonizes annual and herbaceous plants (Chandrasekhar and Sashank 2017, Cossa 2011, Brown et al. 1995). Besides, the *Bemisia tabaci* population differs according to geographic location, fecundity, dispersal behavior, insecticide susceptibility, natural enemy complex, invasive behavior, plant virus transmission, and complement endosymbionts (Brown et al. 1995). It is suspected to be a complex of morphologically discrete indistinguishable species (Tay et al. 2012, Thompson 2011, Dinsdale et al. 2010, Stansly and Naranjo 2010, Gerling and Mayer 1996). There are thirty-seven species in the genus *Bemisia* and the *Bemisia tabaci* is a dominating species with faster development and higher longevity and female: male sex ratios (Zhang et al. 2011). The biological details of the species have been researched as compared to other species (Gangwar and Gangwar 2018).

Bemisia tabaci is a polymorphic species. The variability is not only in its morphology (size and form of nymphs) (Bethke et al. 1991, Mound and Halsey 1978) but also found in the ecological parameters (environment, development, pesticide resistance, virus transmission, natural enemies complex, and variable endosymbionts) affecting it (Al-Zyoud and Sengonca 2004, McKenzie et al. 2004, Xu et al. 2003, Kirk et al. 2000, Horowitz and Ishaaya 1996, Markham et al. 1996, Costa et al. 1995, Bedford et al. 1994, Costa et al. 1993a, Rowland et al. 1991). In terms of reproduction habits, a female *Bemisia tabaci* can oviposit over 500 eggs in an arc pattern on the undersurface of middle canopy leaves (Tsueda et al. 2014). The eggs are attached via a stalk (called the pedicel) which is withdrawn after fertilization (Quaintance and Baker 1913). The other species, *Trialeurodes vaporariorum*, deposited eggs on the upper canopy leaves.

Like the other Sternorrhyncha, the whitefly is known to possess incomplete metamorphosis, going through the egg, nymph, pseudo pupal stage (last instar), and adult stage (Plate 2.1). The first instar nymph (crawler) has a number of setae that are mechano-sensory. It is the only stage where it is mobile (Gill 1990). It moves from the site of hatching to a suitable feeding place and stops further movement. After the identification of suitable feeding locations on the lower surface of the leaf, the crawler loses its legs by the next molt and the nymph becomes sessile. The nymph is oval in shape and does not move during the remaining three instars/ nymphal stages. The fourth instar/final nymph is divided into three parts. In the first part, it is flattened and transparent, and primarily engaged in feeding. In the second stage, it turns opaque with waxy ribbons and continues to feed (Cahill et al. 1995). In the third stage of the 4th instar, the so-called pupal stage nymphs with two red eyes (Plate 2.2) are conspicuous. It is a non-feeding stage from which the adult emerges. Unlike other insects with complete metamorphosis, it continues to suck sap while entering the final instar and subsequently stops feeding, thus entering the pseudo pupal stage. During the last stage in the fourth instar, the nymphs do not feed. There is no difference in the percent eclosion if the pupa is retained on the leaf and allowed to develop into adults or it is removed from the leaf surface and placed on moist filter paper for emergence of adults. The adults did not show any difference between the

Plate 2.1. Whitefly adult.

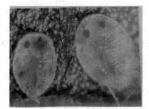

Plate 2.2. Red eye nymphs.

adults emerged from the two kinds of adults except the weight. No variation was recorded with respect to weight of adults in the same species. However, the weight of adults was more in *Bemisia tabaci* than in *Trialeurodes vaporariorum* (Gelman et al. 2007). The adults have wax-covered wings. There is sexual dimorphism in whiteflies. The reproduction is arrhenotokous in which the whiteflies produce haploid males from unfertilized eggs and females from fertilized diploid eggs (Walker et al. 2010). The only exception is *Parabemisia myricae* in which the reproduction is through thelytokous parthenogenesis.

The development of poikilothermic animals is governed by temperature. The total time taken from egg to adult stage is 105 days (15°C) to 14 days (30°C) respectively. It is important to note the variations in fecundity concerning temperature which ranged between 324 eggs (20°C) to 22 eggs (30°C) (Zeshan et al. 2015, Hai Lin et al. 2014, Guo et al. 2013, Xie et al. 2011). Likewise, the value of the intrinsic rate of development governed by temperature is –0.0176 days (15°C) and 0.0989 days (35°C), respectively (Yang and Chi 2006). In all, 11 to 15 generations are recorded in a year; the generation time is presently governed by temperature and is between 18.2 days (35°C) and 81.9 days (15°C) (Yang and Chi 2006).

After emergence, the pale yellowish females lay eggs through parthenogenetic means. The egg hatching is affected at a temperature above 35°C. The most suitable temperature has been identified as 27°C. The effect of climate change on *Bemisia tabaci* was studied in the Netherlands, on the cassava colonizing population of the whitefly (Aregbesola 2018). The development from egg to adult emergence was affected adversely when the temperature reached 28°C and 16°C. The longevity was recorded as 11 days (male) and 19.7 days (female) at 20°C. The peak oviposition/female was 117.5 eggs at 20°C–28°C. Further the abiotic factors such as temperature (maximum and minimum), relative humidity (morning & evening) and sunshine (hrs) exerted profound influence on the development of *Bemisia tabaci*. All these factors contributed upto 57% as revealed by step-wise regression (Jha and Kumar 2017). Similar results were obtained in the studies carried out earlier (Umar et al. 2003).

2.1.2 *Trialeurodes vaporariorum*

This species was first recorded as *Aleurodes vaporariorum* by J.O. Westwood in Europe (Quaintance 1900, Cockerell 1902). However, the origin of this species is sometimes referred to as being in Brazil. This species now inhabits Asia, Africa, North America, South America, Central America, Europe, and Oceania (Anonymous 2019a). In Asia, the countries in which the whitefly has been recorded include China, Bangladesh, Fujian, Hebei, Heilongjiang, Hongkong, Shandong, Shanxi, India, Indonesia, Sulawesi, Iran, Israel, Japan, Jordan, South Korea, Philippines, Singapore, Sri Lanka, Turkey, Uzbekistan, and Yemen (CABI 2020a, Nasruddin and Mound 2016, Erdogan et al. 2008, Chu et al. 2008, Pan et al. 2007, Liu et al. 2003, Zakhidov 2001, AVA 2001, Wang et al. 2000, Yasarakinci and Hincal 1996, Masuda and Kikuchi 1993, Byrne et al. 1990, Kajita 1986). The affected Africa region includes Algeria, Ethiopia, Kenya, Morocco, South Africa, Zimbabwe (CABI 2020a, Byrne et al. 1990). The affected North American regions include Bermuda, Barbados, Belize, Costa Rica, Cuba, Canada (Alberta, British Columbia, Quebec),

Dominican Republic, El Salvador, Guadeloupe, Gautemala, Honduras Jamaica, Martinique, Mexico, Netherlands Antilles, Panama, Puerto Rico, USA (most of the states) (CABI 2020a, Lambert et al. 2003, Lopez and Botto 1995, Osborne and Landa 1992, Steiner 1993, Byrne et al. 1990). Oceania contains Australia, New Zealand, Papua New Guinea and South Australia in which whiteflies are present, While the South American region consists of Argentina, Brazil (Rio Grande, Sao Paulo), Chile, Colombia, Ecuador, Peru, Uruguay and Venezuela. The whiteflies are also present in Europe which covers countries such as Albania, Austria, Belgium, Bosnia and Harzegovina, Czechia, Denmark, Estonia, Fed Republic of Yogoslavia, Finland, France, Germany, Greece, Hungary, Ireland, Italy, Sardinia, Latvia, Lithuania, Malta, Montenegro, Netherlands, Norway, Poland, Azores, Madeira, Portugal, Romania, Russia, Switzerland, Spain, Sweden, Slovenia, Canary Islands, United Kingdom, Serbia, Channel Islands (CABI 2020a).

The life cycle of this species is almost the same as the *Bemisia tabaci*. This whitefly species sucks an enormous amount of sap from the phloem in the lower leaf surface to meet their protein and amino acids requirements. With excessive draining of the sap, the plants become weak and stunted and their growth is retarded, making them look wilted. The whitefly's conical eggs (with the pedicel) are laid on the epidermis in a circular fashion. The adult also continuously rotates itself in a circular pattern for several days and lays eggs at the rate of 2–7 eggs/day on the lower leaf surface. The total fecundity is 534 eggs/female (Lloyd 1922). The first instar nymphs hatch out of the eggs within a week. The nymph is pale green, oval, flat, and scale-like. There are four nymphal instars. Of these, the last instar is referred to as a pupa; in the fourth instar, the nymph initially feeds but soon gets transformed into a pupa and stops feeding. The pupal stage lasts for seven days. The long waxy filaments around the margins of the pseudo pupae are conspicuous. Before the emergence of adults, the red eyes of an adult are apparent in the fourth instar nymph (pseudo pupa). The adults emerge from the pupae through the T-point on the dorsal surface. The adult's longevity is generally around 9–18 days but it could be as long as 30–45 days though it is governed by prevailing temperature, a fact that holds true for other poikilothermic animals as well. The total life cycle of the whitefly is completed in twenty-five days.

The scanty information on weather parameters showed the nil effect of temperature between the range 43–45°C on the sex ratio of *Trialeurodes vaporariorum* (Cui et al. 2008). However, the fecundity of *Trialeurodes vaporariorum* decreased at 43°C and egg hatching was also nil at this temperature. A study was carried out on the biology and morphology of three species of whitefly (*Aleyrodes* sp., *Oxalis corniculata, Bemisia tabaci*) on brinjal (egg plant) and the duration time from egg to adult (males and females) emergence was 24.2 and 28.4, 18.6 and 21.8 and 97.5 and 100.6 days, respectively. Similarly, the male and female ratios were 3.5:1, 4:1, and 3:1 in *Aleyrodes, Oxalis,* and *Bemisia*, respectively. The simulation model (SARAH) based on development components (oviposition rate, adult survival, pre-adult survival, developmental period, and sex ratio) was tested and found suitable to study resistance to antibiotics in Aleyrodidae (Baig et al. 2015). The simulation model was then tested on two species of whiteflies, namely *Bemisia tabaci* and *Trialeurodes vaporariorum*, and found to be successful (Van Giessen et al. 1995). A further study carried out also showed variations in the development parameters of *Trialeurodes vaporariorum* on different hosts with a higher intrinsic rate (rm) on zucchini (rm = 0.12) as compared to tomato (rm = 0.10) (Calvitti and Buttarazzi 1995).

2.1.3 *Trialeurodes abutilonea*

This species is highly polyphagous and prevalent in North, South, Central America, and the Caribbean, devouring 140 plants of thirty-three families. The biology of *Trialeurodes vaporariorum* and *Trialeurodes abutiloneus* (=*Trialeurodes abutilonea*) does not differ from each other. Extensive research work on this species was done as early as the Thirties/late Forties in the twenthieth century (Husain and Trehan 1942, Husain and Trehan 1933).

The damage inflicted by these whiteflies is known to reduce both the quantity and quality of products such as cotton (Butter and Kular 1987). The immature opening of bolls leads to immature lint, which is responsible for production of weak fiber lacking in strength and a reduction in seed germination. It is primarily a pest of crops cultivated in Cuba and the US. The spread of this species has been recorded in Arizona, California, Colorado, Florida, Georgia, Kentucky, Illinois, Kansas, Missouri, Lousiana, Maryland, North Carolina, New Mexico, New York, Utah, Texas, and Virginia in the US. It was earlier known to damage *Abutilon theophrasti*. Being from the category of polyphagous pests, it devours *Euphorbia, Geranium, Hibiscus, Petunia* (including several weed species). It sucks the plant's sap and transmits plant viruses like Abutilon yellows, Diodia Vein Chlorosis and Tomato Chlorosis. It has also been identified in Canada, Brazil, Jamaica, Mexico, El-Salvador, Honduras, Trinidad, Peurto Rico, and Columbia as a pest. It lays its eggs only on lower leaf surfaces (similar to the *Bemisia tabaci*) and the freshly laid eggs are pale yellow in color and glossy in appearance. After 5–6 days, the eggs split longitudinally to give rise to crawlers. The crawlers select a suitable spot for feeding on a leaf. The development of the first three instars is completed in about six days. During the fourth molt, the nymph transforms into a convex-shaped pupa. The pupa is covered with waxy hangings. This waxy material makes the insect look like a box in the pupal stage. The development of the fourth instar is over in about 5–6 days, post which the adult's emergence takes place. The complete life cycle thus takes about 16–20 days. The adult whitefly is either tube-like (male) or spindle-shaped (female), depending on the sex. The pupal cases are slightly bigger in females than in males. The adults of this species are altogether different from the other species of the whitefly. The fore wings of the other species' adults are marked with a zig-zag pattern; these markings are absent in the *Trialeurodes abutilonea*. The central part of the pupal case is conspicuous with dark brown dorsal pigments.

2.1.4 *Tetraleurodes perseae*

It is commonly known as the red-banded whitefly and is native to Latin America. It was first seen in San Diego in 1980 (Nakahara 1995). It is prevalent in the Caribbean, Central America, Florida, and Mexico. It is a secondary pest of avocado. Its biology is simple: the insect passes through the egg, nymphal, pseudo pupal stage, and the adult stage. The adult is also marked with a reddish-brown band which is peculiar to this species. The female lays kidney-shaped, oblong, smooth yellowish eggs on the lower surface of the leaves; the egg stage lasts for eleven days. On hatching, the yellow-brown nymphs turn black. The nymphs are flattened and scale-like. The nymphal stage is roughly fourteen days; the nymph is covered with white wax linings on the body which curls upward. The pseudo pupal stage lasts six days. Adults are 0.8 to 1.2 mm in size and sit with wings folded like a roof around them.

Tetraleurodes perseae Nakahara (Hemiptera: Aleyrodidae) is an exotic whitefly in California that is a minor pest of avocados, *Persea americana* (Lauraceae). Field monitoring from 1997–2002 of avocado orchards in southern California could identify only *Tetraleurodes perseae* species of whitefly on these fruit trees. It is probably univoltine, and adult densities show single distinct peaks each year around August. The only hymenopteran parasitoid found attacking this species of whitefly in California was an aphelinid, *Cales noacki* Howard, and this parasitoid caused mortality of whitefly nymphs between 30 to 100% from February–April. Partial life tables constructed from cohorts of *Tetraleurodes perseae* in the field indicated that survivorship from settled first and second instars to emerged adult whiteflies ranged from 34 to 37%. There were no significant differences in marginal mortality rates by life stage for whitefly cohorts enclosed in sealed mesh bags, open mesh bags, or on unenclosed avocado leaves. Survivorship curves constructed from field phenology data indicated that average egg-to-adult survivorship was around 3.5%, which is substantially lower than that suggested from the life table analyses. Laboratory studies conducted at 25°C on excised avocado leaves indicated that 43–46 days are needed by *Tetraleurodes perseae* to complete develop from an egg to an adult. Demographic analyses of laboratory data indicate that this species has a high reproductive potential with net reproductive rate and intrinsic rate of increase estimates being 21.15 ± 1.39 and 0.07 ± 0.001, respectively. The researchers focused on the influence of weather conditions, mainly temperature, on the whitefly breeding habits, and determined that the species completed at least eight generations in the orchards with the highest altitude, ten generations in the middle altitude, and eleven generations in the low altitude orchard (Hoddle 2013, Hoddle 2006). According to the study univoltine nature in avocado orchards in California, USA was demonstrated. These results are in close conformity with the results obtained earlier by Dowell (1982), who estimated eight generations per year of *Tetraleurodes acaciae* (Hemiptera: Aleyrodidae) and *Calliandra haematocephala* in Florida (a native species of Mexico and California) during the year.

Besides these species, another important species, the castor bean whitefly, *Trialeurodes ricini* (Misra) was also studied under four temperatures (15, 20, 25 and 30°C), on three plant hosts, the castor bean (*Ricinus communis* L.), the papaya (*Carica papaya* L.), and the sweet potato (*Ipomoea batatas* L.) (Abdel-Baky 2006). The effect of temperature was visualized on this species of whitefly. The insect development, oviposition, and generation period of *Trialeurodes ricini* got affected on castor bean plants. At 30°C, the egg hatching, nymphal, development, adult longevity, and generation length were the shortest, followed by 25°C, while these characteristics were longer when the insects were reared at 15°C. The female fecundity was greater at both 30 and 25°C and lower at 15°C in September. The temperature threshold (TH) and thermal accumulative effect (degree-days) were calculated and the laboratory studies demonstrated the relationship between seasonal temperature and insect populations. The study also indicated an increase in the population range between 20–30°C temperatures. Therefore, it is important to take into account the adaption of this insect species to high temperatures with extension in its distribution due to increasing global warming. Concerning host, the castor bean is the most preferred followed by papaya. The sweet potato is however non preferred one. Thus, the host plant species differ significantly concerning egg hatching, nymphal survival, female fecundity, and the duration of the life cycle of *Trialeurodes ricini*.

2.2 Extent of Damage

The insects injurious to the interest of men and known to inflict losses are commonly referred to as pests. According to one estimate, the annual global loss due to animal pests is between 18–20% with a monetary value of $470 billion US dollars (Oerke 2006). In Indian agriculture, the loss was reported around 15.7% with a monetary value of $36 billion US dollars (Sharma et al. 2017). The earlier figures of estimated loss at the global level have been quoted at $200 billion US Dollars in 1993 (Gawel and Barlett 1993). For example, cotton is a crop which is globally cultivated in an area of about 35 million hectares and produces nearly 26 million tons of lint (ICAC 2015). At the same time, it is highly vulnerable to the attack of insect pests, and therefore consumes comparatively more pesticide as compared to other crops.

The whitefly is a polyphagous pest and inflicts great losses in numerous crops. The two species of whitefly that affect many crops are *Bemisia tabaci* or the tobacco/ cotton whitefly and *Trialeurodes vaporariorum* or the glasshouse whitefly. The tobacco whitefly, has been found associated with Leaf Curl Virus disease in chilli but no transmission experiments were conducted (Pruthi and Samuel 1942). The incidence was 15–25% in chilli fields. The cotton whitefly, *Bemisia tabaci* species, Middle East Asia Minor 1 (MEAM 1) and Mediterranean (MED) is prevalent worldwide and has been wrecking havoc in the cultivation of field, vegetables, and ornamental crops both infield and in screen houses (De Barro et al. 2011, Skaljac et al. 2013, Gueguen et al. 2010, De Barro et al. 2003). When present in large numbers, whitefly feeding can affect plant growth, causing distortion, discoloration, yellowing, or silvering of leaves. The study was carried out to assess the extent of silvering of leaves and it was experimentally proved that the feeding of silver whitefly adults upto 20 for 48 hours on zucchini did not express the silvary symptoms (Costa et al. 1993). On the other hand, when immatures are allowed to develop on leaves they could cause silver leaf symptoms. The adult feeding on the other hand are unable to induce this kind of symptoms.

Initially, the whitefly occurs as a sporadic pest before transforming into an epidemic. The change in the damaging status of this insect in different parts of the world is difficult to explain. However, the intervention of humans to get high productivity with excessive use of insecticides and use of transgenics, coupled with favourable climate and fast reproduction potential, have led to a sudden spurt in the pest population. The detailed reasons for this have been discussed elsewhere in this book.

As mentioned earlier, the whitefly sucks the sap from the lower leaf surface and produces yellow spots on an upper surface of the leaf and finally, the plant vigor is reduced and the plants turn yellow, presenting a sickly appearance. Both nymphs and adults are destructive to the crop as both suck sap from the leaves. While feeding/sucking sap from the lower leaf surface, the whiteflies draw a lot of sap and exude a sugary substance called honeydew that falls on the leaves. The leaves become sticky to touch from the upper surface. The leaves soon get covered with black sooty mold (Plate 2.3). The whitefly infestation becomes apparent once the black sooty mold develops on the leaves and completely covers them; however, till then, the infestation remains hidden. Once the leaves are covered with black sooty mold, the plants are unable to carry out photosynthetic activity and manufacture food for the development of a plant. The plants become weak and stunted and unable to bear fruits. The other problem created by sooty mold in ginneries is stickiness of cotton which is mainly caused by honeydew secreted by sucking pests (aphids and whiteflies) (Hequet et al. 2007). The honeydew secreted by whiteflies contains largely the trehalulose type of sugar-a disaccharide while melezitose-a trisaccharide is the main sugar content of honeydew of aphids. The honeydew is mainly a kind of sugar on which sootymold develops responsible for stickiness and to get rid of this problem the producer spends extra money which is a direct loss to the farmer. The worldwide loss due to insect pests in cotton has been reported as 80% and is

Plate 2.3. Damage of whitefly.

the maximum among the crops affected with this pest. Among the insect pests, loss due to sucking pests of cotton has been reported at 36%. For example, in the past, the first severe whitefly infestation was observed in a part of north India (now Pakistan) in the late 1920s and early 1930s (Husain and Trehan 1940), followed by in Sudan and Iran (1950s), El Salvador (1961), Mexico (1962), Brazil (1968), Turkey (1974), Israel (1976), Thailand (1978), Ethiopia (1984), and Arizona and California (1980)(Johnson et al. 1982). The outbreaks of this pest on cotton in India were noticed in different cotton-growing states in India, viz. undivided Punjab (1930–43), Andhra Pradesh (1984–87), Tamil Nadu, Maharashtra and Karnataka (1985–87), Gujarat (1986–87), and north India (1996–2000, 2015) (Chaba 2018). In India, the outbreak during 2015 was devastating and it destroyed 2/3rd of the cotton in the country and about 15 farmers committed suicides due to loss of cotton tonnage (Varma and Bhattacharya 2015). An experimental result demonstrated a higher seed cotton yield of the protected crop (2038 kg/ha) than the unprotected crop (1313 kg/ha) in India (Makawana et al. 2018). In cotton crop, the loss in ginning outturn and seed germination to the extent of 3–5 percent each has been reported (Singh et al. 1983). The loss in yield, ginning outturn and fiber qualities, particularly the maturity of fiber has been apparent in cotton (Butter and Kular 1987, Singh et al. 1983) (Table 2.2). The direct damage is done through sucking the sap by both nymphs and adults. The affected plants look pale and wilted. The excessive drainage of sap causes shedding of fruiting bodies and inflicts heavy loss in tonnage. It is known to diminish the quality of produce (Pollard 1955).

The whitefly is also known as a pest of pulse crops where it acts as a vector of a large number of plant viruses. Therefore, the losses inflicted on these crops are much more as compared to other crops. According to one estimate, the loss in black gram is between 17.4 to 71% (Mansoor-ul-Hassan et al. 1998) depending on location. Further, the loss in urd bean (*Vigna mungo*) has been calculated as 30 to 70% (Marimuthu et al. 1981). The overall loss was determined as 71% in black gram as estimated in another study (Sexena 1983).

Above all, these insects are efficient vectors of plant viruses. The whiteflies are carried through air currents and the viral diseases appear suddenly in the

Table 2.2. The effect of whitefly feeding on seed germination and fiber qualities in cotton (Butter and Kular 1987).

Whitefly Nymphs/ pupae per leaf	Seed germination (%)	Ginning outturn (%)	Fiber length (mm)	Gravimetric Millitex (fineness)	Micronaire 10–6 g/inch (fineness)	Maturity coefficient
0	47.4(43.5)	30.8(38.6)	25.6	121.0	3.08	0.73
20	25.8(30.5)	35.0(36.2)	24.9	128.8	3.27	0.67
40	16.7(25.5)	03.0(33.2)	24.7	135.0	3.43	0.64
60	15.6(22.8)	28.9(42.5)	23.7	137.8	3.50	0.60
CD P = 0.05	(4.7)	(1.9)	-	NS	NS	(0.06)

Parentheses are angular transformations while outside parentheses information are original mean values.

affected areas. The outbreaks of viral diseases in the recent past have been recorded (Navas-Castillo et al. 2014, Pan et al. 2012). They can act as vectors for over sixty plant viruses belonging to the *Geminivirus, Closterovirus, Nepovirus, Carlavirus, Potyvirus,* and *Ipomovirus* genera. The virus group transmitted by whiteflies earlier known as *Geminiviruses* has now been designated as *Begomoviruses* (Brown 2000, Brown 1991). These are agriculturally important causing yield losses in crops of about 20 to 100%. More than a hundred virus diseases are spread by this whitefly, like African Cassava Mosaic Virus, Cotton Leaf Curl Virus, and Tomato Yellow Leaf Curl Virus. The host plants are indirectly damaged by the transmission of more than fifty *Geminiviruses*, for example, Tomato Yellow Leaf Curl Virus (Markham et al. 1996), Tomato Mottle Virus, and Bean Golden Mosaic Virus (McAuslane 2000). *Begomoviruses* are characterized by yellow mosaic appearance, yellow veins, leaf curling, stunting, and vein thickening (for example, Cotton Leaf Curl Virus disease) Tomato crops throughout the world are particularly susceptible to many different *Begomoviruses*, and in most cases exhibit Yellow Leaf Curl Virus symptoms (EPPO 2014, EPPO 2006, Rashmi et al. 2008, Bedford et al. 1994, EPPO 1998). The losses in tomato due to the Tomato Leaf Curl Virus vectored through the whitefly, *Bemisia tabaci*, are enormous. The incidence due to the virus is up to cent percent in tomato in some accounts (Butter and Rataul 1981). As a result of heavy incidence/damage in tomato, the farmers had to dispense with the July crop. This was devastating as earlier the farmers used to cultivate three crops in a year but now the farmers are left with only two crops. Both species of whiteflies *Trialeurodes vaporariorum* and *Bemisia tabaci* causing damage in the form of irregular ripening of fruits in tomatoes has recently been highlighted (Nauen et al. 2014). It is known to cause damage to tomato by sucking sap, secreting honey dew, covering leaves with sooty mold and making plants incapable to carryout photosynthesis. Eventually, these sickly-looking plants shed their leaves almost completely. As a result the plants remain weak and bear lesser and poor quality fruits.

The chapter contains information related to bioecology of whiteflies and the nature and extent of the damage inflicted by them. The data about the insect's generalised biology has been updated with the inclusion of the latest information. The whitefly's growth from egg to adult has been detailed: the eggs and their stalks (pedicel) are laid on the undersurface of the leaves. This is followed by the production of crawlers which move out, to settle on a suitable spot and live a sessile life till their emergence as adults. In this process, the insect passes through four nymphal instars in about two weeks. The fourth instar nymph is termed as a pseudo pupa or red-eye nymph.

The changes in the biology of the whitefly with respect to weather parameters have also been discussed. The ecological conditions were studied taking into consideration the profound influence of changing weather conditions. The two weather parameters affecting the whitefly's development, temperature and relative humidity, have also been discussed. Being a poikilothermic animal, the profound influence of temperature on the life cycle of the whitefly *Bemisia tabaci* is evident. Additionally, the bioecology of three more important species (*Trialeurodes abutilonea, Trialeurodes vaporariorum,* and *Tetraleurodes perseae*) has also been described. The detailed information on the extent and nature of damage inflicted by the whitefly has been presented. To recap, the insect sucks the sap and exudes honeydew on which black sooty mold develops that interferes in the photosynthetic activity of the plant. As a result, the plants look weak due to the draining of sap and reduced photosynthesis. The whitefly is also a vector of several dreaded *Geminiviruses* in nature. The loss in yield due to insect pests is not that alarming in crops but the insect is far more harmful as a vector of viruses. The cent percent loss in some of the crops suffering from viral diseases is common. Resultant crop failures have also been reported at many locations. For example, the premature boll opening in cotton is common. The data presented showed weak fiber and reduced seed germination due to whitefly damage. Another example is the immature ripening of tomato fruits, damaging the quality of produce along with reduction of tonnage.

CHAPTER 3

Taxonomic Status, Biotypes, and Endosymbionts

◇◇◇

3.1 Taxonomic Status

The whitefly, *Bemisia tabaci* being cosmopolitan in distribution thrives on 600 species of plants. It is a complex species undergoing evolutionary change from over a hundred years ago. It contains biotypes and exhibits differences with respect to virus transmission efficiency, rate of development, kind of endosymbiont, host acceptance, and physiological host damage (Oliveira et al. 2001). The spread of this pest is generally associated with the international trade of high-value crops through human involvement, and with the movement of air currents, whereas, within the field, its movement is slow and is limited to adjoining plants, with the disturbance of plants taking place by the agricultural operations carried out regularly by human beings. The rampant increase in this pest's population the world over has necessitated taxonomic studies on this species (Martin and Mound 2019, Hodges and Evans 2005, Sundararaj 2014). The whiteflies in the southern states of the United States of America were identified with the use of an identification guide developed by Hodges and Evans (2005), Sundararaj (2014) did commendable job in whitefly identification in India. A recent study was conducted on the systematic cataloging of Aleyrodidae using host plant and natural enemies data from British Museums (Mound and Halsey 1978). These taxonomists provided pictorial keys to facilitate the identification of whiteflies.

The preliminary studies on whiteflies have indicated the likely presence of species or existence of biotypes or races in this insect species. In the absence of precise detection techniques, the generation of taxonomic data remained scanty. As whiteflies are fast-spreading insects and known to occupy remote locations, taxonomists have been compelled to go far to a large number of locations to collect the whiteflies because of the confusion in the identification of species and biotypes. As a result, several species described earlier had to be revised to eliminate misidentification. *Bemisia tabaci* (Gennadius) is currently believed to be a species complex, with many biotypes, two of which have been described as cryptic species, *Bemisia tabaci* and *Bemisia argentifolii* (Bellows and Perring). Based on these studies, seven groups within the species complex were proposed (Perring 2001). Now, it is known that there are two widespread biotypes: Q (MED) and B (MEAM1) (De Moraes et al. 2017). By 1980, the B-biotype came into the limelight in the Middle East, and it rapidly spread to other areas. In 1984, it was spotted out in Australia (Gunning et al. 1995). This biotype's greater adaptability, coupled with fast migration, caused enormous losses in crop plants and have led to the whitefly, *Bemisia tabaci* being categorized as one of the most devastating pests of crops throughout the globe. Thus, the scientific data generated about this species is largely related to the *Bemisia tabaci* species complex and its endosymbionts. It is vital for developing management strategy of whitefly.

The *Bemisia tabaci* species complex is globally distributed, thus species are named depending on the locations such as the Mediterranean; Middle East-Asia Minor 1; Middle East-Asia Minor 2; Indian Ocean; Asia I; Australia/Indonesia; Australia; China; China 2; Asia II-1; Asia II-2; Asia II-3; Asia II-4; Asia II-5; Asia II-6; Asia II-7; Asia II-8; Italy; sub-Saharan Africa 1 (SSA1); sub-Saharan Africa 2 (SSA2); sub-Saharan Africa 3 (SSA3); sub-Saharan Africa 4 (SSA4); sub-Saharan Africa 5 (SSA5); New World; and Uganda (Mugerwa et al. 2018, Mugerwa et al. 2012). Recently, Asia II-9, Asia II-10, Asia III, and China 3, and Asia I-India and New World 2 have been added. Two new groups were identified using mt CO1 and 16 s tRNA technology (Kanakala and Ghanim 2018). The species of *Bemisia tabaci* throughout the world include Asia 1, India Asia I, Asia I-India, Asia II-1, Asia II-5, Asia II-7, Asia II-8, Asia II-11, Asia II-13, MEAM K, China 3, MEAM1 (India), Asia I, Asia II-1, Asia II-5, Asia II-7, MEAM1 (Pakistan), Asia I, Asia II-1, Asia II-5, China 3 (Bangladesh), Asia II-, South East Asia (Nepal), MED, Asia I (Cambodia), Australia/Indonesia, Asia II-7, Asia II-12 (Indonesia), Asia I, MED, China-2, Asia II-7 (Malaysia), Asia II-5 (Myanmar), Asia I, East Asia (Singapore), MED, MEAM1, Asia I, Asia II-1, Asia II-2, Asia II-3, Asia II-4, Asia II-6, Asia II-7, Asia II-9, Asia II-10, Asia IV, China 1, China-2, China-3, China-4 (China), MED, MEAM1, MEAM2, Asia I, Asia III, Japan-1, Japan-2, Asia II-1, Asia II-6 (Japan), MED, MEAM1, Japan-2 (South Korea), MED, MEAM1, Asia I, Asia III, Asia II-1, Asia II-6, Asia II-7, Middle East (Taiwan), MED, MEAM1 (Egypt), MEAM1 (Iran/Iraq/Israel/Jordon/Kuwait/Saudi Arabia/United Arab Emirates), Asia II-1, MED, MEAM1 (Syria), MED, MEAM1, Asia I (Turkey), MEAM1, Africa (Yemen), MED (Algeria/Burkina Faso), MED, sub-Saharan Africa 1 (Benin), sub-Saharan Africa 1 (Burundi), Cameroon Africa, MED, Sub Saharan Africa 2, sub-Saharan Africa 3, Sub-Saharan Africa 4 (Camron Africa), sub-Saharan Africa 1, sub-Saharan Africa 3 (Democratic Republic of Congo), MED, sub-Saharan Africa 3, sub-Saharan Africa 1 (Ghana) MED, sub-Saharan Africa 1, sub-Saharan Africa 2 (Ivory Coast) Indian Ocean (Madagascar), sub-Saharan Africa 1 (Malawi), sub-Saharan Africa 2 (Mali), Indian Ocean (Mauritius) Spain 1 (Spain) Sub Saharan Africa 1 (Mozambique), MED, sub-Saharan Africa 2 (Nigeria) MED, MEAM1 (Senegal/Morocco) Indian Ocean (Seychelles) MED, New World (Sudan) sub-Saharan Africa 1 (Swaziland) MED, MEAM1, sub-Saharan Africa 1 (South Africa), MED, sub-Saharan Africa 1 (Tanzania) MED, sub-Saharan Africa 3 (Togo), MED, MEAM1, sub-Saharan Africa 2 (Tunisia) Uganda, Indian Ocean, MED, sub-Saharan Africa 2, sub-Saharan Africa 5, sub-Saharan Africa 1 (Uganda), sub-Saharan Africa 9 (Zambia), MED (Zimbabwe). MEAM1 (Mayotte) MED, MEAM1, MEAM2, Indian Ocean Europe (Reunion),

Bosnia and Herzegovina MED (Bosnia) MED, MEAM1 (Croatia/Cyprus/Greece), MED, MEAM1, Indian Ocean, New World 1 (France), MED, MEAM1, Italy 1, Ru (Italy) MED (Netherlands/Czech Republic//Uruguay), MEAM1 (Netherlands Antilles/Cuba/Trinidad and Tobago/Demonian Republic) MED. MEAM1, /MED, sub-Saharan Africa 2, sub-Saharan Africa 3, Italy 2 (Spain). MEAM1 New World, South America (Australia), MEAM1, New World 2, MED (Argentina) New World 2 (Bolivia) MEAM1, MED, New World 1, New World 2 (Brazil), New World 1 (Colombia/Belize), New World 1, MEAM1 North America (Venezuela) MED, MEAM1, New World 1 (USA) MED, MEAM1 North America Canada, New World 1 (Honduras) MED, MEAM1, New World 1 (Mexico) New World 1 (Panama), MED, MEAM1, New World 1 Micronesia, Oceania (Guatemala) Asia II-5 the Greater Antilles, Caribbean (Nauru) MEAM1, New World 1 (Puerto Rico) (Kanakala and Ghanim 2018).

A total of thirty-four morphologically indistinguishable species were identified (Boykin 2014, Lee et al. 2013, Boykin et al. 2012, Dinsdale et al. 2010) MEAM1, also named as *Bemisia argentifolii*. Besides, sub-Saharan Africa 1 (SSA1) and sub-Saharan Africa 2 (SSA2) were considered as the putative species (Legg et al. 2002). Despite this, there was an ambiguity in the taxonomic position in which *Bemisia tabaci* (Gennadius) commonly known as a tobacco/sweet potato/cotton whitefly and earlier named *Bemisia gossypiperda* (Misra & Lamba)/*Bemisia longispina* (Priesner & Hosny)/*Bemisia nigeriensis* (Corbett 1936). The complex is a pack of thirty-seven species of Asian origin (Boykin 2014, Mound and Halsey 1978). The African origin of this pest has been revealed by molecular studies (Campbell et al. 1996). The New World, India/Sudan, and Old World—three new distinct groups have been identified by mitochondrial 16S Ribosomal sub-units (Frohlich and Brown 1994). The B-biotype (silver leaf whitefly, *Bemisia argentifolii*) during the 1980s (Brown et al. 1995a) was recorded as a separate species (Bellows et al. 1994). Like the identification of B-biotype, the population, differences are also recorded in population of West Africa and Uganda with respect to host selection (Navas-Castillo et al. 2011) and vector ability in transmission of plant viruses (Martin 2005, 2004).

The abundance of whitefly on earth is attributed mainly to the monoculture of crops. The explosion of the whitefly population due to human activity is mainly from the crops raised in the screenhouse, particularly vis-à-vis the *Trialeurodes vaporariorum* species. It was introduced in Europe and the United Kingdom (UK) from Mexico by orchardists in 1956. It was further spread from screen houses to open fields and recorded in Finland by 1963. This whitefly species was established in the Mediterranean region and invaded Greece in 1978; the interference of cultivators played a key role in its spread. It continues to multiply without any restriction. The *Bemisia tabaci* biotypes B and Q were also detected in the UK at the quarantine station. A recent study on taxonomic status (Vyskocilova et al. 2019) revealed MED (Q1 and Q2 groups) and ASL (old MED) groups. The studies were further extended to determine fecundity, survival, and the progeny/(offspring). The oviposition of MED whitefly was confined to middle leaves of tobacco and tomato, whereas the adult population laid a large number of eggs on top leaves in pepper and sweet potato plants.

In the last three decades, an effort has also been made to determine the presence of species in the *Bemisia tabaci* complex or races/biotypes. It was mainly the development stages, that is, the puparium, and the adult in the three putative species that were taken into consideration. The determination technique, mt COI, was applied to identify biotypes or strains in *Bemisia tabaci* on a wider scale. In one of the studies carried out to explore biotypes, efforts were made to examine the eighteen populations from North and Central America, the Caribbean, Africa, Middle East, Asia, and Europe (Bedford et al. 1994). These were examined and compared for biological and genetic characteristics to help in identifying the biotypes. On examination, nine populations showed the occurrence of B-types or B-biotypes based on a mating criterion. In all, thirty-six morphological indistinguishable putative species were studied.

Similar collections were made from India/regions such as Asia I, Asia II, and Asia III from Amravati, Delhi, and Ludhiana, respectively (Chaubey et al. 2015) as the Asian continent (India) is the one Dinsdale taken to determine the biotypes in the whitefly population (Lisha et al. 2003). The evaluation of morphogenetic variations of putative species was identified and recorded. Based on the data generated, the 4th instar pupae and adults of the Asia III population were found to be longer than those in the Asia I and Asia II populations. The diversity in different genetic groups (Africa, Asia I, Asia I-India, Asia II 1–12, Asia III, Asia IV, Asia V, Australia, Australia/Indonesia, China 1–5, Indian Ocean, Ru, Sim Middle East Asia Minor I-II (MEAMI-II), Mediterranean (MED), MEAM K, New World 1–2, Japan 1–2, Uganda, Italy 1) has been recorded (Tay et al. 2012, Dinsdale et al. 2010, Zang et al. 2006, Simon et al. 2003, Frohlich et al. 1999). Also, Sub Saharan Africa 1–5 were noted (Hu et al. 2018, Roopa et al. 2015, De Barro et al. 2011, Dinsdale et al. 2010, Boykin et al. 2007).

Studies were carried out in south India taking into consideration two strains of *Bemisia tabaci*: the sweet potato strain and the cassava strain. The other factors that were considered were the development of the insect on different hosts, the efficiency of virus transmission, and the assessment of esterases (Lisha et al. 2003). Both strains of *Bemisia tabaci* were reared on different hosts such as cassava, sweet potato, egg plant and tobacco and further tested by Lisha et al. (2003). Of these two strains, the sweet potato-reared strain of whitefly failed to breed on cassava (*Manihot esculenta*) and the other cassava strain failed to develop on sweet potato. But all the three hosts, sweet potato (*Ipomoea batatas*), eggplant (*Solanum melongena*), and tobacco (*Nicotiana tabacum*), were the common hosts of both strains in nature. Further, the sweet potato whitefly strain did not transmit the Cassava Mosaic Virus to the cassava while the cassava strain population successfully transmitted the Cassava Mosaic Virus from cassava to cassava through routine tests. It was therefore demonstrated that two biotypes in *Bemisia tabaci* exist in the population. The variations were also recorded in an esterases band formation pattern with respect to the two populations. The results of the study of the two populations revealed drastic variations with respect to breeding on common hosts, virus transmission, and the nature of the esterases bands. These variations strongly support the existence of two strains/ biotypes in the population of *Bemisia tabaci* in India (Lisha et al. 2003). However, these strains were not given any specific names. Another study carried out in Andhra Pradesh (India) demonstrated the presence of B-biotype in *Bemisia tabaci* for the first time (Chandrashekar and Shashank 2017). The host range is highly variable: *Bemisia tabaci* devours avocado, banana, cabbage, capsicum, cassava, cauliflower, citrus fruits, coconut, cotton, eggplant, garlic, guava, legumes, mango, mustard, onion, peach, pepper, radish, squash, soybean, tomato, and tobacco (Palaniswami et al. 2001).

The cassava is an important host of whiteflies as well as a source of food for 1,500 million people in Africa and of one-fifth population of the world (apart from Africa). It is also a favourite food of the people of Asia and South America. *Bemisia tabaci* is available in two clusters (Ug1 and Ug 2) on cassava. A phylogenetic analysis of population of adults and 4th instar nymphs with mtCOI revealed the presence of the Ug1 cluster on cassava and non-cassava adjoining crops. No Ug 2 cluster of whitefly was detected on any plant species including cassava (Sseruwagi et al. 2006). The crop is vulnerable to two devastating whitefly-borne viral diseases (that is, the Cassava Mosaic virus and Cassava Brown Streak Virus) (Simon et al. 2006). These diseases are limiting factors in the successful cultivation of cassava in this region.

As in Asia, whitefly collections were done in Africa to study biotypes based on the importance of the pest in this region (Were et al. 2007). The population study of *Bemisia tabaci* (MEAM1) demonstrated eight (Asia I, Asia I, Asia II-1, Asia II-5, Asia II-6, Asia II-7, Indian Ocean, and Australia (Dickey et al. 2013), morphologically indistinguishable cryptic species, with the application of Mitochondrial Cytochrome oxidase 1. Besides, other studies reported thirty-seven biotypes using the mitochondrial COI gene and the Nuclear ITSI gene (Boykin and De Barro 2014, Dinsdale et al. 2010, Boykin et al. 2007, Frohlich et al. 1999). It is already known that *Bemisia tabaci* is a complex of twenty-four morphologically indistinguishable species that rules out the presence of biotype after naming B-biotype as *Bemisia argentifolii* (De Barro et al. 2011). It was identified using mitochondrial cytochrome oxidase 1 (mtCO1). The first large invasion of the B-biotype in 1980 occurred in MEAM. The Middle East Asia Minor 1 (MEAM1) from its MEAM region (Jordan, Iran, Kuwait, Israel, Pakistan Saudi Arabia, United Arab Emirates, Syria, and Yemen), spread over to fifty-four countries viz. American Samoa, Antigua, and Barbados, Argentina, Australia, Austria, Belize, China, Canada, Brazil, Colombia, Cook Island, Costa Rica, Cyprus, Dominican Republic, Egypt, France, Fiji, French Polynesia, Denmark, Germany, Greece, Guatemala, Grenada, Guadeloupe, Guam, Honduras, Italy, Japan India, Martinique, Marshall Island, Mauritius, Mayotte, Mexico, New Caledonia, New Zealand, The Netherlands, Niue, North Mariana Islands, Norway, Panama, Poland, Puerto Rico, Reunion, Saint Kitts, Newis, New Mexico, South Africa, South Korea, Spain, Taiwan, Tobago, Trinidad, Tunisia, USA, Venezuela. The Q-biotype from its origin countries (Algeria, Crete, Croatia, Egypt, France Greece, Italy, Israel, Morocco, Portugal, Spain Sudan, Syria, and Turkey) spreads to ten other countries (Canada, China, Guatemala, Japan, Mexico, The Netherlands, New Zealand, South Korea, Uruguay, USA). The more sensitive modern techniques, namely RAPD-PCR, AFLP, mtCO1 and 16 s ribosomal ITSI microsatellites, CAPS and SCAR, and mtCO1 were used to distinguish differential groups (Ueda and Brown 2006, De Barro et al. 2000).

Bemisia tabaci is also a complex of 40 cryptic species. A study carried out in Bangladesh identified four cryptic species with interspecific variation: Asia I was in abundance, Asia II-1 and Asia II-5 were moderate; however, Asia II-10 was noted in the central region. This study was found useful for learning about the diversity and geographic distribution of cryptic species in Bangladesh (Khatun et al. 2018). In that country, *Bemisia tabaci* was earlier designated as a single species, but due to the study, is now considered as a combination of two sister mitochondrial groups (Q1 and Q2) and ASL species (Old MED group) based on the differential pattern of its oviposition rate, fecundity, off-spring rate, survival (Vyskocilova et al. 2019). The comparison of the two groups based on thirteen host plants belonging to nine families was made. The female whitefly preferred egg-laying on the middle canopy leaves on tobacco and tomato plants while the other group selected top young leaves of pepper and sweet potato for oviposition. It was thus confirmed that the whitefly population exists in groups rather than single species. The buildup of populations of these tiny creatures is enormous as compared to the rest of the fauna owing to certain characteristics such as unique mode of reproduction without fertilization (haplodiploid), wide host range, protected cultivation of crops, and favorable weather conditions (Vyskocilova et al. 2019). The differences were statistically significant for all the four parameters.

In India, a recent study, two groups of *Bemisia tabaci* population were identified using 4,253 (mtCO1) sequences from 82 countries and 1,226 16S/23S rRNA endosymbionts from 32 countries. The study further showed divergence of different endosymbionts sequences within the species complex (Kanakala and Ghanim 2019) which are as listed below

I) *Hamiltonella-1*,

II) *Portiera-2* (P1 and P2),

III) *Arsenophononus-2* (A1 and A2),

IV) *Wolbachia-2* (supergroup O and B),

V) *Cardinium-4* (C1 to C4), and

Vi) *Rickettsia-3* (R1 to R3).

Another study was conducted in India to analyze the genetic variations in host plants in the species prevalent in this region as evident from mitochondrial cytochrome oxidase 1 (mt CO1) sequences. Seventy-one samples (4th instar nymphs and adults) were collected and the result demonstrated four indigenous species (Asia 1, Asia II-7, Asia II-8, MEAM1), and MEAM1-K. The study demonstrated different number of host plants in Asia 1 (17), Asia II-7 (9) and Asia II-8 (5). MEAM-K was a new species in the area, that had a great resemblance (92.6%) to MEAM1. MEAM1 possessed wide host range and produced more honey dew. They gave an indication of new species or existence of biotypes or races in the B-biotype of the *Bemisia tabaci* complex (Roopa et al. 2015).

Instead of biotypes, recent research has focused on exploring the presence and variation of bacterial endosymbionts in complex species of whitefly and noted three species and clades (SSAI, SSA2, and Reunion) in Kenya (Mugerwa et al. 2012).

The *Bemisia tabaci* complex was found to contain *Candidatus Portiera aleyrodidarum* (primary endosymbiont) to enrich its diet (Sloan and Moran 2012, Thao and Baumann 2004). Secondary symbionts that supplemented the host's survival, and the development of whiteflies, were also detected (Baumann et al. 2006, Baumann 2005, Thao and Baumann 2004, Moran and Telang 1998, Baumann and Moran 1985). Amongst the secondary endosymbionts, *Rickettsia* (Gottlieb et al. 2006), *Wolbachia* (Nirgianaki et al. 2003, Zchori-Fein and Brown 2002), *Hamiltonella* and *Arsenophonus* (Sloan and Moran 2012, Thao and Baumann 2004), *Cardinium* (Weeks et al. 2003), *Fritschea* (Everett et al. 2005), *Hemipteriphilus* (Bing et al. 2013) were also present in the population. Another endosymbiont named *Candidatus Fritschea bemisiae* strain *Falk* belonging to the Simikaniaceae family of the bacterium was recorded using sixteen S rRNA from whiteflies (Thao et al. 2003).

All the genetic groups from China were examined and the endosymbionts detected include *Wolbachia: Rickettsia: Arsenophonus: Hamiltonella* and *Cardinium* (MEAM1/MED). Besides these, the other population from China showed the presence of *Hamiltonella* (MEAM1 and MED) (Marubayashi et al. 2014, Zchori-Fein et al. 2014, Bing et al. 2013, Skaljac et al. 2013, Skaljac et al. 2010, Gueguen et al. 2010, Chiel et al. 2007, Hunter et al. 2003, Zchori-Fein et al. 2001); *Rickettsia* (Asia II, Asia II-7, China 1 and MEAM1) (Bing et al. 2013a), and *Arsenophonus, Cardinium, Rickettsia,* and *Wolbachia* (Africa/China/India) (Ghosh et al. 2015, Bing et al. 2013, Singh et al. 2012). *Hamiltonella* and *Fritschea* were however absent from the population. The infection of secondary symbionts of *Hamiltonella* and *Rickettsia* (MEAM1) and *Hamiltonella* in Q1 (Gueguen et al. 2010), *Cardinium* (MEAM1) and *Cardinium, Fritschea* and *Wolbachia* (MED) (Gorsane et al. 2011) were detected. The possible reason for such differences could be due to variations in host plants, differential use of pesticides, or something wrong with the data recording.

The diversity of endosymbionts in *Bemisia tabaci* with 16s rRNA gene sequences of 298 accessions in thirty-two countries of six continents (Asia Africa, North America, South America, Europe, and Australia) was studied. The findings detected *Portiera aleyrodidarum* in a cryptic species of *Bemisia* (MED, MEAM1, Asia I, Asia II-1, Asia II-6, Asia II-7, China-1, Japan, Sub-Saharan Africa)/other countries (China, Japan, France, Australia, Brazil, India, South Africa, Tanzania, Malawi, Uganda, Nigeria, USA, Mexico, Israel, Pakistan). Of these, *Rickettsia* has been recorded in 7 species/19 countries (MED, MEAM1, Asia II-1, Asia II-3, Asia II-7, China-1 and sub-Saharan Africa)/(China, Japan, Israel, Burkina, Montenegro, Croatia, South Korea, Sudan, Brazil, Tunisia, Reunion, Antilles, India, Bangladesh, Tanzania, Malawi, Uganda, Nigeria, and Pakistan). Besides, *Cardinium* (MED, MEAM1, Asia I, Asia II-1, Asia II-3, Asia II-6, Asia II-7, Japan, Sub-Saharan Africa, and New World). *Arsenophonus* (MED, Asia I, Asia II-1, Asia II-3, Asia II-6, Asia II-7, Sub-Saharan Africa and New World), *Fritschea/Hamiltonella* (MED, MEAM1 and New World 2), and *Hemipteriphilus* (China 1-China) were also found in populations of whitefly. The endosymbionts namely, *Portiera aleyrodidarum, Rickettsia, Wolbachia, Hamiltonella, Arsenophonus, Cardinium,* and *Fritschea* were present in the MED population. However, the MEAM1 population contains all the above endosymbionts, except *Arsenophonus* and *Hemipteriphilus*. Interestingly, the presence of *Fritschea* and *Hamiltonella* was restricted to MED, MEAM1, and New World species, and were not associated with *Bemisia*'s new complex species. *Portiera aleyrodidarum* lot contained two major Clades viz. P1 and P2. The selected *Portiera* share was > 95% in one lot from Japan, Brazil, and sub-Saharan Africa, Australia. *Portiera* share from another lot from China had < 95% share. The percentage share of *Rickettsia* was divided among three clade/groups which are as under:

- Clade R1, *Rickettsia* (invasive and indigenous) is present in Australia, China, Japan, Israel, and sub-Saharan African countries,
- Clade R2: *Rickettsia* is restricted only to the sub-Saharan Africa species, and
- Clade R3: *Rickettsia* has only been isolated from the original indigenous population from sub-Saharan Africa, Asia II-7 and Asia II-1 from China and India.

Hamiltonella was recorded from the two invasive species (MED and MEAM1) of one major cluster (H1) from Brazil, Japan, the USA, China, and South Korea. *Wolbachia*, grouped into sixteen supergroups named A–Q inhabiting a wide range of arthropods and filarial nematodes (Glowska et al. 2015). In our analysis, two supergroups B and O were recorded in *Bemisia tabaci Wolbachia* sequences. The O supergroup (MED and Asian species) from China and Japan; and the B supergroup (MED and MEAM1 invasive species) from Asia, sub-Saharan Africa, China 1, Japan, and Australia, and in indigenous species were noted. Both supergroups were collected from the MED species from China. *Cardinium* is a complex of four major clusters (C1–C4), where;

- C1 is an invasive MED species from China;
- C2 cluster is from the Asian/MEAM1/sub-Saharan African species from four different regions (India, China, sub-Saharan Africa, and the US);
- C3 has also been recorded (invasive MEAM1 species from USA);
- C4 has been isolated from MED/MEAM1/Japan/sub-Saharan Africa/Asian/New World species from ten countries such as China, Japan, Sudan, Spain, South Korea, Morocco, France, Cameroon, Uruguay, and the USA. These isolates belong to the three Clades (C1, C2, and C4) present in China. The isolates that appeared in C4 include MED, MEAM1, and Asia II-1 and Asia II-6. Clade C2 had only indigenous species (Asia II 1, Asia II-7, and sub-Saharan Africa).

The analysis of *Arsenophonus* species revealed greater diversity in whitefly *Bemisia tabaci* complex and it helps to explore more insight that is helpful to understand the ecology (Mouton et al. 2012). The endosymbiont with greater diversity showed variations in populations and therefore is grouped into A1 and A2. The endosymbiont samples from Beni-Sweif, Ismailia, Kalyobia, El-Fayoum, Tanta, and Kafr El-Sheikh contained 80–90% endosymbionts while Banha had 30–40% population of endosymbionts as recorded through PCR-RAPD in Egypt (Fahmy and Abou-Ali 2015).

- Clade A1 is from an indigenous Asian species from China, while
- Clade A2 is further divided into sub-clades (a to k). Amongst these clades, the variations in the Chinese *Arsenophonus* strains have two sub-clades viz. (i) A2e and (ii) A2h.

The three indigenous sub-clades are named

I) A2 a

II) A2 f

III) A2

Similarly, strains in the sub-invasive species (MEAM1 and MED) and New World 2 species were recorded in Brazil. In contrast, *Fritschea* in MED, MEAM1, and New World 2 while *Hemipteriphilus* in China 1 (China) were detected. The "cryptic species" were used to distinguish the whitefly population with the application of a modern technique (3.5% mitochondrial cytochrome oxidase gene sequence divergence) causing twenty-four *Bemisia tabaci* cryptic species to be determined and nominated. The genetic divergence showed forty-two distinct groups. The newly described morphologically indistinguishable species, including Africa, Asia I, Asia I-India, Asia II 1–12, Asia III, Asia IV, Asia V, Australia, Australia/Indonesia, China 1–5, Indian Ocean, Ru, Middle East Asia Minor I-II (MEAM), Mediterranean (MED), MEAM K, New World 1–2, Japan 1–2, Uganda, Italy 1, and sub-Saharan Africa 1 5 have been currently delimited at the global level (Hu et al. 2018, Roopa et al. 2015, Skaljac et al. 2017, Firdaus et al. 2013, De Barro et al. 2011, Dinsdale et al. 2010, Boykin et al. 2007). Two *Portiera* groups (P1 and P2) were identified in this analysis: The molecular sequences belonging to the P1 group of the population from Australia, Brazil, China, and Japan and the P2 group of the population from China, India, Japan, and sub-Sharan African were studied. The four groups of *Cardinium* (C1–C4) demonstrated by the study are given as below:

- C1 is from China,
- C2 is from China, India, and sub-Saharan Africa,
- C3 is from USA, Cameroon, China, France, Japan, South Korea, Spain, Sudan, and Uruguay, and
- C4 is a product from the whole world.

In *Wolbachia*:

- B is from Australia, China, Japan, India, sub-Saharan Africa, and South Korea countries and
- C is from China and Japan subgroups.

Based on this analysis, > 40 cryptic species of *Bemisia tabaci* and endosymbionts grouped in the same clades were recorded. The presence of *Hamiltonella* was noted in all infecting species. The occurrence of a horizontal transfer between two species of whiteflies (*Bemisia tabaci* and *Trialeurodes vaporariorum*) within the same plant host and endosymbiont species was observed (Skaljac et al. 2013, Skaljac et al. 2010).

3.2 Biotypes

As the insects continue to develop on resistant varieties of crop plants, it can be inferred that this insect population is capable of feeding, developing, and surviving on plants already resistant to other populations of the same species (Kogan 1982). Of the *Bemisia tabaci* complex the species (MEAM1-B biotype MED-Q biotype and MW 2 biotype) is already known. Many morphological indistinguishable species are available and efforts are being made to identify biotypes taking into consideration endosymbionts, pesticide resistance host plants, and viral transmission using latest molecular techniques. Intensive cropping, irrigated monoculture with high inputs of fertilizers and pesticides in 1905 caused the production of biotypes in whiteflies (Brown et al. 1995).

The insect has a worldwide distribution, occupying most geographic regions. A systematic study at the desired level is still lacking. The variations in population buildup of whiteflies led the researchers to explore the concept of the presence of races. The occurrence of host races or strains in population led to the naming of biotypes by 1950 while working with *Bemisia tabaci*. However, little emphasis was given to explore the mechanism in operation through the molecular and biochemical pathways (De Barro 2007, Brown et al. 1995). Initially, the parameters of the virus-vector relationship and a definite association of whitefly with the host were considered as components to identify a differential pattern of virus transmission level. The different host plants are known to alter the parameters of the biology of whitefly. Besides the host range, other characters such as allelochemicals, the attraction of parasitoids, the presence of endosymbionts for the enrichment of sap, and level of susceptibility/resistance to insecticides are taken to describe races of whitefly. By taking into account these features, the biotypes of *Bemisia tabaci* namely B and Q were identified (Horowitz and Ishaaya 2014). The B-biotype of *Bemisia tabaci* was designated as separate species, *Bemisia argentifolii*, based on the samples of whiteflies collected from California and Florida (Bellows et al. 1994) and there is a great genetic diversity in species of whiteflies (Sundararaj 2014). The identification of whitefly adults is problematic but helpful in pupae. To avoid misidentifications and erroneous naming of new species, the Polymerase chain reaction (PCR) fragment amplified from the mitochondrial cytochrome oxidase 1 (mtCO1) gene was considered appropriate, as done in Colombia (Ovalle et al. 2014). The genetic diversity was studied using molecular markers. Of these markers, Amplification and Polymorphic Analysis of Restriction Enzyme Digestion of mitochondrial Cytochrome Oxidase-1 gene was considered most appropriate to measure the degree of variability (Martin 2004). Besides this, cloning, sequencing phylogenetic analysis of mitochondrial 16S rRNA gene and Rapid Amplified Polymorphic DNA—PCR Rapid PCR analysis are the molecular techniques. The whiteflies are separated from other members through the utilization of evolutionary eighteen S rDNA sequences (Campbell et al. 1994, Campbell et al. 1993). To differentiate species, the measurement of the length of

DNA nucleotides is taken into account. The exceptionally long length between 2200–2500 bp in this group (whiteflies) has been recorded against 1900–1925 bp (aphids/psyllids/mealybugs). With the use of this technique, five putative species namely SSA-9, SSA-10, SSA-1, SSA-12, and SSA-13 were identified. Sub Saharan Africa (Burkino Faso) is known for the presence of three biotypes viz An (sub Saharan non silver leafing), AsL (sub Saharan Africa silver leafing) and Q (without any information). Of these Q contains three sugroups (Q1, Q2 and Q3).

As mentioned earlier, *Bemisia tabaci* is a complex of many species or biotypes. It has eleven genetic groups and these groups contain thirty-four morphologically indistinguishable species. These biotypes were identified based on morphogenetic characteristics of adult and nymphal stages with a microscope. The characteristics in 4th instar nymphs taken into consideration are the vasiform orifice, operculum, lingula, and length of margins of wax, including the width of the tracheal comb. The adult morphogenetic characters include compound eyes, antennal segments, vasiform orifice, tarsus, and genitals. The biotypes were also identified taking into consideration the differential host range, pesticide resistance, endosymbiont, diversity and extent of virus transmission, etc. The studies carried out to identify these biotypes using Polyacrylamide Gel Electrophoresis (PAGE) indicated the occurrence of 3 (Ludhiana) and 4 morphs (Abohar/Bathinda/Faridkot) in Punjab, India (Singh et al. 1998). The differential response in pesticide resistance confirmed the presence of biotypes. The biotypes B and Q are the predominant ones and these are widespread. Similar studies carried out in China identified six biotypes (JHJ1, JHJ2, JHJ3, and Cv) including B and Q (Wan et al. 2009). Efforts were made at various places to know the presence of biotypes in many more regions. Until 2010 cryptic species of *Bemisa tabaci* were 24 only but subsequently 15 new species were added after thorough identification with phylogenetic analysis making a total of 39 species. The study confirmed the presence of one indigenous (NW2) and two invasive species (MEAM1, MED) in Argentina (Alemandri et al. 2015). Of these MEAM1 species was confined to melon, MED to tomato, and MEAM1 was present both on beans and melon (Watanabe et al. 2019). Based on the new classification, the species MEAM1, NW, and NW2 were recently investigated in Brazil (Marubayashi et al. 2013) in addition to *Bemisia tuberculata,* and *Trialeurodes vaporariorum.* Of these, *Bemisia tuberculata* was found in Sao Paulo, Mato Grosso, and Rio de Janeiro states of Brazil (Alonso et al. 2012, Rabello et al. 2008) while *Trialeurodes vaporariorum,* only in Sao Paulo (Lourencao et al. 2008). It was the only report from Brazil which indicated that the B-biotype of *Bemisia tabaci* is not a complex of biotypes/groups/races. It is rather a species in itself. Biotypes An and BR (Invasive), and B (introduced) are already known (Queiroz et al. 2016). After the invasion of MEAM1 in Brazil and four years after it, 1,237 samples were collected to make an assessment of the species. The major species found was MEAM1 in north, north-western, and central Brazil whereas, MED was located only in five states in the south-western region. The MEAM1 was found associated with a virus but MED had nothing to do with a virus. The other species from the new world were not detected from Brazil. The intensity of endosymbionts in MEAM1 was variable in two populations.

Another study on the taxonomic position demonstrated that *Bemisia tabaci* is a complex of thirty-one species (Lee et al. 2013): Asia I, Asia II-1, Asia II-2, Asia II-3, Asia II-4, Asia II-5, Asia II-6, Asia II-7, Asia II-8, Australia, Australia/ Indonesia, China1, China2, Italy, Mediterranean, Middle East Asia Minor 1, Middle East Asia Minor 2, Indian Ocean, New World, Sub Saharan Africa 1, Sub Saharan Africa 2, Sub Saharan Africa 3, Sub Saharan Africa 4, Uganda, Asia I-India, Asia III, China 3, Asia II-9, Asia II-10, and New World 2, including the ones already described (Dinsdale et al. 2010, Ueda et al. 2008, Chowda-Reddy et al. 2012, Alemandri et al. 2012, Hu et al. 2011). The species Asia II 1, Asia II 2, Asia II 3, Asia II 4, Asia II 5, Asia II 6, Asia II 7, Asia II 8 and species Asia I, Asia III, Australia/ Austria, Indonesia, China 1, China 2, China 3 were found as clusters. It is, therefore, demonstrated that *Bemisia tabaci* is a complex of thirty-one species rather than a single species as revealed by molecular methods like allozyme and mtCO1.

The biotypes were mainly studied using RAPD-PCR, and mit CO1 techniques. As mentioned earlier, there are two widespread whitefly biotypes such as Q (MED) and B (MEAM1). Of these two biotypes, Q is resistant to most potential insecticides. The B-biotype is wide spread and recorded in north Africa (Egypt, Mauritius, Reunion), west Africa (Tunisia), China (Fujian, Hebel, Shandong, Xinjiang), India (Gujarat, Karnataka, Maharashtra), Iran, Israel, Japan, Honshu, Jordon, Pakistan, South Korea, Taiwan, Turkey, and Yemen. It is also recorded from Austria, Cyprus, Denmark, France, Spain, Canary Islands, Germany, Greece, Italy, Cicily, Netherlands, Norway, Poland, Portugal, Antigua and Barbuda, Belize, Canada, Nova Scotia, Costa Rica, Dominican, Grenada, Guadeloupe, Guatemala, Honduras, Martinique, Mexico, Nicaragua, Panama, Puerto Rico, Saint Kitts, Newis and Tobago, and USA (eleven states). Besides all these areas, it has also been detected in Oceania (American Samoa), Australia (New South Wales/Queensland), Tasmania, Western Australia, Cook Islands, Guam, Fiji, Marshall Islands, and South America (Jameis Reo grande endorse, Bahe Golas/Marshall Minas Gesals Parade, Pernambuco/Rio De Sao Paulo, Colombia, Venezuela). The biotype-Q is also widespread and was recorded in Africa, South Africa, Asia, China, Guangdong, Jiansu, Shandong, Tibet, Israel, Japan, Ryukyu Islands, South Korea, Turkey, Greece, Crete, Italy, Ontario, Costa Rica, Mexico, USA (about twenty states). The United states in which Q-biotype is present includes Alabama, Arizona, Connecticut, Florida, Georgia, Hawaii, Ohio, Pennsylvania, Texas, New York, New Hampshire, New Jersey, Indiana, Michigan, Illinois, Kentucky, Lousiana, Maryland, New Hampshire, and New Mexico. South America is also afflicted with this biotype and the affected countries include Argentina, Brazil, and Uruguay. It has not been recorded in Ireland, United Kingdom, Finland, and Sweden (Kiriticos et al. 2020, Cuthbertson and Vanninen 2015).

The study carried out in Africa demonstrated biotypes in Africa 1 (Cassava and eggplant) and Africa 2 (polyphagous) based on the association of host (Burban et al. 1992), and Nauru (Taiwan) and sixteen biotypes namely B, Q, An, M, B2, L, H, A, C, N, K, R, E, S, J, P in Pakistan (Hsieh et al. 2006, Saeed et al. 2012). *Bemisia tabaci* species preferred cassava/beans/sweet potato/weeds as their hosts in Africa. The populations observed are Africa 1–5 (SSA1–5), SSA1–5, Uganda 3 (SSA-6), Africa (SSA7), East Africa 1, Mediterranean (MED), MEAM1 and Indian Ocean

(IO) (Legg et al. 2002, Berry et al. 2004, Gueguen et al. 2010). The other species such as SSA1 and SSA2 were designated as a vector of cassava viruses including New World from US (Barbosa et al. 2014, Marubayashi et al. 2013, Legg 2010, Legg and Fauquet 2004) and Asia species (Asia 1–4, Japan, China 1–3) (Legg et al. 2002, Ellango et al. 2015).

The different genetic groups of *Bemisia tabaci* have been identified in India and these include Asia I, Asia II-1, Asia II-5, Asia II-7, Asia-8, Asia II-11, China 3, MEAMI-B (Ellango et al. 2015). The distribution of endosymbionts has been given. Of these species, SSA10 and SSA11 were clustered with the New World species. Great diversity in the population of *Bemisia tabaci* has been obtained in Kenya while working with whitefly vectors of viral diseases of cassava (Cassava Brown Streak Virus and Cassava Mosaic Virus) using mitochondrial Cytochrome Oxidase 1 (mtC1-DNA) (Manani et al. 2017). *Bemisia tabaci* showed a lot of diversity in its population. The phylogenetic analysis indicated five clades in the populations of Sub-Sahara-Africa-1 and Sub-Sahara-Africa-2. Similarly, BCDE Clades were identified from the western region and Nyanza regions. Besides this, the B Clade were pinpointed on the coastal region and Nyanza areas. Based on geographic location, the species have been named as Mediterranean Middle East Asia Minor1; Middle East Asia Minor2; Indian Ocean/Asia 1; Australia/Indonesia; China; China-2; Asia II-1; Asia II-2; Asia II-3; Asia II-4; Asia II-5; Asia II-6; Asia II-7; Asia II-8; Italy; Sub-Saharan Africa-3; Sub-Saharan Africa-4; Sub-Saharan Africa-5; New world, Uganda, etc. The worldwide emerging species are MEAM-1 (B-biotype = *Bemisia argentifolii*) and MED (*Bemisia tabaci*; Q-biotype). These have now been spread and posed a serious threat to crop cultivation (Brown et al. 1995). These two biotypes namely B and Q have recently invaded the Western North Pacific Region (China, Korea, Japan and Taiwan) playing havoc with the cultivation of crops (Hsieh et al. 2007). The modern technique for quick identification of biotypes of whitefly (namely, RAPID PCR has been put in to use) (De Barro and Driver 1997). Furthermore, the silver whitefly, *Bemisia argentifolii* population found on poinsettia, sweet potato, cotton, and tobacco plants were tested for biotypes and variations in banding pattern of esterases was apparent in the populations of South America, USA, Africa, Middle East, and the Caribbean (Costa et al. 1993, Brown et al. 1995).

Allozyme markers/esterases through protein variability were also applied to distinguish the biotypes in whiteflies in another study (Costa and Brown 1991). The susceptibility to pesticides is another component through which the variability can be seen. The existence of biotypes (B and Q-biotypes) has been detected in southeast China based on the pesticide resistance. The result of the study indicated variations in the level of susceptibility to different pesticides. The moderate to the high level of resistance to imidacloprid and thiamethoxam, medium to high to alphamethrin, low to medium to fipronil, low to spinosad has been reported (Wang et al. 2010). The whiteflies were practically immune to abamectin pesticides. The other factor on the bases of which biotypes are distinguished is the presence of endosymbiont species meant for the enrichment of sap which lacks amino acids and proteins (De Moraes et al. 2018). The study confirmed the presence of endosymbionts viz *Hamiltonella* and *Rickettsia* in MEAM1 through the analysis of mtCO1 as against the presence of *Hamiltonella*,

Rickettsia, Wolbachia, and *Arsenophonus* in other species of whitefly. Further, the mitochondrial oxidase subunit (mtCO1) analyses have also indicated the presence of endosymbionts (Surapathrudo and Ghanim 2015, Boykin and De Barro 2014).

In the Brazilian state Bahia (Bondar 1967), forty-three species of whitefly were separated based on the presence of endosymbionts, both obligatory (*Portiera aleyrodidarum*) and facultative (*Arsenophonus, Wolbachia, Cardinium, Fritshea, Rickettsia, Orentia-like organism*) (Bing et al. 2013). The endosymbionts were helpful in increasing the host suitability of natural enemies (Hadjistylli et al. 2016, Dalton 2006), to enhance tolerance to high temperature (Brumin et al. 2011) to facilitate virus transmission (Gottlieb et al. 2010, Rana et al. 2012), and to increase susceptibility to pesticides (Kontsedalov et al. 2008).

The information gathered from China reported endosymbionts, namely *Portiera* Aleyrodidae (Primary) and *Rickettsia* and *Hamiltonella* (Secondary) from MEAM1 *Bemisia tabaci* (Shi et al. 2018). The *Rickettsia* endosymbionts are reported from bacteriocyte cells, wax glands, and collateral glands in the body. To develop, there is a significant increase of *Portiera* from 2nd to 4th instar nymphs and adult stage while secondary endosymbionts (*Arsenophonus*) development is in the adult stage.

Based on an analysis through mtCO1, five whitefly species namely, sweet potato whitefly, *Bemisia tabaci* B-biotype (MEAM1), Greenhouse whitefly, *Trialeurodes vaporariorum* (A-biotype commonly called New World 2), Acacia whitefly, *Tetraleurodes acaciae* and *Bemisia tuberculata* from cassava. The RNA sequence revealed that a higher presence of *Hamiltonella* and *Rickettsia* was present in all populations of MEAM1, while *Cardinium* was in three populations, and MEAM and NW2 had *Fritschea* and *Arsenophonus.* Both *Fritchea* and *Arsenophonus* endosymbionts were also present in *Tetraleurodes acaciae* and *Bemisia tuberculata* (Marubayashi et al. 2014). Of these, MEAM1 was the only species in north, and north-western and central Brazil, while MED in states from south-east to the south. The low genetic diversity of *Hamiltonella* and *Rickettsia* was recorded. It can be concluded from the available work carried out in different regions and enough literature that three species MEAM1 (the Middle East, Asian Minor 1), MED (Mediterranean) and NW (New world) are widespread through the world.

B-biotype of *Bemisia tabaci* is not a complex of biotypes/groups/races. It is rather a species in itself. To recap, the biotypes are the morphologically indistinguishable population of *Bemisia tabaci* but differ in other biological traits such as host range, host adaptability, and ability as the vector of the virus. As per the available studies, the biotypes of whiteflies are B-biotype (Synonym of Q-biotype), Biotype MEAM1-A in India (related to MEAM1), in addition to B and B2 (both related to MEAM1), biotypes Q, J, and L (Sub-Saharan Africa), biotype-A, biotype-C, biotypes D, F (Africa/MEAM1). The biotypes B and Cv of *Bemisia tabaci* were compared for variation in biological parameters. Both these biotypes showed differences in biological traits. These traits were the efficient, low-cost, effective, and reliable solution to the current problem of species identification (Hebert et al. 2003).

B-biotype preferred vegetable crops while the attraction was great towards ornamentals by biotype Cv in China (Qiu et al. 2009). Another study in China carried out with RAPD-PCR for the identification of three species of whitefly *Bemisia tabaci* showed B-biotype had three bright DNA bands between 250–600 bp, the Q biotype only has one bright band at 300 bp, Cv with no band between 250–500 bp, both Cv and Q had two bands at 750 and 1000 bp and B had one band at 1000 bp (Qiu et al. 2011). It was further inferred that B-biotype and Q-biotype are closer than B and Cv or Q and Cv. In all 6 biotypes namely B, Q, Cv, ZHJ-1, ZHJ-2, and ZHJ-3 are present in China (Li and Yan 2013). In Taiwan, 4 species are present include *Trialeurodes vaporariorum, Aleurodicus diapersus, Bemisia argentifolii,* and *Paraleyrodes bondari.* Furthermore, in all six biotyes (A, B, Q, An, S and Nauru) have been recognized in Taiwan (De Barro et al. 1998).

Genomics has facilitated the symbiotic relationship between the endosymbiont and the host animal (Moya et al. 2008). In the comparison of two biotypes of *Bemisia, Rickettsia* may thus be an excellent model to study the evolution and transmission pathways.

3.3 Endosymbionts

Endosymbionts are the mediators of ecological traits in the host insect, whiteflies, and help in enriching a diet deficient in essential amino acids, parasite resistance and heat tolerance (Feldhaar 2011). There are 1,200,000 total insect species in all, and almost 50% are harboring endosymbionts. These endosymbionts are present in bacteriocytes and live in a mutualistic relationship (Kikuchi 2009). Symbiotic relationships among animals and microorganisms are common in nature. The obligate association of *Buchnera aphidicola* in aphids contributes towards the fitness of aphids. The aphid, *Acyrthosiphon pisum,* harbors bacterium responsible for the increased vulnerability of the hymenopteran parasitoid, *Aphidius ervi.* The interaction between a symbiotic bacterium and a host's natural enemy provides a mechanism for the persistence and spread of symbiotic bacteria (Oliver et al. 2003). The bacterial inhabitant of insects is antagonistic and in symbiotic relationships with the latter. Besides the insects, mites are also recorded as ectoparasites feeding on insect hemolymph. They also act as vectors of *Spiroplasma poulsonii* (a male-killing endosymbiont of Drosophila) from *Drosophila nebulosa* females to *Drosophila willistoni* (Jaenike et al. 2007).

The symbionts are found in special cells but their role remained hidden for some time on account of the non-availability of detection methods (Dale and Moran 2006). With the introduction of PCR, however, their role has been defined (Baumann and Moran 1997). These endosymbionts in insects are being utilized for correcting the nutritional deficiency in sap-sucking insects (Dadd 1985). These endosymbionts are well known in aphids and are also present in whiteflies (Clark et al. 2010, Clark et al. 1992).

Of the symbiotic bacteria, the endosymbionts present in *Bemisia tabaci* are a well worked-out organism. As *Bemisia tabaci* is a complex of numerous species, it is expected to contain genetically distinct groups/biotypes (Liu et al. 2012, Tay et al. 2012, De Barro et al. 2011). While working with the complex species *Bemisia,* the concept of biotypes/species was explored. The biotypes of *Bemisia tabaci,* I.e., B and Q were identified through the PCR based on the presence of endosymbionts. The study carried out in China demonstrated the existence of two major groups of Q-biotype depending upon the presence of secondary endosymbionts (Chu

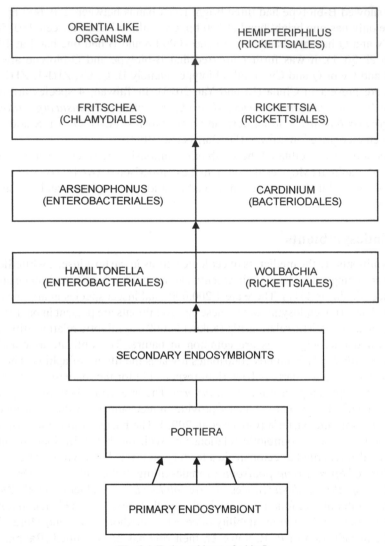

Fig. 3.1 Endosymbionts of whitefly.

et al. 2011). The endosymbionts are of two categories (Fig. 3.1) primary (*Portiera* Aleyrodidae—to enrich the food in the form of amino acid and carotenoids) and secondary to enhance the fitness of species and, to confer resistance to pesticides, enhance virus transmission, tolerance to high temperature, etc. *Rickettsia* like *Wolbachia* is a reproductive manipulator, associated with male-killing, parthenogenesis, and fertility. This model is helpful in study of non-pathogenic *Rickettsia* to advance our understanding in such situations (Perlman et al. 2006). The bacterial endosymbionts influence biology (Gottlieb et al. 2008) while the endosymbiont *Portiera aleyrodidarum* (Sloan and Moran 2012, Thao and Baumann 2004) enriches the diet. The facultative secondary endosymbionts, *Arsenophonus*

(Enterobacteriales), *Hamiltonella* (Enterobacteriales) (Moran et al. 2005), *Fritschea* (Chlamydiales) (Everett et al. 2005), *Cardinium* (Bacteroidetes) (Weeks et al. 2003), *Rickettsia* (Rickettsiales) (Gottlieb et al. 2006), *Wolbachia* (Rickettsiales) (Nirgianaki et al. 2003, Zchori-Fein and Brown 2002), and *Orientia-like organism* (Bing et al. 2013) are not essential.

The whitefly contains a primary endosymbiont named *Portiera aleyrodidarum*. The secondary endosymbionts are *Hamiltonella* (Enterobacteriales) (Thao et al. 2004, Moran et al. 2005) *Wolbachia, Rickettsiales* (Nirgianaki et al. 2003, Zchori-Fein and Brown 2002) *Cardinium,* Bacteroidetes (Weeks et al. 2003) and *Fritschea* (Chlamydiales) (Everett et al. 2005) and *Arsenophonus* (Enterobacteriales) in *Bemisia tabaci.* Also, the secondary endosymbionts namely, *Hamiltonella, Rickettsia, Orentia Like Organism* (OLO), *Fritschea, Arsenophonus, Wolbachia, Cardinium* are present in all species of whiteflies. The whiteflies prevalent in Israel did not contain *Cardinium* and *Fritschea* but did contain all the other endosymbionts (Bing et al. 2012, Gottlieb et al. 2008).

After a thorough investigation, the presence of the primary endosymbiotic bacteria, *Portiera,* in all individuals of whiteflies has been documented (Thao and Baumann 2004), which enriches its food in nutrients. With respect to secondary endosymbionts, differential combinations of facultative endosymbionts were identified. Further, the variations in endosymbionts as noted in *Wolbachia* that exist in two subgroups A and B in whiteflies were also recorded (Ahmed et al. 2010a). The differential patterns of facultative endosymbionts demonstrate insect biology (reproduction, survival, and fecundity), pesticide resistance, and vector ability.

Of these endosymbionts, *Wolbachia* exists in two subgroups (A and B) in whiteflies as well as in the natural enemies of whiteflies (Ahmed et al. 2010 a). In a study conducted by Ruan et al. (2006), the two biotypes of *Bemisia tabaci* were fed with three antibiotics, tetracycline, ampicillin trihydrate, and rifampicin, to evaluate the fitness of their offspring feeding on cotton. These three antibiotics did not affect *Portiera aleyrodidarum* but eliminated the secondary endosymbionts. In the B-biotype, the tetracycline or ampicillin trihydrate accelerated the survival while rifampicin reduced the development of the offspring. In the non-B ZHJ-1 population, the adults treated with tetracycline or ampicillin trihydrate also showed acceleration in the development of their offspring but without any corresponding effect on their survival. However, the treatment of adults with rifampicin significantly reduced both the development and survival of offspring. It can be both negative and positive.

Whiteflies are sap-sucking insects that harbor prokaryotic primary endosymbionts (P-endosymbionts) within specialized cells located in their body. Whiteflies also harbor S-endosymbionts (Enterobacteriaceae). These insects feed on a poor-quality diet deficient in mineral elements, which is subsequently improved upon by symbionts (primary endosymbiont-*Portiera aleyrodidarum*) present in the whitefly. The endosymbionts are present in specialized cells called bacteriocytes/ bromosomes. There is a common sequence of 16 S rDNA in *Bemisia tabaci* when present in the form of *Bemisia argentifolii, Siphoninus phillyreae,* or *Trialeurodes vaporariorum.* The presence of primary bacteria is mandatory in all species of whiteflies (Thao and Baumann 2004).

Based on the presence of the P-endosymbiont/hosts, whiteflies could be grouped into at least five clusters. The major subdivision was between the subfamilies Aleyrodinae and Aleurodicinae. Unlike the P-endosymbionts of many other insects, the P-endosymbionts of whiteflies were related to *Pseudomonas* and possibly to the P-endosymbionts of psyllids (Thao and Baumann 2004). *Candidatus Portiera aleyrodidarum,* a primary endosymbiont, consists of an extremely reduced genome responsible for the enrichment of a diet deficient in amino acid and sulfur (Santos-Garcia et al. 2015, Santos-Garcia et al. 2014). The Tomato Leaf Curl Virus (ToLCV) disease is managed through both chitosan and *Pseudomonas* sp. Chitosan enhances the biocontrol efficacy of *Pseudomonas* sp. against ToLCV (Mishra et al. 2014).

The whitefly *Bemisia tabaci* contains seven endosymbionts present in bacteriocytes and found scattered throughout the entire abdomen, wax glands, and collateral glands. These novel infection patterns help to uncover the function of *Rickettsia* in the hosts of the whitefly (Shi et al. 2018). Specific endosymbionts are present in closely related species of whiteflies while the endosymbionts are different in different species of whiteflies. The endosymbionts are known to perform functions beneficial to the host. These functions include:

- Primary endosymbionts enrich the poor diet into a nutrient-rich diet. *Cardinium* and *Arsenophonus* enhance the fitness of the host (Santos-Garcia et al. 2012). The symbionts called *Candidatus Sulcia muelleri* (Bacteroidetes), and *Candidatus Baumannia cicadellinicola* (γ-Proteobacteria), not related to whiteflies produce essential amino acids as well as many vitamins. While the phenomenon of co-symbiont also occurs in insects, it has not been demonstrated in whiteflies (McCutcheon et al. 2009). The whitefly *Bemisia tabaci* (Gennadius) (Hemiptera: Aleyrodidae) harbors several bacterial symbionts. Among the secondary (facultative) symbionts, *Hamiltonella* has a high prevalence and the previous reports indicated increased fitness of whitefly. The data suggest that *Hamiltonella* may play a previously unrecognized role as a nutritional mutualist (Su et al. 2014). The endosymbionts also take care of composition/infection frequencies (Thao and Baumann 2004).

- The stage-specific location of endosymbionts helps in the detection of symbionts (Marubayashi et al. 2014, Zchori-Fein et al. 2014, Skaljac et al. 2010).

- The interference of endosymbionts in host physiology/ecology/reproduction is occurring (Su et al. 2014, Chiel et al. 2007, Zchori-Fein and Perhnam 2004, Hunter et al. 2003, Zchori–Fein et al. 2001).

- Fast evolution is brought about by these symbionts (Himler et al. 2011). It is thus important to trace the evolutionary history of insects taking into consideration the presence of different endosymbionts and their density (Ghanim et al. 2010).

- The thermotolerance of the whitefly is dependent on endosymbionts (Brumin et al. 2011). *Rickettsia* is known to enhance the ability of whiteflies to tolerate high temperatures.

- The resistance to insecticides is brought about by endosymbionts (Kontsedalov et al. 2008). The significant influence on the susceptibility of whitefly to thiamethoxam, imidacloprid, pyriproxyfen, and spiromesifen was recorded taking into consideration the different combinations of endosymbionts (*Rickettsia-Arsenophonus and Wolbachia-Arsenophonus*), present in the whitefly. The high density of symbionts is known to enhance the ability of whitefly (biotype B and Q) to detoxify the earlier mentioned chemicals (Ghanim and Kontsedalov 2009).

• The host fitness is enhanced by endosymbionts (Chiel et al. 2009) reflected through the rapid growth, better survival and higher proportion of females in a generation (Himler et al. 2011) In addition the premoult development in whiteflies was significantly faster with symbionts.

• Endosymbionts protect the host against the attack of pathogens, parasites, encoding toxins and virulence factors (Hendry et al. 2014).

• The vector virus transmission is successfully performed through *Arsenophonus* symbionts (Rana et al. 2012, Gottlieb et al. 2008). It has been demonstrated in a well-worked virus, Cotton Leaf Curl Virus vectored by *Bemisia tabaci* Q-biotype and *Trialeurodes vaporariorum* (Ghanim 2014, Rana et al. 2012).

The symbiont, *Cardinium hertigii* is known to cause cytoplasmic incompatibility (Weeks et al. 2003). However, bacterium-free whiteflies are better transmitters of the Cassava Mosaic Virus (the Uganda strain). Plus, there is a higher emergence of bacterium-free adult whiteflies as compared with viruliferous whiteflies (Ghosh et al. 2018). This is enough indication to report that endosymbionts can be beneficial as well as detrimental to the host. *Bemisia tabaci* (Hemiptera: Aleyrodidae) harbors *Portiera* as primary and *Hamiltonella, Arsenophonus, Cardinium, Wolbachia, Rickettsia,* and *Fritschea* as secondary endosymbionts (as mentioned earlier) in sixty-one field populations; and the variations could be explained by biotypes, sexes, host plants and geographical locations of these field populations (Pan et al. 2012b).

The horizontal and vertical transmission/spread of endosymbionts takes place in nature (Ahmad et al. 2015, Ahmed et al. 2013, Caspi-Fluger et al. 2011, Vautrin and Vavre 2009, Gottlieb et al. 2008). So much so, the horizontal transmission of the endosymbiont, *Wolbachia* was confirmed through spiders/parasitoid wasp in addition to vertical transmission (Li et al. 2017, Ahmed et al. 2015, Baldo et al. 2008). The horizontal transmission of *Rickettsia* via cotton, basil, and nightshade in phloem through insects/fleas (Li et al. 2019, Hirunkanokpun et al. 2011) has been demonstrated (Caspi-Fluger et al. 2011). Phylogenetic studies have implicated the frequent horizontal transmission of *Wolbachia* among arthropod host lineages. However, the ecological routes for such lateral transfer are poorly known (Sintupachee et al. 2006). Further horizontal transmission of *Cardinium* is also reported between different phloem sap-feeding insect species through plants (Gonella et al. 2015). The phylogenetic tools used can be assessed through web application through CIPRES and it is free of cost to the public (Miller et al. 2015, Miller et al. 2010).

The diversity of bacteria being harbored by whiteflies is large. Thus both the categories of bacteria (culturable and endosymbionts) were isolated in the studies carried out on *Aleurocanthus woglumi* (Pandey et al. 2013). The bacteria species were culturable on the media under laboratory conditions. All fourteen culturable genera (*Arthrobacter, Bacillus, Brevibacillus, Bhargava, Curtobacterium, Kocuria, Listeria, Micrococcus, Paenibacillus, Pantoea, Rhodococcus, Rummellibacillus,* and S*taphylococcus*) were isolated from the citrus black fly, *Aleurocanthus woglumi* (Pandey et al. 2013). *Portiera* and other endosymbionts like *Cardinium, Wolbachia, Rickettsia,* and *Arsenophonus* along with *Bacillus, Enterobacter, Paracoccus,* and *Acinetobacter* were detected. The phylogenetic analysis of 16S rDNA sequences of *Cardinium, Wolbachia, Rickettsia,* and *Arsenophonus* showed separate clusters. The results suggest that the presence of large number of species of bacterial endosymbionts in the *Bemisia tabaci* makes it more complex than the whitefly host harbouring single species of endosymbiont (Singh et al. 2012). The bacterial isolation will go a long way in helping to identify cultural media. The bacteria present in the body of insects are transmissible through transovarial transmission but the bacteria present in leafhopper are transmissible via plant feeding (Purcell et al. 1994). The spread of bacterial endosymbionts through whiteflies is vertical, that is, with the complete transmission of the bacteriocyte into the eggs and the progeny (Costa et al. 1996).

The great population of whitefly had a higher concentration of *Wolbachia* and *Hamiltonella* but *Arsenophonus, Cardinium,* and *Rickettsia* were absent. *Wolbachia* strains (two) were identified and the two strains of the endosymbiont were found together in the same population but not in the host individual of multilocus strain (Tsagkarakou et al. 2012). The analysis of genome same sequences of *Portiera* and *Hamiltonella* in the *Bemisia tabaci* showed a reduction in *Portiera* (357 kb), without affecting essential amino-acids and carotenoids and the genome of *Hamiltonella* not only provided vitamins and cofactors, but also completed the missing steps of some of the pathways of *Portiera* (Rao et al. 2015). MEAM1 harbors *"Candidatus Portiera aleyrodidarum"* and, *"Candidatus Hamiltonella defensa"* and *Rickettsia* sp., whereas MED has only *"Ca. Portiera Aleyrodidae"* and *"Ca. Hamiltonella defense."* Both *"Ca. Portiera aleyrodidarum"* and *"Ca. Hamiltonella defensa"* are intracellular endosymbionts in the bacteriomes, and *Rickettsia* sp. is found scattered in the body. After cold treatment at 5 or 10°C or heat treatment at 35 or 40°C for 24 h, respectively, the infection rates of all symbionts had not significantly decreased based on diagnosis PCR (Shan et al. 2014, Shan et al. 2016). Antibiotics rifampicin selectively eliminated certain symbionts from Middle East-Asia Minor 1 *"Candidatus Portiera aleyrodidarum"* and two secondary symbionts *"Candidatus Hamiltonella defensa"* and *Rickettsia*. Neither the primary nor the secondary symbionts were completely depleted in the adults (F0) that fed for 48 h on a diet treated with rifampicin at concentrations of 1–100 µg/ml. However, both the primary and secondary symbionts were nearly completely depleted in the offspring (F1) of the rifampicin-treated adults.

The prevalence of the primary endosymbiont *Portiera aleyrodidarum* and secondary endosymbionts (*Arsenophonus* and *Wolbachia*) in the two invasive biotypes (B and Q) and one indigenous biotype (Cv) in China was determined (Ahmed et al. 2010). The bacteria contain carotenoid-synthesizing genes for enriching the diet (Santos-Garcia et al. 2012, Sloan and Moran 2012). The presence of endosymbionts (*Arsenophonus and Hamiltonella*) in three species of whiteflies (*Bemisia tabaci, Trialeurodes vaporariorum, Siphoninus phillyreae*) in southeast Europe was demonstrated (Skaljac et al. 2017). *Rickettsia* was present only in *Bemisia tabaci* while *Hamiltonella* was present in both the *Bemisia tabaci* and *Siphoninus phillyreae* populations of whiteflies. *Fritschea* was noted in all the three species of whitefly.

In a study by Zhu et al. (2017), the dissection of the host and filtration method was followed to extract the bacteria (Zhu et al. 2017). The cotton whitefly is also a vector of the Tomato Yellow Leaf Curl Virus and the presence of *Rickettsia* endosymbiont increases the retention of the virus in the whitefly vector (Kliot et al. 2014). The study was carried out with three species, that is, *Bemisia tabaci*, *Trialeurodes vaporariorum*, and *Siphoninus phillyreae* using the modern technique, mitochondrial cytochrome oxidase 1 (mtCO1). With the application of this technique on *Bemisia tabaci*, the population is divided into four lots: Q1 (MED), Q2 (Middle East), Q3 (Burkina Faso), and Q4 (Ivory Coast, Burkina Faso, and Cameron). Similar data were recorded in MED (Frohlich et al. 1999, Gottlieb et al. 2008). The whitefly populations, differentiated into thirty-four species, were collected from the states of Sao Paulo, Bahia, Minas, Gerais, and Parana in Brazil for the identification and infection of endosymbionts using the method of sequencing the rRNA gene. Of these 23, 4, 4, 1, and 3 species were identified as belonging to MEAM1, *Trialeurodes vaporariorum*, *Bemisia tuberculata*, NW2, and *Trialeurodes acaciae*, respectively (Marubayashi et al. 2014). In all the above five species the MEAM 1 population was found infected with *Hamiltonella* and *Rickettsia*. Four endosymbionts, namely *Hamiltonella*, *Rickettsia*, *Cardinium*, and *Fritschea* were detected in the *Bemisia tabaci*, MEAM1 and NW2 populations. The species-inhabiting endosymbionts were investigated for their molecular and biological aspects (De Barro et al. 2011, Dinsdale et al. 2010, Alemandri et al. 2012, Firdaus et al. 2013a) and named as MEAM1, NW1, and NW2.

In Israel, the endosymbionts are present in B-biotype (*Hamiltonella*) and Q-biotype (*Wolbachia*; *Arsenophonus*) and both B and Q (*Rickettsia*) (Chiel et al. 2007). Further, *Fritschea* in A-biotype (NW2) from USA (Thao et al. 2003), *Cardinium* in B and Q-biotype from China (Chu et al. 2011), and Q from Croatia (Skaljac et al. 2013, Skaljac et al. 2010) and France were detected. The whitefly, *Trialeurodes vaporariorum*, and *Siphoninus philyreae* from Croatia had *Portiera*, *Hamiltonella*, and *Arsenophonus*.

The primary endosymbiont, *Portiera Aleyrodidae*, and two secondary endosymbionts, *Arsenophonus* and *Cardinium*, are present in a Chinese population of *Bemisia tabaci*. All these three symbionts are missing from MEAM1 and Asia II-1, Asia II-3, and China-1, whereas, *Cardinium* was present in MED and Asia-1, Asia II-3 but rarely present in MEAM1 (Tang et al. 2018). All the seven endosymbionts (*Portiera*, *Hamiltonella*, *Arsenophonus*, *Cardinium*, *Fritschea*, *Wolbachia*, *Rickettsia*) (Fig. 3.1) were present in MEAM1 and MED of *Bemisia tabaci* (Shi et al. 2018). The species of whitefly, *Trialeurodes variabilis*, *Aleurotrachelus socialis*, and *Bemisia tuberculata* all from Colombia contain primary endosymbiont *Portiera* in all the samples but the secondary endosymbioants *Rickettsia*, *Arsenophonus*, and *Wolbachia* were absent in these species (Gomes-Diaz et al. 2019). Of these endosymbionts, *Cardinium* and *Arsenophonus* were recorded for the first time in these species and demonstrated great diversity in Colombia. *Cardinium* was the most common endosymbiont in these species. It was also ascertained that all the biotypes of *Bemisia tabaci* in Oman contain *Hamiltonella* (Al-Shehi and Khan 2013).

The genetic diversity of *Trialeurodes vaporariorum* was studied by analyzing thirty-eight samples from eighteen countries with mt cytochrome oxidase I and

Hamiltonella, Rickettsia, Arsenophonus, Cardinium, Wolbachia, and *Fritschea* were found in most samples of individuals. All the samples examined showed at least one secondary endosymbiont. *Arsenophonus* was present in all individuals while *Wolbachia* and *Cardinium* occurred in low frequency in Greece. *Rickettsia, Hamiltonella,* and *Fritschea* were absent and showed low diversity similar to *Bemisia tabaci* (Kapantaidaki et al. 2015).

The Tomato Leaf Curl Virus (ToLCV) disease is managed through both chitosan and *Pseudomonas* sp. Chitosan enhances the biocontrol efficacy of *Pseudomonas* sp. against ToLCV (Mishra et al. 2014).

This chapter was devoted to expounding on the taxonomic aspect of the whitefly, keeping in mind the existence of biotypes and the role of endosymbiotic bacteria in the determination of biotypes. *Bemisia tabaci* was considered as a complex of many species and it is widespread. It has occupied an important niche in the ecosystems of tropical, subtropical, and temperate zones throughout the world. The chapter also highlighted the identification of biotypes of whiteflies in different countries or important regions. It is a known fact that *Bemisia tabaci* has two biotypes based on location namely B-biotype (*Bemisia argentifolii*) (MEAM1) and Q-biotype (MED). However, the information gathered on the differential response of whitefly from varied locations showed the presence of more biotypes taking into consideration, host plants, response to pesticide, presence of endosymbionts, vector ability virus transmission, etc. The modern and highly sensitive techniques such as RAPD-PCR, AFLP, mtCO1 and 16 s ribosomal ITSI microsatellites, CAPS and SCAR, help researchers to differentiate genetic groups based on region, and thus, led to the new biotype (NW-new world) being determined in Brazil. Likewise, many biotypes yet to be named have been determined. The identification of endosymbionts and their presence in terms of species and extent of endosymbionts were also analyzed. The detailed information on the presence of the primary endosymbiont in all whitefly populations and varying presence of secondary symbionts was given. The symbionts and their role have been discussed in detail. The presence of *Portiera, Hamiltonella, Cardinium Arsenophonus, Fritschea, Rickettsia* has been analyzed to determine the biotypes.

CHAPTER 4

Population Sampling and Economic Thresholds

◇◇

4.1 Population Sampling

To study the whitefly from various perspectives, sampling is critical for drawing definite conclusions. The sampling is done to utilize the data for initiating sprays of pesticides or for the release of natural enemies or any other purpose. It is important to know where the sampling is to be done and which stage should be sampled. The distribution of the population within the plant canopy should be known to scientists engaged in sampling of whiteflies. A faulty sampling procedure forfeits the very objective of such investigations. The data pertaining to population sampling, biology and varietal reaction of the whitefly were generated (Chandi and Kular 2020, Butter and Vir 1990). It is mandatory to record the data as per the requirement to develop new methodology and to make refinements in the existing sampling procedures. The sampling methods are different in different crop agroecosystems depending upon the objective of the study; that is, whether the sampling population is chosen to study the economic threshold, the release of natural enemies, or checking the spread of virus diseases. In this monograph, an effort has been made to enlarge upon the efficient methods to estimate the population of whiteflies. To develop methodology due consideration is given to the stage of the insect, part of the plant, and plant canopy.

It is a known fact that the abundance of whitefly in a crop is governed by certain important characteristic features. It has a preference for some characteristics in a crop. To record a population of whiteflies, it is desirable to sample the population from susceptible varieties to have a sufficient population. The physical characteristics such as hairiness, leaf shape, and the internal pH of sap were found to be correlated with whitefly population. The behavior of whiteflies determines the preference/non-preference of whiteflies as these creatures prefer hairy leaved varieties (Kular and Butter 1999), thickened leaf lamina (Butter and Vir 1989) with low content of phosphorus, tannins, and phenols (Butter et al. 1992) in the crops. The whiteflies are generally attracted to yellow color, but the visual color cues determine the settling of the whitefly. The adult flight is governed by green color and perceive signals of 550 nm green light (Issacs et al. 1999). The sampling of eggs from the lower surface of upper canopy leaves is suggested due to the abundance of eggs found there.

The egg counting is recommended under a microscope using 3 cut leaf discs (each 1 sq. cm areas). For counting the adults, three fully developed leaves from the upper canopy are appropriate under field conditions. However, for nymphal counts, nymphs from three fully opened leaves from the middle canopy of cotton are agreed upon. The sampling is generally done by two methods viz. yellow sticky traps (Lu et al. 2012) or adult inspection (Butter and Vir 1990). The adult counts are better in the field while traps are useful in screen houses. The traps are mainly exploited for control of whiteflies rather than monitoring population. The sampling can be done by inspecting nymphs and eggs on the leaves. The counting of numbers/adults is more reliable as compared to sticky traps in the early hours of morning (Ellsworth et al. 1995). In all, at least thirty leaves from the 5th node from the terminals should be selected from four quarters. For counting points, there should be a 10–15 feet distance between two points. It is also essential to count the whitefly population from about thirty plants. Further, the leaves carrying three or more adults should be considered infested while the leaves supporting less than three adults be considered uninfected. By utilizing this technique, the economic threshold of five adults/leaf on cotton and three adults/leaf on melon was determined (Palumbo et al. 1984).

There have been studies (mentioned below) that were carried out in this vein to precisely sample the population of whiteflies. The vertical distribution of whiteflies within the plant and in the whole field has been studied (Von Arx et al. 1983). The adults, eggs and first instar nymphs of whitefly are predominantly found on top the leaves, the second and third instar nymphs on the middle canopy leaves, while the 4th instar red-eye nymphs are found on the older canopy leaves. The percentage-wise population of eggs on the top, middle, and lower canopy leaves was 51.5, 46.7, and 1.8 eggs, respectively (Husain and Trehan 1940). A similar pattern was witnessed in Punjab, India on several genotypes of cotton (Butter and Vir 1990). This observation was further endorsed by the findings in the central (73.4%) and southern (86.2%) parts of India where the fecundity was higher on the top (Naik and Lingappa 1992, Rao et al. 1991). Another study also confirmed the whitefly's preference for egg-laying (73.4%) on the top canopy leaves (Rao et al. 1991). In another on top canopy, egg-laying to the maximum extent (86.2%) was also reported. Another experiment conducted in south India (Naik and Lingappa 1992), wherein the nymphs/pupae were found confined to the middle canopy and eggs on top canopy leaves. In this context, sampling of the eggs and adults from the upper canopy of the cotton plant is advocated, whereas the counting of nymphs from the middle canopy leaves is recommended, considering the abundance of the population (Mohd Rasdi et al. 2009, Butter and Vir 1990).

4.2 Sticky Traps/Application of Remote Sensing

Remote sensing is a technology through which the damage of insect pests can be located/detected in a vast area or a special belt of the crop without landing on the ground. It can also help to detect an infestation of insects, particularly whiteflies. Broad information with respect to affected crops can be gathered within the shortest period. However, it is quite cumbersome to separate the damage of whitefly from other insects of the sucking category. The sensor-based systems are put to use to earmark the affected areas. The methods through which detection and mapping can be done with sensors and other information technology are Global Position Systems (GPS) and Geographic Information Systems (GIS). In remote sensing, the various techniques applied are spectroradiometers, satellite imaging systems of moderate/ high resolution, areal photographic camera, and airborne digital multispectrum and hygrospectrum imaging systems (Yang and Everitt 2011). Traps are desirable to monitor the population of adult whiteflies for surveillance and to trap out the initial population of adults both in the screen houses and in the open fields.

The whitefly has attained pest status in Sudan, Iran, El-Salvador Mexico, Brazil, Turkey, Israel, Thailand, USA, India, and Pakistan. Therefore, numerous studies have been made in these countries for the sampling of whiteflies. The relative abundance of whiteflies have been estimated to be found in the canopy position (Ozgur and Sekeroglu 1986, Butter and Vir 1990, El-Khidir 1965). The method to trap a population of whitefly adults through the yellow sticky traps did not prove highly effective under field conditions; however, it was considered as one of the tactics used to manage a population of whiteflies under screenhouse conditions (Zanik et al. 2018, Lu et al. 2012). Despite shortcomings, the method is considered as the best method for monitoring of the population. As mentioned earlier, whiteflies are known for their attraction to the color yellow and this behavior has been exploited by researchers. The traps are available in different shapes and sizes and are locally prepared by smearing the adhesive gum on plastic sheets, plates or petri dishes. Such traps can be reused after washing with detergents. However, the adhesive gum needs replacement after ten days. Catching adults is comparatively more successful in poor or patchy crops as compared to open fallow fields or dense crops.

In north India, ordinary circular tins of a capacity of 2 kg are used to prepare traps, the inner side of the tin is coated with castor oil (Dhawan and Simwat 1997). Work is already in progress in south India to determine the modalities for efficient surveillance regarding a suitable design use of sticky gel and the suitable height and orientation of traps. The rectangular traps prepared in the south India are considered better than the ones used in the north, there are still efforts being made to improve their design. Experiments are being conducted to determine the appropriate placement of traps in a cropped field. The traps placed horizontally on the crop canopy facing the sky are useful to trap the flying whitefly population (Gerling and Horowitz 1984). The study further advocated the placement of a yellow-colored trap facing the sky to trap flying whitefly adults. Positive correlation (r = 0.7275) between the whitefly adults trapped on sticky traps and egg-laying on plants recorded manually was found. The economic thresholds for the need-based application of pesticides in screen houses have also been determined using this technology. Accordingly, the economic thresholds for a young crop (0.5 whitefly adults/sticky card/day) and the mature crop (2 adults/sticky card/day) in the screen house have been determined (Alston 2007).

Considering the level of the population the appropriate tactic of management should be resorted to for tackling the menace of whiteflies. To make appropriate decisions for pesticide applications, reliable sampling is a must as different whitefly species behave differently during oviposition. While in the other species egg-laying is on the lower surface of leaves or the top canopy leaves, the nymphal population of whiteflies is located on the middle canopy leaves. The other two species (*Aleurothrixus floccosus* and *Dialeurodes citri*) also prefer fully expanded leaves of citrus. Also, while sticky traps are successful in certain conditions; in a particular study, they did not accurately reflect the population levels in melon fields (Lu et al. 2012). The trap catches made near the sprayed fields are inflated after the sprays. In fact, the visual observations of adults on leaves provide a more accurate and practical method as the population is evenly spread in the melon terminal and fields. The sampling of adults during early morning hours within 2 hours of sunrise is desirable (Palumbo et al. 1984). The population of three adults per leaf is recommended. In all, 200 leaves in a field (50 from each quadrant) are appropriate. This estimation is based on a 200-leaf samples; reduction of the sample size will reduce the accuracy of the estimation. Three whiteflies per leaf are the action level for whiteflies on spring melons.

4.3 Economic Thresholds

The economic threshold is defined as the population density at which control measures are applied to prevent the population from reaching the economic injury level (EIL). The economic injury level is the lowest population density at which the control measures should be undertaken to prevent the population from inflicting economic damage. The economic damage is the amount of damage that justifies the cost of control operations. The threshold levels are liable to change with the change in crop variety, etc. The economic threshold is generally based on the population of whitefly (egg/nymph/adult) and damage symptoms. With respect to the stage of the pest, the egg stage is more precise but is time-consuming and cumbersome, thus not considered for determining the threshold. The threshold levels available for whiteflies/different species of whiteflies in different crops are worked out and presented for the application of control tactics (Table 4.1: Economic thresholds). Economic thresholds are essential for the judicious use of pesticides. These have been worked out for most crop pests and whiteflies are the target insects for which thresholds have been worked out in most situations (Vieira et al. 2013, Pedigo et al. 1986). However, economic thresholds are dynamic and subject to change depending on the variations in the prices of inputs used in control operations and the selling prices of commodities (Seiter 2018). Thus, there is a need to refine the thresholds that have already been worked out. The mathematical equation to calculate the threshold is available. Accordingly, the given equation is as under:

$$EIL = CVIDK$$

where **(C)** is the cost of control measures; **(V)** is the value of the commodity; **(I)** is the injury based on the population density (defoliation/damaged fruit); **(D)** is the economic damage; and **(K)** is the proportion of reduced injury levels with control measures. The control manager should be well versed in the terminology of economic thresholds. The economic threshold of whiteflies has been determined for important crops vulnerable to the attack of whiteflies namely cotton, cassava, soybean, tomato, and pulse crops. The economic thresholds levels are determined for different purposes like an application of insecticide, release of natural enemies, and application of insect growth regulators. For example, in cotton, the threshold

Table 4.1. Economic thresholds of whiteflies in different crops.

Whitefly species	Economic threshold	Crop	Country	Reference
Bemisia argentifolii	0.5 adults/7.6 cm² leaf area (Texas) 3 adults/leaf (Arizona)	Cantaloupe	USA, Texas/Arizona	Riley and Palumbo 1995
Bemisia tabaci	8–10 adults/leaf	Cotton	India (Tamil Nadu)	Natarajan et al. 1986 Sundaramurthy 1992
Bemisia tabaci	40% infested leaves (presence of atleast 3 nymphs/adults)	Cotton	USA, California	Goodell et al. 2015
Bemisia tabaci	6–8 adults/leaf	Cotton	India	Sukhija et al. 1986
Bemisia tabaci	The appearance of sooty mold	Cotton	India	Anonymous 1986
Bemisia tabaci	20-immatures (nymphs/pupae)	Cotton	India	Butter and Kular 1986
Bemisia tabaci	5 adults/leaf	Cotton	Pakistan	Safdar et al. 2019
Bemisia tabaci	1-large nymph/Disc plus (3–5-adults/leaf)	Cotton	USA Arizona	Ellsworth et al. 1996
Bemisia tabaci *Trialeurodes* *abutilonea*	0.5 adults/Sticky card/day (young crop) 2 adults/sticky card/day (mature crop)	Cotton	USA Utah	Alston 2007
Bemisia tabaci	3–5 large nymphs/disc	Cotton	USA, Arizona	Ellsworth et al. 1996
Bemisia tabaci	10 adults/leaf 5–10 nymphs/adults per Leaf	Cotton	India, Andhra Pradesh	Reddy and Krishnamurthy 1989 Anonymous 2004
Bemisia tabaci	5.9–15.2 adults/leaf or 6.1–9.8 eggs/cm or 1.7–4.7 nymphs/cm²/leaf	Cotton	USA, Arizona	Naranjo et al. 1996
Trialeurodes abutilonea	50 nymphs/leaf on 5th node (before boll opening) or 25 nymphs/leaf on 5th node (after boll opening)	Cotton	USA, Texas	Anonymous 2020
Trialeurodes perseae	3 or more adults covering > 40% leaves of 5th node	Cotton	USA, Florida	Nakahara 1995
Trialeurodes vaporariorum	50% of young leaves show honeydew on the upper surface	Cotton	USA, Florida	Funderburk et al. 2019
Bemisia argentifolii (IGR application)	3 or more adults covering 40% of leaves of 5th node with the presence of nymphs	Cotton	USA, Texas	Anonymous 2020
Bemisia tabaci	100–200 adults/week on sticky traps	Cotton	Israel	Melamed-Madjar et al. 1982

Table 4.1 Contd. ...

...Table 4.1 Contd.

Whitefly species	Economic threshold	Crop	Country	Reference
Trialeurodes vaporariorum	6.2 adults/leaf	Cucumber	Korea	Jeon et al. 2009
Bemisia tabaci	2–3 adults/Leaf	Melon (Spring)	USA, Arizona	Palumbo et al. 1984
Dialeuropora decempunctata	20-adults/leaf	Mulberry	India, West Bengal	Bandyopadhay et al. 2002
Bemisia tabaci	10-nymphs/leaflet or presence of sooty mold	Soybean	Brazil	Vieira et al. 2013
Bemisia tabaci	136.31–+26.60 nymphs per leaf	Soybean	Brazil	Vieira et al. 2013
Bemisia tabaci (B-biotype)	4-adults/leaf	Sweet Potato	USA, California	Anonymous 2019
Bemisia tabaci	3.25 adults or 4 nymphs/per leaf	Tomato	USA	Perring et al. 2018

of 6–8 whitefly, *Bemisia tabaci*, or 20 nymphs/leaves is for initiating sprays of insecticide against cotton whitefly (Sukhija et al. 1986, Butter and Kular 1986). The adult counting is feasible up to 1000 hours in the morning as the adults are not very active in the morning hours. This counting procedure is appropriate for surveillance study and the timing of control operations. The economic threshold level based on symptoms of damage has also been worked out and it is much easier to follow and is accepted at the farmers' level. The egg and nymphal counts are taken from small areas on the leaf. The eggs and nymphs are equally distributed in all the four quarters of the leaf. It is cumbersome to take the whole leaf for making counts of immatures and is time-consuming as well. The application of growth regulators based on nymphs (0.1–1.0 nymphs per leaf disc) adults (3–5 adults per leaf) is followed in cotton. The tomato crop, in general, does not suffer due to the attack of whitefly, if they are not acting as a vector of plant viruses. The economic threshold on tomato has been determined based on the population of nymphs (5 nymphs/ 10 leaflets) and adults (1 adult/plant and 1 adult/leaflet) (Schuster and Smith 2015). The tomato crop rarely suffers from the attack of whiteflies (Schuster 2004). The uneven ripening of fruit sometimes inflicts loss in the quality of produce and there is a need to tackle this problem. The desired level of whitefly population is yet to be determined to estimate the loss in quality of produce in tomatoes (Nauen et al. 2014). For the practice of recording data on adults of whiteflies, the majority of records are obtained normally through the sticky traps placed about 10 cm above the canopy of the plant. Additionally, sampling helps prevent the excessive use of pesticides as the thresholds are available for cotton, soybean, pulses and cassava crops. With the availability of the threshold, excessive use of pesticides can be mitigated. The judicious use of insecticide is helpful in preventing the ill-effects of pesticide. In case, the population of vector whitefly is to be recorded, it will not serve any purpose because the threshold level is as low as a single insect in vectors. For the control of the whitefly vector, it is desirable to spray crops with pesticides on the appearance

of the vector in the field. The pesticide for vector control should have a knockdown and kill the whitefly within 15–30 minutes as the vector whitefly requires the same amount of time to acquire/inoculate the viruses in the phloem.

The chapter engages with important information pertaining to the sampling of whiteflies. As of now, it is not clear whether adult, nymph or egg count is the most appropriate sample unit, they are the most important factors for obtaining dependable results through sampling. The compiled information suggested that the adults, eggs and nymphs can be sampled through inspection. It is more appropriate than the sampling done through sticky traps. Within the plant, the population of different stages of whitefly is done depending upon the abundance of population. The eggs and adult counts from the upper canopy upto 5th node from the top and nymphs from the middle in cotton are appropriate. For need based application of pesticides the economic thresholds available for different crops are presented. The ways and means are also suggested to count the population of whitefly depending upon the purpose of population estimation.

CHAPTER 5

Pesticides

Resistance and Resurgence

◇◇◇

5.1 Pesticides

It is an established fact that 20,000 insect pests species are known to destroy one third of food production annually throughout the globe and that amounts to 10,000 million dollars in India (Khan et al. 2015). Several pesticides belonging to different groups were introduced in the market and liberally applied against insect pests to get quick relief (Kranthi 2007). The dependence on pesticides in India was comparatively higher than on any other place in the world over. According to one report on consumption of pesticide from India showed about 30% of the total quantity was against bollworms (Prasad et al. 2009). Among the bollworms, *Helicoverpa armigera* was the dominant species in cotton crop ecosystem that developed resistance to most pesticides. The balance in the insect population in nature is maintained by natural enemies as long as this process is free from interference by human beings. However, the interference of man cannot be avoided in crop agroecosystem as the insect pests inflict enormous losses and their management becomes mandatory. In such a situation, the application of pesticides cannot be avoided. The broad spectrum pesticides were available at that time. As these pesticides mostly have a broad target spectrum, they eliminate the pest's natural enemies while helping with the management of the target pest species. Thus, the continuous use of toxic insecticides causes the mortality of natural enemies. With the result that the resurgence of insect pests becomes apparent, the crop yields start dwindling down, and the indiscriminate use of pesticides becomes common. This leads to the continuous and excessive use of recommended insecticides but the overdoses, increased rounds of spray, indiscriminate use without label claim, unwanted mixtures, repeated use of a single group of pesticides are the practices liberally taken up by the farmer against the recommendation. The large scale of use of such malpractices erodes the very principles of pesticide application, turning this normally effective anti-pest strategy into a mockery.

Pest resurgence is a common occurrence in crop ecosystems. The reasons for the resurgence in the whitefly population are many but the predominant one is pesticide-induced resistance. To get rid of resistance problem, biopesticides were introduced. It includes repellents, deterrents or toxicants, insect growth regulators, chemosterilants, attractants, volatile chemicals and biocontrol agents. The cotton crop ecosystem is quite complex and has thus been taken up by scientists to assess the situation and explore the causes behind the alarming increase in the population of whiteflies. Cotton crop remains the more or less the largest consumer of pesticides throughout the globe. The pyrethroids were introduced for use against pests world-wide in the 1970s. But being unstable, their persistence was short and more number of sprays were mandatory to manage pests. It was not a viable preposition. The pyrethroids were made photostable with the long persistence and again introduced in nineteen eighties. The indiscriminate and injudicious use of synthetic pyrethroids (cypermethrin, deltamethrin, permethrin, alphamethrin, and fenvalerate) was first initiated in cotton and led to the widespread resurgence of whiteflies (Anonymous 1989). In the absence of the natural enemy (due to pesticides) complex, the resurgence process accelerated even faster. The heavy pesticide use on cotton in Punjab, India, continued unabated. Following this, abrupt changes in the insect pest fauna were experienced in the cotton crop ecosystem there. Similar trends were noted at many other regions/locations.

For example, in India, the state of Punjab used to harvest around 10–12 lakh bales of cotton from an area of about 7 lakh hectares. Amongst the bollworms, *Helicoverpa armigera*—a key pest of cotton—developed resistance to most pesticides recommended for use on cotton, that is, those based on chlorinated hydrocarbons (endosulfan), organophosphates (triazophos, monocrotophos, quinalphos, phosphamidon dimethoate, phosalone, and carbamates (carbaryl)). Crop failures became common in the region. The farmers generally followed the pattern of 6–8 rounds of spray which soon reached 10–12 rounds of spray. The farmers also resorted to the use of un-recommended mixtures of pesticides. In the year 1981, a new group of potent pesticides called synthetic pyrethroids was recommended by the state of Punjab but their availability was made late in the season during 1983 (the year of crop failure due to the attack of *Helicoverpa armigera*). The farmers just realized one bale of lint/ha (one bale = 170 kg) which was highly unprofitable. This situation had a devastating impact on the farmers, who were demoralized and started thinking of abandoning cotton cultivation in the region. Despite the efforts of the department of Agriculture, the area under cotton cultivation dwindled down to around 5 lakh hectares. However, with the introduction of pyrethroids, the farmers were able to realize a reasonably good yield for about a decade, till the early nineties. And then an outbreak of whitefly *Bemisia tabaci* and mealy bugs during this period was experienced. After the strenuous efforts of scientists, the mealybug was brought under control. But there was no weapon to tackle the menace of the whitefly. It is important to mention at this point that hitherto whiteflies remained a minor pest till the middle of the twentieth century and the research work on this pest was scanty. For example, the initial research work carried out by Husain and Trehan in 1930s and 1940s was subsequently abandoned with the termination of schemes. The first major outbreak of whiteflies was only experienced in the late 1970s/early 1980s. The use of potent pesticides like fenvalerate, cypermethrin, alpha-cypermethrin, deltamethrin, and permethrin, etc. against the bollworms had caused a large-scale resurgence of sucking pests, especially the whiteflies (Dhawan and Saini 1998) throughout the country. In the absence of tolerant varieties of cotton and potent pesticides, the losses continued. Meanwhile, the neonicotinoids came to the rescue and provided relief. Soon these pesticides faded away due to the development of resistance in insects and the danger they posed to the beekeeping industry (Henry et al. 2012). Subsequently, insect growth regulators (IGR) were introduced to effectively control this pest. However, the insect pests have developed resistance to these chitin inhibitors too. The build up of whiteflies continued unabated and there was an enormous spread of whitefly. It has developed resistance to many chemicals belonging to different groups of pesticides with the result,there was a resurgence in pest population. The resistance is slightly different from resurgence. Here, the pesticide no longer effectively controls a high number (> 90%) of individuals in an insect population. In the year 2002, Bt cottons were introduced and proved effective against bollworms. With the introduction of Bt, pesticide use got reduced up to 40%. However, by removing the protection of pesticides, the debilitating impact of sucking pests was experienced. Further, the Bt cottons were also declared as more suitable for harboring a higher population of whiteflies as compared to conventional varieties of cotton (Atta et al. 2015).

The resurgence of whiteflies has been attributed to many other insecticides. For example, the resurgence in the whitefly population due to synthetic pyrethroids has been reported by Butter and Kular (1999). The synthetic pyrethroids caused the mortality of natural mortality factors. During this period, new molecules too were used for the management of whitefly and those included alphamethrin, flucythrinate, and fluvalinate (Butter and Sukhija 1987). The excessive use of these new pyrethroids has also fallen in the same category (Butter et al. 1989). To manage the whitefly menace and other insect pests in the system, several new pesticides were evaluated and recommended for use on cotton, which included heptenophos, acephate, phosphamidon, endosulfan, quinalphos, ethion, methyl parathion, triazophos, amitraz, and oxydemeton methyl. Besides, new groups of pesticides called, neonicotinoids and insect growth regulators were tested and these have shown great promise against whiteflies (Kular and Butter 1991, Sukhija et al. 1989). Most of these have been included in the recommendation (Table 5.1). Furthermore, the excessive use of some of these pesticides (monocrotophos and acephate) late in the season has led to an increase in the whitefly population (Dhawan et al. 2000). The use of systemic pesticides like orthene (acephate), metasystox (oxydemeton methyl), dimecron (phosphamidon), monocil (monocrotophs), zolone (phosalone), hostathion (triazophos), etc. have added to the succulency of cotton crop late in the season; causing the build-up of population and the outcome was the resurgence of whitefly (Table 5.2). Some of these pesticides recommended/advocated (for example, triazophos) to manage whiteflies on cotton also caused a large scale resurgence of mites (*Hemitarsonemus latus*) (Sukhija et al. 1989). To overcome this problem of mite infestation, the restricted use of the pesticide has been advocated in cotton (Butter et al. 1989).

An increase in the population of whiteflies in countries like Egypt, Sudan, Syria, Thailand, Israel, Zimbabwe, China, Pakistan, and India is attributed to the increased use of synthetic pyrethroids. In India, the over dependence on synthetic pyrethroids

Table 5.1. Pesticides evaluated/recommended against whiteflies.

Whitefly (Species)	Chemical name	Class of chemical	Location (State/country)	Source(s)
Bemisia tabaci/ Tomato	Imidacloprid & Profenophos Cypermethrin	Neonicotinoid	India (Bihar)	Jha and Kumar 2017
Bemisia tabaci	Acetamiprid	Neonicotinoid	Israel, Saudi Arabia	Horowitz et al. 1998, Wafaa 2011
Bemisia tabaci/ Brinjal	Thiacloprid	Neonicotinoid	India (Uttar Pradesh)	Kumar et al. 2017
Bemisia tabaci & Trialeurodes vaporariorum/ General Crops	Nitenpyram	Neonicotinoid	China	Liang et al. 2012
Bemisia tabaci	Thiamethoxam	Neonicotinoid	Saudi Arabia	Wafaa 2011
Bemisia tabaci	Clothianidin	Neonicotinoid	China	Fang et al. 2018
Bemisia tabaci Trialeurodes vaporariorum	Spiromesifen	Ketoenols	Germany	Bretschneider et al. 2003
Bemisia tabaci; Trialeurodes vaporariorum	Buperfezin (IGR)	Cyclic Ketoenols	Germany	Bretschneider et al. 2003
Bemisia tabaci: Trialeurodes vaporariorum	Azadirachtin	IGR	Thailand	Kumar and Poehling 2007
Bemisia tabaci	Nuvaluran Pyriproxyfen Chlorfluazuron Teflubenzuron	Insect Growth Regulators (IGR)	Israel	Ishaaya et al. 2003
Trialeurodes vaporariorum	Flonicamid	Pyridinecarboxamide	USA	Morita et al. 2014
Bemisia tabaci Trialeurodes vaporariorum	S-Kinoprene	Juvenile hormone mimic	USA (Florida)	Anonymous 2019
Bemisia tabaci	Buprofezin Lufenuron	Benzoylureas (IGR)	Pakistan	Gogi et al. 2006
Bemisia tabaci Trialeurodes argentifolii	Fenopropathrin	Pyrethroid	USA (Florida)	Hoelmer et al. 1991
Bemisia tabaci	Dinotefuran	Neonicotinoid	China	Qu et al. 2017
Bemisia tabaci	Cyfluthrin	Pyrethroid	India (Andhra Pradesh)	Deepak et al. 2017

Table 5.1 Contd. ...

...Table 5.1 Contd.

Whitefly (Species)	Chemical name	Class of chemical	Location (State/country)	Source(s)
Bemisia tabaci, and other Species	M-pede/Dis-X/Aza-DIRECT/Azatrol/ MoltsX/Mix	Botanical Insecticide (Azadirachtin) (NSKE)	India & World Wide	Jha and Kumar 2017
Bemisia tabaci/ tomato	Mineral oils	Physical suffocation of adults	India (Punjab)	Butter and Rataul 1973
Bemisia tabaci and other Species	Insecticidal soaps (M-pede)	Washing of away of protective layer and breaking of cell walls	USA (Florida)	Butler et al. 1993
Trialeurodes vaporariorum Bemisia tabaci/ Tomato	Chlorantraniliprole	Anthranilic diamides	China	Fang et al. 2018

Table 5.2. Pesticide-induced resurgence in whiteflies.

Pesticide	Class	Reaction	References
Chlorpyriphos	Organophosphate	Resistance	Singh et al. 1998
Acephate	Organophosphates	Resurgence	Singh et al. 1998, Prasad et al. 1993
Monocrotophos	Organophosphates	Resurgence	Singh et al. 1998, Prasad et al. 1993 Dhawan et al. 2000 Basu 1986
Quinalphos	Organophosphates	Resistance	Prasad et al. 1993, Singh et al. 1998
*Triazophos	Organophosphates	Resistance	Prasad et al. 1993
Dimethoate	Organophosphates	Resistance	Prasad et al. 1993
Phosphamidon	Organophosphates	Resistance	Singh et al. 1998
Phosalone	Organophosphates	Resistance	Prasad et al. 1993
Carbaryl	Carbamates	Resistance	Prasad et al. 1993
Fenvalerate	Synthetic pyrethroids	Resurgence	Dhawan et al. 2000, Basu 1986
Cypermethrin	Synthetic pyrethroids	Resurgence	Dhawan et al. 2000
Deltamethrin	Synthetic pyrethroids	Resurgence	Singh et al. 1998, Dhawan et al. 2000 Basu 1986
Fluvalinate	Synthetic pyrethroids	Resurgence	Dhawan et al. 2000, Basu 1986
Fenpropathrin	Synthetic pyrethroids	Resurgence	Dhawan et al. 2000, Basu 1986
Endosulphan	Chlorinated	Resistance	Prasad et al. 1993
Imidacloprid	Neonicotinoids	Resurgence	Ghosal and Chaterjee 2018
Dinotefuron; Clothianidin	Neonicotinoids	Resurgence	Ghosal and Chaterjee 2018

* Resurgence of *Bemisia tabaci/Hemitarsonemus latus* in cotton (Butter et al. 1997, Butter et al. 1989).

for control over bollworms increased the problem of whitefly resurgence in cotton. Jayaraj (1987) observed an outbreak of whitefly with twelve rounds of application of organophosphates and synthetic pyrethroids. The resurgence of whitefly with repeated application of fenopropathrin, fenvalerate, cypermethrin, deltamethrin, and fluvalinate was apparent (Shelka et al. 1987, Natarajan and Surulivelu 1986, Reddy et al. 1986b). The farmers resorted to mixtures of synthetic pyrethroids and organophosphates. Regu and coworkers (1990) concluded that the tank mixing of chlorpyriphos with cypermethrin and fenvalerate resulted in a heavy build-up of the whitefly population and the build-up increased with the increase in sprays.

In Punjab, a heavy buildup of whitefly was observed with the repeated use of fenvalerate, cypermethrin, and deltamethrin (Butter and Kular 1999, Dhawan and Simwat 1997). However, alphacypermethrin did not cause whitefly resurgence (Dhawan and Saini 1998). The resurgence ratio was maximum after the sixth spray and was much higher in cypermethrin applied @ 100 g a.i./ha than in other treatments. The regular liberal use of pesticides was prevalent in cotton cultivation, which paved the way for flaring up of sucking pests on cotton in Punjab while managing the bollworm complex of cotton. The sucking pests (including whiteflies) remained exposed to pesticides and this led to resistance in pests and the resurgence of sucking pests (Butter and Kular 1999). Additionally, an experiment with tank mixture of Humic acid (@0.1%) and fenvalerate (@50 gm a.i./ha) was conducted on cotton, leading to a significantly higher cotton yield (1914 kg/ha) as opposed to the 1041 kg/ha in control (Butter and Kular 1999). Both these products showed positive results in preliminary experimentation. However, a detailed study is needed on these two products vis-a-vis Humic acid and deltaphos and their use against whiteflies.

An earlier study made elsewhere in India also reported the resurgence of whitefly attacks in cotton cultivation (Sundaramurthy 1992), while in another instance, similar information pertaining to the repeated use of synthetic pyrethroids, cypermethrin, and deltamethrin against bollworms on cotton in Thailand caused the resurgence of whitefly, *Bemisia tabaci* (Gennadius) (Wanghoonkong 1981). The continuous use of pyrethroids to control tobacco budworm and pink bollworm in south India was also instrumental in whitefly resurgence. The unique system of parthenogenetic reproduction (haplodiploid/arrhenotokous) without males has made the reproduction process of the whiteflies fast as compared to other insect fauna. California has already experienced an increase in the population of whiteflies (Johnson et al. 1982).

5.2 Pest Build-up

The pest build-up in a crop ecosystem occurs unabated due to the following factors. These factors are related to climate and varieties and cultivation of crops.

Climate Related Factors:

- Effective monitoring and timely surveillance was almost lacking in the system on account of insect status as the whitefly was considered a minor pest. The pest also appears in a cotton crop late in the season, when the crop is already faced with resurgence and pesticide resistance problems.

- The climate change with global warming over the years was in favor of the buildup of this pest. The temperature has risen by 5–6°F over the last forty years throughout the globe.
- The highly favorable warmer weather for whiteflies is still continuing to occur and winters are becoming shorter and milder. Additionally, the early termination of monsoon is responsible for an increased dry spell during the crop season which is always in favor of whitefly multiplication. The cultivation of cotton under rainfed conditions too inflicts heavy damage on the crop as it is highly favorable for the whitefly in nature.

Pesticide Related Factors:

- The pest build-up was mainly due to pesticide-induced resistance and resurgence. Additionally, the unique and fast system of parthenogenetic reproduction (haplodiploid/arrhenotokous) is inbuilt in whiteflies which is also responsible for heavy build-up (Walker et al. 2010).
- The excessive and indiscriminate use of pesticides on several crops has led to resurgence of whiteflies (Dhawan and Saini 1991). Their natural enemies were wiped out from the crop ecosystem and the whitefly increased in the absence of natural mortality factors. In one of the studies carried out in India, researchers saw the resurgence of the mite population in triazophos-treated cotton in the field after four sprays while this chemical in question controlled the mite population in the laboratory (Butter et al. 1997, Dhooria and Butter 1990, Butter et al. 1989). The heavy build up of whitefly in cotton could be attributed to the mortality of natural mortality factors.
- The whiteflies did not confine themselves to cotton but continued to increase on other crops such as vegetables (brinjal, tomato, cucurbits, etc.), pulse crops, soybean, and many more throughout the year. The repeated use of the same pesticides without label claim continued throughout the year. The repeated exposure of whitefly to almost the same insecticides in this period enhanced the magnitude of pesticide resistance and build-up of the whitefly population.
- The hidden infestations of whiteflies due to their sessile nature which leads them to rest on the underside of the leaf surfaces that remain free from direct contact to insecticides till the appearance of black sooty mold. In such a situation, the build-up continues unchecked.
- The most important factor is the development of resistance in whiteflies to very potent chemicals like insect growth regulators and neonicotinoids, etc. (Kumar et al. 2017). The availability of safe and selective insecticides with recommendation/valid permit (label claim) were lacking, Efforts are made to identify the safe and selective pesticides with different modes of action to mitigate the whitefly problem.
- The effective spray technology to be used on whiteflies was deficient. As mentioned earlier, the sessile nature of the pest on the lower leaf surface helps it evade contact with pesticides. The spray techniques for whitefly and other pests

remained the same despite the great differences in the behavior of pests in a crop ecosystem.

- The polyphagous nature of whitefly has contributed towards its quick build-up on weed hosts act as reservoir hosts/overwintering hosts. These hosts ensure the regular supply of food to whiteflies throughout the year (Atta et al. 2015). In addition, many favorable or preferred crops harbor whiteflies in the vicinity of the main crop or follow the crop shortly after.
- The international trade based on high-value crops through human involvement also helps in the fast spread of this pest.

Varietal Factors:

- The cultivation of Bt kinds of cotton without a pesticide umbrella or substantial reduction in pesticide use has created conditions congenial for the build-up of whiteflies. The introduction of Bt cotton with the Bt gene did not require pesticide sprays to manage the dreaded bollworms. Thus, the pesticide quantity on cotton was reduced drastically but it has added to the build-up of sucking pests. The increase in the build-up of the population of sucking pests continued unchecked, including whiteflies (Krishna and Qaim 2012). As a result, the multiplication of whiteflies continued to increase unchecked in cotton in the absence of the pesticide umbrella. Further, Bt cottons are more favorable to the build-up of whitefly populations than the conventional varieties (Maharshi et al. 2017).
- It is the generalized practice of cultivation of cotton in the citrus orchards in the traditional areas of south western districts of cotton in Punjab (India) that further adds to the build-up of the population. The enclosure-like structures create congenial conditions for the heavy build-up of whiteflies. Unfortunately, it leads to the creation of a great store of a whitefly population from where the pest spreads to other crop systems.
- The greenhouse cultivation of crops is favorable for whiteflies and the technology is on the increase, creating the possibility for more infestations. It has become more important as farmers have resorted to organic farming without the use of pesticides. The whitefly is basically a pest of screenhouse and the whitefly buildup is many times more in such a situation.
- The continuous cultivation of highly susceptible, hairy and broad leaf cotton varieties was a common feature in all the cotton-growing areas as these varieties were advocated to check the menace of cotton jassid—a key pest of cotton in Punjab. However, the monoculture of highly pubescent cotton varieties was the common practice in Punjab but unfortunately these varieties were highly susceptible to whitefly (Butter and Vir 1989). It is, therefore, important to develop varieties resistant to both the cotton leafhopper and whitefly species of insect pests.
- The cultivation of preferred hosts/crops (brinjal, okra, cucurbits, melons, potato, beans, cape gooseberry, etc.) in or around or near the target crops further boosts the whiteflies. The cultivation of okra as an inter-crop followed in the cotton belt of Punjab is common. The cultivation of preferred hosts like brinjal throughout

the year has added to the whitefly population as the brinjal is the most favored host of the whitefly (Butter and Rataul 1977).

• The long migration and wide and fast spread of pest through air currents is an important factor to make the whitefly attain pest status. The factor is more critical in areas where whitefly is acting as vector of dreadly plant viruses. The recent introduction of Cotton Leaf Curl Virus. (Begomovirus) vectored by whitefly has recently been introduced in India which is worrisome.

• The unchecked growth of weeds such as *Euphorbia* sp., *Solanum* sp., *Lantana camera, Hibiscus esculentus, Ipomoea cardofana, Duranta* sp., *Abutilon grandifolium,* etc. help the whitefly to multiply. These hosts help the whitefly in the carryover of pest from weeds to cotton and vice versa (Atta et al. 2015).

Cultivation practices:

• Stray cotton plants/ratooning of cotton is not a regular feature but does add to a population increase of whitefly at some locations. At the end of the cotton season, after the last picking of cotton, the cotton sprouts provide palatable and soft substratum which whiteflies feed on and thus leads to a further increase in their numbers.

• Staggered sowing provides a soft substratum to whiteflies over a longer period of time that is ideal for the build-up of whitefly. Cotton sown in the first week of August suffers more than the crop planted in the first week of September in Sudan. Similarly, the cotton sowing done in May suffers heavily from the vagaries of whiteflies in north India.

• The large scale and prolonged use of synthetic pyrethroids is responsible for the resurgence of whiteflies (Butter and Kular 1999). It has been demonstrated that taking fenvalerate as representative of the synthetic pyrethroids used on cotton, the mortality of natural enemies of whitefly due to synthetic pyrethroids could be the reason for the large-scale resurgence of whitefly (Table 5.2)

• Patchy cultivation of cotton on sandy soils poor in organic matter paved the way for a further build-up of whiteflies (Husain and Trehan 1942).

5.3 Pest Resurgence

A major limiting factor in the successful cultivation of crops throughout the globe is the attack of insect pests. The whitefly, being a polyphagous pest, devours a large number of crops. The whitefly, *Bemisia tabaci* Genn. (Hemiptera: Aleyrodidae) now enjoys the status of key pest and is rated as the most devastating insect species (Bennett et al. 2004, Viscarret et al. 2003, Brown et al. 1995, Bedford et al. 1994, Perring et al. 1993, Byrne and Bellows 1991). Outbreaks of this pest had been experienced in Sudan/Middle East/USA (Dittrich et al. 1985, Johnson et al. 1982) and in tropical/and subtropical countries (Australia, Canada, Japan, and The Netherlands) in 1970s/80s (Gerling and Mayer 1996). These incidents are enough to indicate the devastating impact of the whiteflies. The whitefly is responsible for severe damage amounting to hundreds of millions of dollars annually (Menn 1996, Naranjo et al. 1996). Damage is caused due to sap-sucking and exuding of honeydew

(Schuster et al. 1996), physiological disorders caused in the host (Yokomi et al. 1990), and the transmission of *Geminiviruses* (Idris and Brown 2004, Markham et al. 1996, Bedford et al. 1994).

The outbreaks of whitefly are attributed to climatic factors (Byrne et al. 1992, Jayaraj et al. 1986), cultivation practices (Byrne et al. 1992), and the indiscriminate use of broad-spectrum insecticides-induced resistance (Prabhaker et al. 1989, Dittrich et al. 1985, Prabhaker et al. 1985), and elimination of natural enemies due to use of toxic insecticides (Eveleens 1983). At many locations, the resurgence of sucking pests has been attributed to the indiscriminate use of synthetic pyrethroids (Butter and Kular 1999, Patil et al. 1990, Jayaraj and Regupathy 1986, Jayaraj et al. 1986), and through homozygosis (change in the physiology and behavior of the pest) (Dutcher 2007, Abdullah et al. 2006). Based on experience, it appears that the outbreaks are unlikely to stop shortly. The serious attack of whitefly in USA (Prabhaker et al. 1985), Israel (Perry 1985, Byrne et al. 1994), Guatemala, Sudan, Turkey (Dittrich et al. 1990a), Central America, Ethiopia, Mexico, Peru (Martinez-Carrillo 2006), Cyprus, UK, Yemen (Byrne and Devonshire 1993), Pakistan (Ahmad et al. 2001, Ahmad et al. 2000, Ahmad et al. 1999, Cahill et al. 1995, Cahill et al. 1994), and India (Butter and Kular 1999, Patil et al. 1990, Jayaraj et al. 1986) have been demonstrated. The resistance ratio of three insecticides, organophosphate (triazophos), cyclodiene (endosulfan), and neonicotinoid (imidacloprid) were estimated from six states in India, namely, Guntur (Andhra Pradesh), Coimbatore (Tamil Nadu), Kolar (Karnataka), Ludhiana and Bathinda (Punjab), Sri Ganganagar (Rajasthan) and Sirsa (Haryana) using the standard method (Henderson and Tilton 1955, Dutcher 2007):

$$\text{Resurgence (\%)} = \left(1 - \frac{(T_a \times C_b)}{(T_b \times C_a)}\right) \times 100$$

where:

Ta – Whitefly population on treated leaf disc after treatment (%)
Cb – Whitefly population control leaf disc before treatment (%)
Tb – Whitefly population on treated leaf disc before treatment
Ca – Whitefly population on control leaf disc after treatment

An attempt to mitigate the build-up of whitefly in cotton using Humic Acid (HA) and deltaphos (a mixture of synthetic pyrethroid–deltamethrin and orgaophophate-triazophos) showed positive results. It is also important that these measures be tested on other crops. However, thorough investigations are needed in this direction.

5.4 Pest Resistance

The whiteflies have developed resistance to many pesticides even though these pesticides have not been used entirely for the management of whiteflies. The pesticide use was mainly for the control of key pests in a crop agroecosystem. The whitefly has developed resistance to BHC, endosulphan, dimethoate phosalone, acephate, monocrotophos, quinalphos, and carbaryl (Prasad et al. 1993). The resistance to pesticides in many other insect species has also been recorded. The whitefly has developed resistance to insect growth regulators (IGR, S) (Buprofezin-Telus) and Neonicotinoids (Imidacloprid-Marathon/Meri Acetamiprid-Trista Thiomethoxam-Flagship) (Kumar et al. 2017) (Table 5.3). It is a fact that whitefly biotypes B and Q now possess a high degree of resistance to IGR (Pyriproxyfen-Distance). The whiteflies are now practically immune to this growth regulator (Horowitz et al. 1994). The major reasons are mainly due to the occurrence of the haplodiploid type of reproduction in whiteflies which involves the quick selection and fixation of genes. This development of resistance is the occurrence of a continuous breeding cycle on the treated substratum (Denholm et al. 2008).

High-value crops are host plants of whiteflies and are also traded internationally. These factors are thus responsible for the rapid spread and multiplication of whiteflies. To mitigate the growing menace of a heavy build-up of whitefly, readymade mixtures of insecticides such as deltaphos are available and should be tried in cotton crop to control whiteflies and bollworms. On testing, the deltaphos has shown great promise against both the insect species (Butter et al. 1992c).

As can be seen from examples in various instances earlier, cotton is extremely essential to highlight the case of pesticide resistance. Cotton, a long duration crop (4–6 months), requires a protection umbrella usually for a long period; the pest problems it faces are generally quite complex. The pesticide consumption on cotton is the highest as compared to other crops. According to one report, the cotton crop consumes over 80% of the pesticide in India and more than 50% through the world over. In Punjab, India, the cotton crop was highly vulnerable to the attack of insect pests (Fig. 5.1). Initially, the cotton bollworms (*Helicoverpa armigera/Pectinophora gossypiella/Earias insulana* and *Earias vittella*) and cotton jassid (*Amrasca bigutulla bigutulla*) were the key pests but the whitefly along with aphid were categorized as minor pests appearing late in the crop season. Nobody would ever dream that whiteflies from the category of sucking pests would be so devastating with a tag of a key pest of cotton. Being a pest of minor importance, scientists did not pay much attention to this pest. Now the whitefly *Bemisia tabaci*, being a polyphagous, cosmopolitan, and invasive cryptic species, is considered as the most devastating insect, possessing the capability to develop resistance to pesticides. The reports of the development of resistance in insects to pesticides (DDT) have been made public. The first case of the development of resistance in insects (housefly) to pesticides was reported about pests affecting public health. The pests of crops followed suit and a large number of published reports highlighting the development of resistance in insect pests was later made available. Simultaneously, the reports of development of resistance in whitefly to more than forty insecticides have been documented (Naveen et al. 2017). The earlier study illustrated that the hazard rate/efficacy of the insecticide is not dependent on the dose-effect but also on mortality time. Thus, the log-rank test was better than the IRAC method as reported by scientists while studying imidacloprid against *Bemisia tabaci* (Naveen et al. 2012).

Bemisia tabaci possess unique genetic qualities linked with insecticide resistance, that is, efficient detoxification mechanism and virus transmission. These characteristics led the whitefly to be a highly invasive polyphagous pest and efficient vector of plant pathogens. Further, these characteristics helped in resolving

Table 5.3. Pesticides resistance in whiteflies/biotypes.

SNo.	Insecticide	Location	Remarks/Level of resistance	Reference
1	Imidacloprid, Thiamethoxam	Brazil	Neonicotenoids	Silva et al. 2009
2	Deltamethrin, Bifenthrin, Chlorpyriphos, Dimethoate, Acetamiprid, Thiamethoxam, Endosulfan, Pymetrozine	Burkina Faso	Resistance to insecticide is due to the presence of invasive species	Houndete et al. 2010
3	Sulprophos, Fenthion, Malathion, Parathion, DDT, Permethrin	California (USA)	*Bemisia tabaci*	Prabhaker et al. 1988
4	Imidacloprid, Thiamethoxam, Acetamiprid	China	Q-biotype	Luo et al. 2010
5	Imidacloprid, Thiamethoxam, Alpha-cypermethrin, Fipronil, Spinosad	China	B and Q, biotypes	Wang et al. 2009
6	Synthetic Pyrethroids, Neonicotinoids	China	> 1000 fold resistance (SPs)	Ma et al. 2007
7	Imidacloprid, Thiamethoxam, Bifenthrin	Cyprus	Carboxyl esterase enzyme (imidacloprid)	Vassiliou et al. 2011
8	Carbosulfan, Aldicarb, Profenofos, Cypermethrin, Lambda-cyhalothrin	Egypt	All populations were susceptible to pyriproxyfen	Kady and Devine 2003
9	Imidacloprid	Germany	B- and Q-biotypes monooxygenases	Nauen and Denholm 2005
10	Spiromesifen	Germany	*Trialeurodes vaporariorum*	Karatolos et al. 2012
11	Neonicotinoids	Greece	*Trialeurodes vaporariorum*	Pappas et al. 2013
12	α-cypermethrin, Bifenthrin, Primiphos-methyl, Endosulfan, Imidacloprid	Greece/Crete	High-level resistance	Roditakis et al. 2005
13	Imidacloprid	Guatemala	B-biotype	Byrne et al. 2003
14	Cypermethrin, Endosulfan, Chlorpyriphos	India	*Bemisia tabaci* Low level (Chlorpyriphos)	Kranthi et al. 2002
15	Imidacloprid	Iran	Detoxification by cytochrome 450 monooxycorbogenase	Basij et al. 2016
16	Organophosphates	Israel	Acetylcholinesterase overexpression of carboxylesterase (B-biotype)	Alon et al. 2008
17	Pyriproxyfen	Israel	B- and Q-biotypes	Horowitz et al. 2005

Table 5.3 Contd. ...

...Table 5.3 Contd.

SNo.	Insecticide	Location	Remarks/Level of resistance	Reference
18	Imidacloprid	Israel	Cytochrome P450s monooxygenases, B- and Q-biotypes	Karunkar et al. 2008
19	Methidathion	Israel	Sweet potato Whitefly	Bloch and Wool 1994
20	Profenofos, Cypermethrin, Imidacloprid, Diafenthiuron	Malaysia	B- and Q-biotypes	Shadmany et al. 2015
21	Profenofos, Triazophos, Parathion-methyl Ethion	Pakistan	Low to high resistance	Ahmad et al. 2010
22	Dimethoate/Endosulfan Mixture and Mixture of Methomyl and Amitraj	Sudan	*Bemisia tabaci*	Ahmed et al. 1987
23	Monocrotophos, Carbofuran	Sudan, Turkey, Guatemala, Nicaragua	Mechanism of resistance is variable in four populations	Dittrich et al. 1990a
24	Imidacloprid, Pymetrozine	Turkey	Overexpression of Cytochrome P450 monooxygenase detoxification	Nauen et al. 2015
25	Bifenthrin, Fenopropathrin, Formothion, Triazophos, Buprofezin	Turkey	No resistance (IGR); Four strains (tested)	Erdogan et al. 2008
26	Imidacloprid, Acetamiprid, Thiacloprid, Thiamethoxam	Turkey	*Bemisia tabaci*	Satar et al. 2018
27	Neonicotinoids, Pymetrozine	UK	Overexpression of cytochrome P450 monooxygenase detoxification	Gorman et al. 2010
28	Profenofos, Cypermethrin	UK	Presence of multi resistance in all the three collections	Cahill et al. 1995
29	DDT, Malathion, Resmethrin	UK	*Trialeurodes vaporariorum*	Wardlow et al. 1976
30	Neonicotinoids	USA	B-biotype	Castle and Prabhaker 2013
31	Acetamiprid, Dinotefuran, Imidacloprid, Thiamethoxam	USA	Insecticide resistance	Prabhaker et al. 2005
32	Pyriproxyfen, Buprofezin, Mixture of Fenopropathrin plus Acephate, Imidacloprid, Thiamethoxam, Acetamiprid	USA	Q-biotype in New world	Dennehy et al. 2010
33	Neonicotinoids	USA	*Bemisia* B-biotype	Schuster et al. 2010

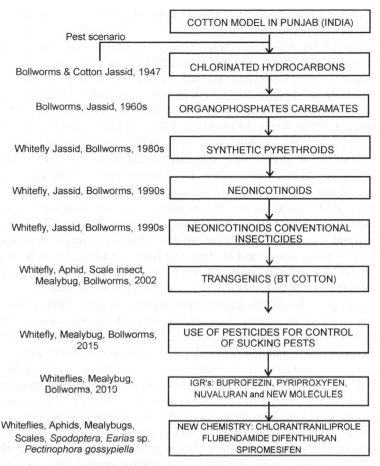

Fig. 5.1. Changing scenario of pests & pesticide use in punjab, India.

species complex and genetic information to manage the whiteflies and the spread of viruses (Chen et al. 2016). The whitefly *Bemisia tabaci* studied in the Indian subcontinent showed significant resistance in imidacloprid and thiamethoxam, monocrotophos and cypermethrin among the Asia-I, and cypermethrin, deltamethrin, and imidacloprid Asia-II-1 group. With the introduction of Bt kinds of cotton in India in 2002, the bollworms were brought under control but the sucking pests started increasing greatly in number from the year 2002. As a result, the pesticide usage had risen from 2374 MT in 2006 to 6372 MT in 2011 onwards to check these sucking pests (Gutierrez et al. 2015). Neonicotinoids pesticides continued to be used to manage whiteflies and cotton jassid and other minor sap-sucking pests in India (Kranthi 2012, Sethi and Dilawari 2008). These insecticides were also sprayed indiscriminately and injudiciously and the problem of resistance got further accentuated (Kranthi et al. 2002a, Kranthi et al. 2001).

Resistance in whiteflies to insecticides became widespread and occurred in various regions around the globe, including Brazil (Silva et al. 2009), Burkina Faso

(Houndete et al. 2010), China (Luo et al. 2010, Wang et al. 2010), Colombia (Cardona et al. 2001), Cyprus (Vassiliou et al. 2011), Egypt (Kady et al. 2003), Germany (Nauen and Denholm 2005, Alon et al. 2006), Greece (Roditakis et al. 2005), Guatemala (Byrne et al. 2003), India (Kranthi et al. 2001), Iran (Basij et al. 2016), Israel (Horowitz et al. 2002, Horowitz et al. 2005, Horowitz et al. 2008, Kontsedalov et al. 2012, Horowitz and Ishaaya 2014), Italy (Nauen et al. 2002), Malaysia (Shadmany et al. 2015), Nicaragua (Dittrich et al. 1990), Pakistan (Ahmad et al. 2015), Spain (Gorman et al. 2010), Sudan (Dittrich et al. 1990), Turkey (Erdogan et al. 2008), and USA (Cahill et al. 1995, Denholm et al. 1998, Prabhaker et al. 2014, Dennehy et al. 2010, Schuster et al. 2010, Castle et al. 2013, Prabhaker et al. 2014). The insect's resistance to conventional insecticides and a mixture of synthetic pyrethroids and conventional ones was recorded. The resistance caused by such combinations led to the outbreak of this insect species. The efficacy of conventional insecticides and new molecules was compared. The order of toxicity to whitefly was ethion > triazophos > acetamiprid > imidacloprid > thiamethoxam. Among different locations, ethion was found to be most toxic and imidacloprid was least toxic. The new insecticide, dinotefuran was found to be most effective followed by diafenthiuron, spiromesifen, and flonicamid (Singh 2017). No pesticide resistance was recorded in profenophos, ethion, triazophos, bifenthrin, and imidacloprid (Singh et al. 2001a). The detailed information collected on pesticide resistance showed widespread resistance to all categories of pesticides in organophosphates, chlorinated hydrocarbons, carbamates, synthetic pyrethroids, and new molecules with a different mode of action (Sharma et al. 2020). The resistance ratios with respect to insecticides were 7.2 in BHC (Prasad et al. 1993), 11.5 in DDT (Wardlow et al. 1972), 60 in endosulfan (Ahmed et al. 1987, Perry 1985), acephate 31.7 in acephate (Singh et al. 2001a), 10 in chlorpyriphos (Byrne et al. 1994), between 24–31 in dichlorvos (Wardlow et al. 1972), 236–311 in dimethoate (Dittrich et al. 1985), 6–100 in fenthion (Wardlow et al. 1972), 7.1 in malathion (Prabhaker et al. 1985), 750 in methamidophos (Cahill et al. 1994), and 41–78 in methyl parathion (Watve et al. 1976). Further, the resistance ratios more than > 1000 in profenophos (Cahill et al. 1996), 150–200 in monocrotophos (Dittrich et al. 1985), 9–92 in quinalphos (Dittrich et al. 1985), 38–54 in sulprofos (Prabhaker et al. 1985), 23–310 in triazophos (Erdogan et al. 2008), 20–310 in formothion (Erdogan et al. 2008), 2500 in cyfluthrin (Cahill et al. 1996), 170 in cypermethrin (Dittrich et al. 1990), 81–124 in deltamethrin (He et al. 2007), 315 in fenpropathrin (Prabhaker et al. 1985), 3.5 in fenvalerate (Prabhaker et al. 1985), 12–29 in permethrin (Prabhaker et al. 1985), 40 in imidacloprid (Prabhaker et al. 1985), 47 in buprofezin (Cahill et al. 1996), 10–32 in acetamiprid (He et al. 2007), 50–164 in thiamethoxam (Vassiliiou et al. 2004), and 6500 in pyriproxyfen (Devine et al. 1999) were recorded. The decision to include pesticides in IPM can be extracted from the list depending upon the level of resistance.

　　The resistance to traditional OPs, pyrethroids, and carbamates in bollworms *Helicoverpa armigera* (Hübner) is already known. The resistance in whitefly to pyrethroids was due to mutation of two para type voltage in gated sodium channel gene and in organophosphates due to mutation in acetylcholinesterase enzyme gene (Roditakis et al. 2006, Mouton et al. 2015). The information is quite useful to develop IPM. An other study carried with alfa-cypermethrin, imidacloprid and primiphos methyl demonstrated differences in all the three insecticides. The resistance in imidacloprid was less established and less persistent than the primiphos methyl. Low level of resistance in primiphos methyl rules out the possibility of cross resistance with organophosphates and synthetic pyrethroids (Roditakis et al. 2008). The resistance in *Bemisia tabaci* (Q-biotype) was recorded in almost all the insecticide groups except pyridaben and spiromesifen, which were highly effective against this insect in Spain (Fernandez et al. 2009). Investigations on insecticide resistance were done in five Mediterranean countries (Spain, France, Morocco/Greece/Tunisia) using molecular tests. A high level of resistance was noted in pyrethroids and organophosphates. Based on the differential reaction, three distinct groups (France, Spain, Morocco/Greece/Tunisia) were identified in *Tetraleurodes perseae* on avocado (Gauthier et al. 2014). In *Bemisia tabaci,* neonicotinoid cross-resistance (imidacloprid) in Q-biotypes from Spain and B-biotype from Israel was recorded (Rauch and Nauen 2003). The resistance to synthetic pyrethroids on account of increased ester highdrolysis and modification of target of acetylcholinistrase enzyme in organophosphates group of pesticides was confirmed in B-biotype (Byrne et al. 2000). The resistance to pesticides in cotton whiteflies was expected to develop at a much faster rate than the bollworms, the reasons for which are already elucidated. Efforts are being made to manage this resistance in whiteflies with various measures including use of refugia (Camere et al. 2009). The insecticide groups like organophosophates, neonicotinoids, carbamates, chlorinated hydrocarbons, and new molecules are used depending on the situation; that is, the selection of the pesticide is made depending on the situation, considering the resistance to a particular group. Care is taken to create a plant refuge to dissolve the resistance by allowing interbreeding between the population in the refuge area and the target areas to preserve the susceptible population with susceptible genes in the target crop area. To mitigate resistance, the refuge strategy is being advocated for a large number of insect species, including the whitefly (Carriere et al. 2012, Crowder and Carriere 2009). Keeping the refugia in resistant management programs is helpful as it significantly delays the appearance of resistance.

The other factor taken into consideration is the target stage of the pest. For example, if the availability of target stage (eggs) is in abundance, the pesticide must be selected from the growth regulators. The alternate use of pesticides and tank mixing is done carefully. The correct dose and application intervals are important and well taken care of to mitigate the resistance. The selective use of pesticides is important to preserve beneficials and is also an important area that is looked into while resorting to pesticide use. A field strain of *Bemisia tabaci* (Gennadius) in California, showed a reduction in resistance level to malathion, methyl parathion, sulprofos, permethrin, and DDT with the addition of synergists (Prabhaker et al. 1988). The operation was well organized by the insecticide resistance action committee to check further resistance in whiteflies.

Efforts are being made to examine the metabolic resistance in which the whiteflies arc unable to detoxify the new chemical, able to penetrate via the skin, and to modify the behavior of whiteflies so that these insects may not avoid the toxicant, enhance penetration of the pesticide, and control detoxification of pesticides. The use of insect-specific, short single-stranded DNA insecticides, in the future is likely as the intelligent insecticides developed with the interaction of scientists and manufacturers would offer safe, affordable, and effective control of the whitefly (Oberemok et al. 2015).

The resistance to pesticides in cotton whitefly was expected to develop at a much faster rate than in the bollworms (El-Latif and Subrahmanyam 2010, Srinivas et al. 2004, Kranthi et al. 2002, Kranthi et al. 2001, Armes et al. 1996). Cahill et al. (1996) reported resistance to monocrotophos and other organophosphate insecticides in *Bemisia tabaci* strains from the USA, Central America, Europe, Pakistan, Sudan, and Israel. However, high resistance levels to monocrotophos during 1992–1996 were lowered considerably by 2000 when the use of the product for whitefly control in Pakistan was reduced (Ahmad et al. 2002). Field strains of *Bemisia tabaci* collected from twenty-two cotton-growing districts across India exhibited a high level of resistance to methomyl and monocrotophos and moderate resistance to cypermethrin (Kranthi et al. 2002a).

Two putative species of *Bemisia tabaci* are the main focus of attention currently: the Middle East-Asia Minor I (that is MEAM1 or B-biotype) and the Mediterranean (MED or Q-biotype) species of the complex, which account for both of the major global invasion events recorded for the group (De Barro et al. 2011, De Barro and Ahmed 2011), particularly in the Nearctic and Western Palearctic regions (Panini et al. 2017, Li et al. 2012, Castle et al. 2010, Fernández et al. 2009, Roditakis et al. 2005). However, insecticide resistance studies on whiteflies are quickly increasing in the Eastern Palearctic and Indo-Malay regions, likely due to the enhanced problems detected with the current spreading of the MED species in China and India (Naveen et al. 2017, Zheng et al. 2017, Wang et al. 2017a, Basit et al. 2013, Yuan et al. 2012, Basit et al. 2011).

Three studies on whitefly resistance to insecticides in Neotropical America, two of them from Central America (Santillan-Ortega et al. 2011, Byrne et al. 2003), and a single study from South America (Silva et al. 2009) are reporting low to moderate levels of resistance to the neonicotinoids, imidacloprid, and thiamethoxam. High levels of lambda-cyhalothrin resistance in undetermined biotypes have been recorded in China (He et al. 2007). Further, the resistance to neonicotinoids is also frequent in populations of MED (or biotype Q) (Zheng et al. 2017, Wang et al. 2017b, Yuan et al. 2012), and a moderate level of resistance has been reported in Egypt and Israel in undetermined putative species (He et al. 2007, Kady and Devine 2003, Ahmad et al. 2002).

The occurrence of resistance to seven insecticides among Brazilian populations of putative whitefly species such as Middle East-Asia Minor I (MEAM1) (*Bemisia tabaci* B-biotype) was surveyed. The likelihood of the control failure of five insecticides registered for this species was also determined. Resistance was detected to the insecticides azadirachtin, spiromesifen, lambda-cyhalothrin, diafenthiuron, and imidacloprid. In contrast, the likelihood of control failure was low for diafenthiuron, and was high ($> 100\times$) to very high levels ($> 1000\times$) in all the rest of them and the overall efficacy was particularly low ($< 60\%$) and the control failure likelihood was high ($> 25\%$) and frequent (70%) for the bioinsecticide azadirachtin, followed by spiromesifen, and lambda-cyhalothrin (Dangelo et al. 2018). Among the neonicotinoids, thiamethoxam is still found effective against two species of whitefly (*Bemisia tabaci* and *Trialeurodes vaporariorum*) (Santillan-Ortega et al. 2011). It was the resistance to azadirachtin among MEAM1 Brazilian whiteflies which posed a serious threat to organic cultivation in the country. The whitefly management was, however, totally dependent on this product in vegetable crops (MAPA 2017).

Low chlorantraniliprole resistance among whiteflies was detected and demonstrated in populations of MEAM1 and MED from the United States and China (Xie et al. 2014, Caballero et al. 2013, Li et al. 2012). In West Africa, some field populations showed a significant loss of susceptibility to pyrethroids such as deltamethrin [resistance ratio (RR) 3–5] and bifenthrin (RR 4–36), to organophosphates (OPs) such as dimethoate (RR 8–15) and chlorpyrifos (RR 5–7) and neonicotinoids such as acetamiprid (RR 7–8) and thiamethoxam (RR 3–7). *Bemisia tabaci* was also reported resistant to pymetrozine (RR 3–18) and endosulfan (RR 14–30) (Houndete et al. 2010). In Pakistan, cotton and sunflower collections form four locations had similar resistance levels in *Bemisia tabaci* (Basit et al. 2013). The Vehari collections showed higher resistance to acetamiprid, thiacloprid, and nitenpyram compared with those of others. Average resistance ratios for acetamiprid, thiacloprid, and nitenpyram ranged from 5- to 13-, 4- to 8-, and 9- to 13-fold, respectively. The Multan and Vehari collections also exhibited moderate levels (9- to 16-fold) of resistance to buprofezin (Basit et al. 2013). Six populations of *Bemisia tabaci* Gennadius Q-biotype in Spain had low to moderate levels of resistance to azadirachtin (0.2- to 7-fold), buprofezin (11- to 59-fold), imidacloprid (1- to 15-fold), methomyl (3- to 55-fold), pyridaben (0.9- to 9-fold), pyriproxyfen (0.7- to 15-fold), and spiromesifen (1- to 7-fold) when compared with a contemporary Spanish Q-biotype reference population (LC50 = 2.7, 8.7, 15.2, 19.9, 0.34, 20.9, and 1.1 mg L^{-1}, respectively) (Fernandez et al. 2009). The report from Multan (Pakistan) also demonstrated the high level of resistance in dimethoate and deltamethrin (Ahmad et al. 2002). A moderate level of resistance in whitefly was apparent in acephate, monocrotophos, and dimethoate lambda-cyhalothrin, and bifenthrin. Another report indicating the development of resistance in whitefly to more than forty insecticides strands was presented (Naveen et al. 2017). Insecticide resistance mechanisms were assigned to mutations of the target protein, decreasing the affinity and increasing the detoxification of esterases enzymes, cytochrome P45-dependent monooxygenases, or glutathione S-transferase. The likelihood of resistance development to novel insecticides is predicted. The farmers of Burdwan, West Bengal (India), are already using imidacloprid, methyl parathion, dichlorvos, and phorate indiscriminately, without any recommendation, and without the necessary precautions (Banerjee et al. 2014).

5.5 Selective Use of Insecticides

The development of DDT in 1935 was known as the golden era in pesticide history. It was followed by organophosphates, carbamates, and synthetic pyrethroids. These pesticides could not sustain on account of the development of resistance in insects and safety to human beings. Subsequently insecticides namely neonicotinoids and insect growth regulators were developed and which was once again designated as the 'Golden Era' in pesticide history (Casida and Quistad 1998).

5.5.1 *Neonicotinoids and New Molecules*

The insecticides with systemic and translaminar action, pronounced residue, and a unique mode of action namely Imidacloprid, acetamiprid, thiamethoxam, thiacloprid/ clothianidin, and dinotefuran were identified (Elbert et al. 2008). The chemicals were highly potent against sucking insect pests and target specific, safe for non-target organisms, and least destructive to the environment. On account of these qualities, these insecticides were considered highly suitable candidates in integrated pest management (Jeschke et al. 2010). However, their toxicity to honey bees put a question mark on the prolonged use on cotton (Henry et al. 2012).

Flonicamid: Flonicamid (pyridine carboxamide) a novel systemic compound with a unique mode of action and has contact activity. Flonicamid effectively manages the pests, the whiteflies, and the leafhoppers in the cotton agroecosystem.

Pymetrozine: It is a systemic translaminar pesticide of benzoyl phenyl urea used as a soil drench at sowing and found effective against sucking pests (Ishaaya et al. 2003). Azomethrine is a pyridine-a chitin inhibitor (Fuog et al. 1998) and flonicamid (pyridine carboxamide), a novel systemic and contact compound with a unique mode of action are effective against whiteflies.

Insect Growth Regulators: It is a category that includes buprofezin, nuvaluron, and pyriproxyfen. Buprofezin is known to control the whitefly *Bemisia tabaci* through the suppression of embryogenesis and progeny formation while pyriproxyfen, being a mimic natural hormone, disrupts growth. It is highly useful against young instars and eggs of whiteflies and prevents younger instars from maturation into adults. Pyriproxyfen is a stomach and contact poison (skin penetration) which disrupts egg-laying and egg-hatching and prevents the young ones from attaining the adult stage. It does not act against adult whitefly.

Spiromesifen/Spirotetramat: These molecules are Ketones derivatives of Tetronic acids (Spiromesifen) and Tetramic acid (Spirotetramat) and are used on cotton and tea. The major benefit of these products is especially promising against all stages of the pest, whitefly (*Bemisia* spp.), on vegetables and fruits. Their safety towards natural enemies of this pest make these an important component of integrated pest management (IPM). Besides, these products offer excellent control of the management of the pesticide-resistant population of whiteflies. However, the resistance to spiromesifen and spirotetramat from the whitefly has been reported, and therefore, it is advocated for use only against susceptible populations of whitefly (Bielza et al. 2019). Another pesticide (Spyridon) effective against sucking pests and mites is likely to be available during the current year. It hails from the category of spiromesifen and spirotetramat insecticides (Muchhlebach et al. 2020).

Diafenthiuron: It is a pro-insecticide and acts on the energy-producing enzymes in the mitochondria after activation and causes immediate paralysis of the pest. The complete mortality occurs in 3–4 days and thus, it is highly suitable for IPM.

Diamides: Flubendiamide and chlorantraniliprole are effective against tissue borers and sucking pests (Sattelle et al. 2008, Lahm et al. 2009). Of these, chlorantraniliprole (Cyazypyr) inhibits lipid synthesis and creates problems in the development of eggs. It does not allow the whiteflies to resurge. Chlorantraniliprole is harmless to the parasitoid wasp species and is a highly suitable candidate in IPM. *Spodoptera littoralis, Helicoverpa armigera* (Hübner), and cotton whitefly nymphs of *Bemisia tabaci* are also manageable with this medium.

With regard to application of pesticides, the oil application through an electric machine for thorough coverage is significantly better than all other conventional and boom sprayers typically used in the industry. The use of chemicals is always considered better due to their effectiveness and low cost in pest management (Singh et al. 2000). The insect growth regulators like buprofezin 2-tert-butylamine-3-isopropyl-5-phenyl-3, 4, 5, 6-tetrahydro-2H-1, 3, 5-thiadiazole-4-one have been tried and provide good control over the pest. Applaud (IGR) has been advocated as a promising molecule as it showed great toxicity against the greenhouse whitefly *Trialeurodes vaporariorum* (Westwood) in all its developmental stages. The larval mortality (LC50) against the first, second, third, early, and middle fourth instar with applaud was 10.659, 0.482, 2.45, 11.6, and 15.7 ppm, respectively. Buprofezin gave significantly promising results as compared to chinomethionat, methidathion, and fenvalerate but did not cause the desired mortality of eggs, pupae, and adults (Yasui et al. 1985). The use of the conventional insecticide, triazophos, against whiteflies was effective but the repeated sprays of triazophos on cotton caused the resurgence of mites (*Hemitarsonemus latus*). It caused large scale mortality of natural enemies on cotton (Butter et al. 1989, Butter et al. 1997). However, the same insecticide gave promising control of *Bemisia tabaci* under laboratory conditions (Dhooria and Butter 1990). Under the laboratory conditions, the natural enemies were absent, thus, triazophos did control whiteflies in cotton.

This chapter highlights the use of homemade remedies and insecticides of plant origin to protect the crops against the invasion of herbivores. The introduction of DDT proved a panacea for all ills. Looking at its success, carbamates, chlorinated hydrocarbons, organophosphates were introduced. Soon these pesticides started posing serious problems of insecticide resistance, toxic residues, and the resurgence of non-target pests; all the ill-effects of pesticides were highlighted in the book, *Silent Spring* by Rachel Carson, creating awareness about these problems. Another group of pesticides, synthetic pyrethroids, was introduced which caused the resurgence of whiteflies. The neonicotinoids were introduced to check the menace of sucking pests. The large-scale resistance to neonicotinoids in whiteflies was another worrisome point. In the year 2002, Bt kinds of cotton were introduced to check the menace of bollworm. It gave a further boost to the whitefly build-up. The reasons for the build-up of whiteflies were mainly the parthenogenetic reproduction without males (haplodiploid arrhenotokous) due to their polyphagous nature, the cultivation of susceptible germplasm, a suitable climate, and the introduction of Bt cotton, to name a few. Currently, there are new molecules available, namely, spiromesifen, spirotetramat, chlorantraniliprole, flubendazole, difenthiuron, flonicamid, buprofezin, etc. which have a different mode of action and are safe for non-target animal fauna. These have been advocated to manage whiteflies in cotton in addition to the pesticide resistant management tactics recommended in Arizona and Georgia which have been detailed in this chapter. To get rid of resistance problem, biopesticides were introduced. It includes repellents, deterrents, toxicants, Insect growth regulators, chemosterilants, attractants, volatile chemicals and biopesticides/biocontrol agents.

CHAPTER 6
Host Plant Resistance

The evolutionary aspect of host plant interaction goes back about 350 million years ago to when man learned to protect plants against the invasion of herbivores. Around 1000 years ago, the art of selection of plants was based on fruit quality produce (sweet and robust fruit) as it was easy to select and recognize the trait. As indicated, man selected, for example, big-sized sweet berries and popularized the cultivation of such groups of plants. At that time man was ignorant about the science of plant breeding or genes controlling the size and flavor traits of fruits. It was the only technique the farmers knew to bring about the improvement in plants with respect to desired traits (Aharoni et al. 2004). The high-yielding but input-responsive berries were further distributed for large-scale cultivation. Soon the desired category of fruit plants occupied substantial areas, becoming monoculture cultivation of that fruit. With the adoption of the monoculture of sweet berries, the farmers started facing multiple plant-protection problems. The more complex insect pest problems forced farmers to look for more effective management techniques. Thus, with an emphasis on science, the Winter Majestic variety of apple resistant to *Eriosoma lanigerum* was developed in 1831. Before that, a wheat variety resistant to Hessian fly, *Mayetiola destructor,* was developed and introduced in the USA. To develop a resistant variety, major consideration was given to understanding the components of plant resistance, which was divided into three different categories. Plant resistance is defined as the relative amount of heritable qualities possessed by the plant that influences the ultimate degree of damage done by the insect (Painter 1951). Its three components are namely, non-preference (antixenosis), antibiosis, and tolerance. Of these, non-preference has been identified as the response of insects towards the characteristics of the plant that make the plant unattractive to insect for feeding, oviposition, and shelter. The antibiosis refers to the adverse effect of the plant on the biology of insects, while the tolerance is the ability of plants to resist and withstand the damage by a pest population that is enough to severely destroy the susceptible plant. Additionally, terms such as ecological/pseudo resistance/observed resistance terms were coined to describe the plant resistance. This resistance has two categories: the first one is host evasion in which the destructive stage of the pest and susceptible stage of the crop do not synchronize. As a result, some of the plants in the field show a resistant reaction despite being susceptible to the insect pest. It is generally referred to as escape mechanism or host evasion. The other category in ecological terms is pest avoidance, in which the plants escape in nature due to lack of infestation or injury owing to transitory circumstances. In this case, the escaped plants are genetically susceptible but remain healthy as these escape the attack of whiteflies in nature. Host evasion and host avoidance are terms that represent susceptible plants which behave in a resistant manner due to the conditions referred above. The research work on plant resistance in whitefly was initiated at a slow pace as per the necessity or importance of whiteflies. It was generally considered as a minor pest on various crops and did not inflict serious damage to host crops. It was considered as a more important pest of crops in situations where it acted as a vector of many dreaded plant viruses (Butter 2018). Recently, the whitefly has attained a key pest status and is playing havoc with the cultivation of crops. The research work on this pest which hither to remained scanty has now been taken up on war footing. Despite the non-availability of germplasm with resistant genes, some headway on cotton and cassava, vegetable crops, pulses, and soybean, etc. have been made in this direction during the last two decades.

6.1 Screening Techniques

The screening of germplasm has not been taken into account as the relevant techniques were lacking. During the search for exploring resistant genes, efforts were made to initiate and strengthened the research work on whiteflies. The strenuous efforts initiated in this direction have led to the development of a suitable methodology to screen the germplasm in important crops. Screening techniques are now available in some important crops such as cassava, cotton, soybean, black and green gram.

Cassava: The cassava is an important host and is a source of food for 1,500 million people in Africa and of one-fifth of the world's population beyond Africa (Cossa 2011, Were et al. 2007, Sseruwagi et al. 2006). The crop is vulnerable to two viral diseases: the Cassava Mosaic Virus and the Cassava Brown Streak Virus; both of these are vectored by the whitefly (Simon et al. 2006). As early as 1986, efforts were made to screen germplasm in cassava with the techniques developed till then (Berlinger 1986). As a prerequisite, efforts were made to identify resistant varieties in cassava based on a study of the population of whitefly and damage symptoms in Colombia (Berlinger 1986). Six scale techniques, each based on the pest population and damage symptoms, were developed and were soon applied to screen varieties of cassava resistant to whiteflies (Berlinger 1986). The scale, based on the whitefly population attacking the cassava was developed to screen the germplasm. It is as under:

Grade	Population (Nymphs/Pupae) per leaf
Grade I	Nil Whitefly
Grade II	1–200,
Grade III	201–500,
Grade IV	501–2000,
Grade V	2001–4000
Grade VI	> 4000

The technique based on the damage symptoms was developed. It was a 6-point scale, the details of which are as under:

Grade	Symptoms
Grade I	No damage
Grade II	The young leaf green but slightly flaccid
Grade III	The young leaf with slight curling,
Grade IV	The curling and yellow mottling of apical leaves,
Grade V	The curling and yellowing with sooty mold on leaves
Grade VI	Leaf necrosis with sooty mold

Cotton: The screening of cotton germplasm was initiated at different centers. Detailed investigations were made to determine the bases of resistance. The morphological, biochemical, and nutritional bases of resistance to cotton whitefly have been identified. These characters were identified and incorporated into different varieties of cotton. The work on the development of resistant varieties was never taken up seriously as the whiteflies were included in the category of minor pest in most crops. The strenuous efforts were made to develop technology for the management of whiteflies once this insect has attained the status of a key pest specially in cotton crop. Earlier the pesticides were targeted against cotton bollworms complex. The whitefly was also treated with the chemicals as it could not escape the pesticide exposure. However, exposure to pesticides caused multifarious changes in our agroecosystem. The development of synthetic pyrethroids as potent chemicals gained momentum and these chemicals were introduced to control the bollworm complex the world over. These new chemicals did a commendable job to contain bollworms. However, the introduction of pyrethroids proved highly damaging for the cotton crop due to changes in pest fauna. The build-up of sucking pests, particularly whiteflies, was so alarming that outbreaks became common and the insect attained a major pest status. It gained more importance because of its vector ability. Thus, characteristics imparting resistance to whitefly in the crop are now being identified and incorporated in the high yielding varieties of crops. A great deal of research has been focused on the development of resistant varieties of cotton and cassava crops. The screening of germplasm work was taken up in cotton and cassava crops initially. In cotton, cotton bollworms and cotton jassid were the key pests and thus, resistant/tolerant cultivars to these key pests were also introduced. These cultivars kept the key pests, such as cotton jassid, under check. However, soon, screening techniques based on young and mature cotton crops were developed against whitefly. Now the screening work against whitefly is extended to many other crops. The popularization of screenhouse cultivation of crops has further added to the importance of creating plant resistance to whiteflies. This is because this insect has increased beyond proportion and it requires immediate control in the screenhouse. The detail of these screening techniques each based on five grades has been presented. The technique involving 2-leaf stage with 200 whiteflies/plant is as follows (Kular and Butter 1996):

Grade	Symptoms
Grade I	Healthy cotton leaf
Grade II	The conspicuous yellow spots on upper leaf surface

Grade III   The complete coverage of the leaf with yellow spots
Grade IV   The puckering of leaf lamina
Grade V   Severe puckering of the lamina with leaf drooping

A technique with a 4-leaf stage (release of 200 whitefly adults) was developed to screen cotton germplasm (Kular and Butter 1996). The time taken to develop the symptoms (1–5 grades) were noted. The symptoms were divided into five grades, the details of which are as noted under:

Grade   Symptoms
Grade I   Healthy leaves
Grade II   The presence of yellow spots on middle canopy leaves
Grade III   The presence of honeydew on lower canopy leaves
Grade IV   The complete blackening of lower canopy leaves
Grade V   Severe blackening with drooping of leaves

In this technique, the plants at attaining the age of the 4-leaf stage, 200 whiteflies were released and allowed to feed and multiply up to 20 days. The time taken for the appearance of different grades was recorded. Thus, those cotton varieties which did not attain Grade III—symptoms within twenty days were regarded as resistant cultivars. Similarly the time taken for the appearance of grade-III, grade-IV and grade-V was also recorded.

Another technique based on mature crops for the screening of cotton germplasm has been developed and advocated for screening. In this technique, the raising of cotton germplasm in a glasshouse is recommended. The whiteflies adults @ 2000/90 plants are released on thirty-day-old plants. The whiteflies are allowed to feed and multiply on cotton up to 45–51 days (Kular et al. 1995). The feeding symptoms of variable intensity expressed as grades were taken into account. The following grades were considered for categories of germplasm based on comparative resistance. The grades were as listed below:

Grade   Symptoms
Grade I   Healthy plant
Grade II   The presence of chlorotic spots on the upper surface of middle canopy leaves
Grade III   The presence of honeydew on the upper leaf surface with stickiness
Grade IV   The partial coverage of the lower canopy leaves with sooty mold
Grade V   The complete coverage of the lower canopy leaves with drooping of leaves

Observations of the appearance of different injury grades were recorded in a period of fifty-one days. The Grade II–injury symptoms appeared within 45–51 days and these plants were classified as resistant. The resistant genotypes did not show symptoms beyond Grade II injuries. The susceptible genotypes, however, attained higher grades within this period. This method was followed to screen large germplasm in cotton.

Pulses: Apart from cotton and cassava, screening techniques were developed for pulses. The screening techniques are also available in Punjab (India) for varieties of pulse crops to protect against heavy damage inflicted by whiteflies, particularly on the beans of both African and Asian origin. For black gram, five grades (based on symptoms) were recognized (Taggar and Gill 2011, Taggar et al. 2013). In this technique, the population of whitefly@100 whiteflies at 3rd trifoliate leaves was released and allowed to multiply for up to six weeks. The population records were made at weekly intervals and continued up to six weeks. The grades were recorded. The data were converted into scores through the calculation of the Whitefly Resistance Index (WRI).

Grade	Symptoms
Grade I	Healthy leaves
Grade II	The presence of yellow spots
Grade III	The Initiation of sooty mold
Grade IV	The severe blackening of leaves
Grade V	The drying up of leaves

The method of calculation of WRI is explained as under.

$$\text{Whitefly Resistant Index (WRI)} = \frac{G1 \times P1 + G2 \times P2 + G3 \times P3 + G4 \times P4 + G5 \times P5}{P1 + P2}$$

The Whitefly Resistance Index (WRI) takes into account the number of leaf injury grades (G) and the number of plants (P) in that particular grade to develop a sound method of screening the germplasm in beans in Punjab (India) against the whitefly. It is presented as follows (Taggar and Gill 2016):

Grade	Symptom	Score-Category	Category/Reaction
Grade I	Healthy	> 1.0	Resistant
Grade II	Yellow chlorotic spots	1.01–1.50	Moderately Resistant
Grade III	The Initiation of black sooty mold	1.51–2.50	Moderately Susceptible
Grade IV	The severe blackening of leaves	2.51–3	Susceptible
Grade V	The drying up of leaves	> 3.51	Highly Susceptible

Also, the population-based criteria for black gram resistant to whitefly was advocated which are as follows (Taggar and Gill 2016):

Grade	Whitefly Population (Nymphs/Pupae)
Grade I	0–10
Grade II	10.1–20
Grade III	20.1–30
Grade IV	30.1–40
Grade V	40.0–50

In this method, the population counts (nymphs and pupae) were made.

Soybean: The method to screen the varieties/cultivars soybean resistant to whitefly in Turkey is also available, the detail of which are as follows (Gulluoglu et al. 2010):

Grade	Population (Eggs plus immatures per 2.85 cm² leaf area)	Reaction
Grade I	< 10	Highly Resistant
Grade II	11–20	Resistant
Grade III	21– 35	Moderately Resistant
Grade IV	36–50	Susceptible
Grade V	> 51	Highly Susceptible

6.2 Bases of Resistance

The mechanism of the plant's resistance to whiteflies has three components viz non-preference (antixenosis), antibiosis, and tolerance. Of these components, the non-preference is defined as the component in which the characteristics of the plant make the plant unattractive to the whitefly for shelter, feeding, and oviposition. Whereas, in antibiosis, the plant affects the biology of the feeding herbivore and reduces the fecundity, longevity, and offspring and increases the mortality. The third component of resistance is tolerance in which the plant growth compensates for the insect attack despite its susceptibility. It has been seen that the whitefly is affected by external and physical factors of the leaf lamina namely hairiness, pubescene, glandular trichomes, leaf shape, internal and chemical features like pH of leaf sap in different plant species (Berlinger 1986).

Cotton: In cotton, generally, three kinds of traits are apparent viz okra type, characteristic features of cotton leaves for preference and non-preference of the crop, glabrous versus non-glabrous nature of leaves determining the rate of oviposition, and lastly, the biochemicals interference in the development of various stages (Miyazaki et al. 2013). The earlier categorization of characteristic features was into internal and external traits (Berlinger 1986). The positive correlation between hair density/density of stellate trichomes and leaf area and the whitefly population (*Bemisia tabaci/ Bemisia argentifolii*) (adult and nymph) was obtained in cotton (Chu et al. 2003, Chu et al. 2002, Bashir et al. 2001, Chu et al. 2001, Chu et al. 2000, Chu et al. 1999, Aheer 1999, Chu et al. 1998). Further, in cotton, the glandular trichome (Yu et al. 2010) and IV *Solanum habrochaites* accessions LD 1777 synthesize sesquiterpenes (7-zingiberene) and R-Curcumene, which impart resistance against the whitefly (Bleeker et al. 2012). The only remedy to mitigate the crop loss due to whiteflies is through the development of resistant varieties of crops. But it is an arduous task to develop resistant varieties of plants due to the non-availability of resistant sources. Despite these difficulties, some headway has been made in this direction. The bases of resistance such as morphological (Butter and Vir 1989), biochemical (Butter et al. 1992), and nutritional (Butter et al. 1996) to cotton whitefly were identified and utilized. Two techniques for screening cotton germplasm at the seedling stage and during the mature crop have been put forward to complete this long overdue task to identify resistant cotton germplasm (Kular and Butter 1996, Kular et al. 1995). By deploying these techniques, promising crop lines were selected, and resistant sources were identified. These lines were further tested and incorporated into our traditional breeding programs. In all, forty morphological characters in cotton contributing towards resistance have been identified (leaf hair density, leaf thickness, hair length, angle of insertion of leaf hair, and several gossypol glands) (Table 6.1). Various physiochemical plant characters, namely, toughness, leaf length/width, number of leaves/plant, plant height, etc. were identified for their role in resistance in addition to biochemical compounds viz. phenols, tannins, gossypol (Deb et al. 2015). Of these, two characteristics viz. hair density and thickness of leaf contributed to the extent of 92% as revealed by step-wise regression (Butter and Vir 1989). Glabrous (without pubescence) and thinner leaves contributed to resistance to whitefly in cotton. To determine the morphological resistance, step-wise regression using eggs, nymphs, and adults as dependable variables and plant characters as independent variables were taken in the model:

$$WF(e) = Bo + B1 \ (Hl) + B2 \ (Hd) +$$
$$WF(n) \ B3 \ (Ha) + B4(La) + B5(Lth) +$$
$$WF \ (a) \ B6 \ (GiGi) + B7 \ (GGp) + B8 \ (GiGi) + Ei$$

WF(e), WF(n) and W(a) denote eggs, nymphs, and adults, respectively; Hl (hair length), Hd (hair density), Ha (hair angle), Lth (leaf thickness); GiGi (gossypol glands on the internode), GGp (gossypol glands on petiole), GGl (gossypol glands on leaf lamina), and B6 to B8 are regression coefficients. The correlation coefficient between the whitefly population (eggs, nymphs, and adults) and plant characteristics were worked out. The values of correlation coefficient between the eggs (–0.2027, 0.7176,** 0.6790,** 0.0441, 0.3021, –0.4579, –0.1094, 0.4011); nymphs (–0.2367, 0.7158,** 0.7187,** 0.0613, 0.2475, 0.4191, –0.0737, 0.2752); adults (0.0026, 0.3424, 0.2162, 0.0247, –1320.0975, –0.1436, 0.3085, 0.5253*) and leaf area, leaf thickness, leaf hair, length of leaf hair, angle of leaf hair, gossypol of leaf lamina, petioles and internode, respectively were presented. The density of trichome was found to be positively correlated with oviposition (r = 0.0.94) in a study by Thomas et al. (2014).

Another study was made on cotton that demonstrated the role of hairs or trichomes, thickness, toughness, leaf length/width, number of leaves/plant, plant height, phenols, tannins, and gossypol in conferring resistance against insect pests (Deb et al. 2015). Similarly, the correlation between plant nutrients (nitrogen, phosphorus and potassium) and whitefly population was obtained. Of these nutrients, phosphorus was found negatively correlated with the whitefly population, both at vegetative as well as reproductive stages of cotton crop (Butter et al. 1996). The correlation between adults (–0.5828* and 0.6389*) and nymphs (–0.4352 and –0.6009*), eggs (–0.3633 and –0.3735) with phosphorus at vegetative and reproductive stages, respectively, were found to be negatively correlated. Biochemical bases were worked out which indicated the role of phenols/di-hydroxy-phenols and tannins in imparting resistance against whitefly (Butter et al. 1992). The co-relation coefficient between a population of whitefly and the contents of chemical nutrients was negative. Furthermore, the nutritional parameters like nitrogen, phosphorus, and potash were also examined for their role in conferring resistance against whiteflies under laboratory conditions (Butter et al. 1996). The correlation coefficient between phosphorus and whitefly

Table 6.1. Characteristic features/factors imparting resistance against whiteflies in crop plants.

Category (Type)	Character (Plant/part)	Reaction/ component	Extent/content	Source(s)
Morphological	Thickness of leaf lamina, leaf pubescence with bright green color (eggplant/cotton)	Resistant	Low and thin density of hair	Khan et al. 1993, Mound 1965
Morphological	Pubescence of leaf/ length of hair (cotton)	Resistant	Glabrous leaf/ low hair density and longer hair length	Butter and Vir 1989, Khan et al. 1993, Butler and Henneberry 1984
Morphological	Okra leaf or semi-okra leaves/(cotton)	Resistant	Narrow leaves	Husain and Trehan 1940
Morphological	Gossypol gland (cotton)	Resistant	High density	Butter and Vir 1989
Morphological	Nectaries (cotton)	Tolerant	Nectariless	Anonymous 1989
Morphological	Leaf area (cotton)	Resistant	Lesser leaf area	Butter and Vir 1989
Morphological	Leaf color (cotton)	Non preferred	Red leaf non-preferred	Husain and Trehan 1940
Anatomical	Thickness of leaf/stem/ (cotton)	Susceptible	Thickened leaf lamina	Butter and Vir 1989, Kular and Butter 1999
Anatomical	Thickness of leaf lamina cuticle (cotton)	Resistant	Thinner leaves	Butter and Vir 1989, Kular and Butter 1999
Anatomical	Leaf Iron content (cotton)	Resistant	Low content	Rao et al. 1990
Nutritional	Leaf phosphorus content (cotton)	Resistant	High content	Natarajan and Surulivelu 1986, Butter et al. 1992a
Nutritional	Potassium (cotton)	Resistant? (inconsistant results)	High content	Butter et al. 1992a, Butter et al. 1996
Biochemical	Tannins/(cotton)	Resistant	High content	Butter et al. 1992
Biochemical	Phenols/Dihydroxy phenols (cotton)	Resistant	High content	Butter et al. 1992
Biochemical	pH of sap/(cotton)	Resistant	Acidic pH	Berlinger et al. 1983a
Ecological	Sowing (cotton)	Resistant	Early sowing	Reddy and Rao 1989
Ecological	Early/timely sown (cotton)	Escape/ Tolerant	Early sown	Singh and Butter 1997
Ecological	Irrigation or rainfall (cotton)	Resistant	Heavy rainfall/ irrigation	Husain and Trehan 1940
Biochemical	Flavonols (cotton)	Resistant	High content	Butter et al. 1992

Table 6.1 Contd. ...

...Table 6.1 Contd.

Category (Type)	Character (Plant/part)	Reaction/component	Extent/content	Source(s)
Nutritional	Nutrient content (eggplant)	Resistant	High content of nitrogen/potassium/zinc/copper	Mansoor-ul-Hassan et al. 2000
Morphogenic	Glossy leaf Lamina (collard)	Non-preference	Repellent due to glossy leaf surface	Jackson et al. 2000
Transgenic	Protein from edible Fern Tma12 (cotton)	Resistant	Presence of Tma12 (Lytic polysaccharide monooxygenase)	Shukla et al. 2016, Yadav et al. 2019
Morphological	7.Epizingiberene/glandular trichomes (Tomato)	Repellent/toxic	Semiochemical	Bleeker et al. 2012, Bleeker et al. 2011
Morphological	Leaf width (Sugarcane)	Resistant	Between 12–14 mm	Agarwal 1969
Morphological	Spiny clones/ (Sugarcane)	Resistant	High density of spines	Agarwal 1969
Morphological	Nectaried and gossypol glands	Resistant	Nectariless and glanded cotton	Tcach et al. 2019
Cotton Plant	Body (color)	Tolerant	Red color	Butter and Singh 1993

population was also negative. Based on the promising results, the experiment was extended to the field. The field experiment was planned with the sole purpose to examine the role of phosphorus element in imparting resistance to whitefly. The application of phosphorus @ 30 kg P_2O_5/ha under field condition caused a drastic reduction in the population of adults as well as nymphs of whitefly and enhanced the seed cotton yield significantly. The field experiment was planned and conducted based on feedback received from the farmers regarding the application of Diamonium phosphate to cotton during the active season. The higher content of phosphorus probably made the texture of leaf lamina of cotton unsuitable for oviposition and feeding. The nymphs too failed to develop on highly fertilized cotton with phosphorus (Butter et al. 1992a). Based on biochemical constituents of cotton, the correlation coefficient between total sugars/tannins/flavonols/phenols/alpha dihydroxy phenols and the population of whitefly was negative. The correlation between the nymphal population and tannin content and between egg and tannin were -0.5179 (P = 0.05) and 0.7543) (p = 0.01), respectively. Similarly, negative correlations between total phenolic and dihydroxyphenolic content and the total eggs were obtained, and the value of correlation was -0.6670 (P = 0.05) and -0.5863 (p = 0.01), respectively (Butter et al. 1992).

In cotton, the whitefly population was positively correlated with nitrogen, potassium, zinc, and copper content of the plants (Mansoor-ul-Hasan et al. 2000).

The leaf brightness, red- and green-colored foliage decreased whitefly preference but the blue color did not have a significant effect (Zhang and Liu 2016). Moreover, considering the whole color component of leaves among the eggplant varieties, the degree of blue had a positive effect, and the degree of saturation and brightness reduced whitefly preference.

Eggplant: In whitefly-prone areas, eggplant varieties with thick and highly pubescent moderately brighter green color cultivars are identified as resistant and should be given preference. The study on the role of nutritional and defensive chemicals in leaves like nitrogen, glucose, fructose, sucrose, amino acids, total phenolic compounds, and the influence of moisture content of the eggplant leaves on feeding and ovipositional response, fecundity, and longevity of whitefly were carried out. In all, six eggplant varieties (H149, JSZ, JGL, TLB, DYZ, and QXN) were allowed to be fed on by the whitefly adults of *Bemisia tabaci*. Of these varieties, 'H149' and 'JSZ' had the most eggs. Higher fecundity and longevity, and a lower nymphal development period were recorded on JSZ. The variety of TLB recorded the least abundance of adults and eggs. The highly susceptible variety (JSZ) had higher nitrogen, glucose, amino acids, and lower moisture content (Hassanuzzaman et al. 2018). On the other hand, the resistant one (TLB) had higher total phenolic content. The findings of earlier work indicated a positive correlation between eggs/adults with trichome density, trichome length, leaf lamina, eggplant palatability for feeding (Hasanuzzaman et al. 2016).

Tomato: In tomato, the important characteristics are glandular trichomes and their exudates which determine the abundance of insects. These trichomes are of two types: glandular and non-glandular ones, of which the glandular is responsible for synthesis, storage, and secretion of secondary metabolites whereas non-glandular regulate water use, maintain temperature and photosynthesis, invite pollinators, and above all, act as physical barriers to the movement of insects. The glandular trichome VI *Solanum habrochaites* accessions (PI-134417 and PI 134418) synthesize and store methyl ketones that impart resistance against insects (Yu et al. 2010, Ben-Israel et al. 2009, Antonious and Snyder 2008, Fridman et al. 2005). The glandular trichomes *Solanum habrochaites* sub-species *glabrum* synthesizes methyl ketones and also imparts resistance. Besides, *Solanum pennellii* and *Solanum alapagos* are responsible for the storage of Acyl sugars (Lucatti et al. 2014, Schilmiller et al. 2012, Firdaus et al. 2012). Acyl sugars of wild tomato *Lycopersicon pennellii* alter the settling patterns of the *Bemisia tabaci* (Homoptera: Aleyrodidae) and reduces its ability to oviposition (Liedl et al. 1995). The mechanism of resistance imparted to tomato against root-knot nematode through complimentary genes (Mi-1.2 intact gene and Mi-1.4 complimentary gene) has also been identified (Milligan et al. 1998). The gene Mi-1.2 in resistant nematode varieties also provides resistance to Q-biotype of *Bemisia tabaci* in tomato and these varieties reduced the pest's fecundity significantly (Nombela et al. 2001).

Oilseed Crops (Collard/Seasame/Sunflower) The varieties of collard (*Brassica oleracea*) were screened for their reaction to whitefly. Out of this screening, glossy leaf cultivars namely SC Glaze, SC Landrace and Green Glaze were found highly resistant to whitefly (Jackson et al. 2000). It was a non preference type of resistance.

Another study conducted in Brazil on screening of Green Collard (*Brassica oleracea* var *acephola*) demonstrated highest level of antixenosis in VE and J genotypes of collard (Domingos et al. 2018). The studies carried out in Pakistan identified PR-14-2 variety of sesame (*Sesamum indicum*) resistant to whitefly, *Bemisia tabaci* (Wadhero et al. 1998). In Venezuela, 2 more cultivars (Fonucla and UCLA-1) were identified which were rich in foliar leaf acidity and this was identified as antibiosis category of resistance (Laurentin et al. 2003). Another oilseed crop (Sunflower-*Helianthus annuus*) was evaluated for resistance and PN 2KS, SH 3322 and IBD-2KS genotypes were identified as resistant (Aslam and Misbah-ul-Haq 2003).

6.3 Resistant Varieties

Cassava: Cassava is a food crop as well as a rich source of starch. It is a staple food of the Colombia region in South America and there is a strong need to protect this food crop against insects. Agronomic characteristics and vegetative propagation, drought tolerance, staggered sowing, prolonged growth cycle, and intercropping are responsible for changes in pest fauna (Bellotti et al. 2012, Bellotti 2008). The screening of germplasm against whiteflies has been undertaken at the International Centre for Tropical Agriculture, Colombia with particular reference to cassava. The host plant's sources of resistance against whiteflies are very rare. In spite of this, the varieties, namely, SSA, SSA7, Nase 1, Nase 2, Nase 3 (Migyera) inc., and Fiona were developed as resistant/tolerant to whiteflies. In all, 5000 clones of cassava were collected and tested at this international center and one clone, ME Cu-72 (72.5%), was identified as resistant to the whitefly *Aleurotrachelus socialis* (Carabali et al. 2010, Bellotti and Arias 2001) (Table 6.2). The development of nymphs was significantly reduced and maximum mortality was obtained on this cultivar. The antibiosis type of resistance was recorded in this cultivar. The whitefly's feeding on the cassava crop was established and highest longevity (5.6 days), oviposition rate (2.6 eggs/female per 2 days), and shortest rate of development (44.4 days) were obtained (Carabali et al. 2005), indicating the establishment on Manihot/Euphorbia/Jatropha. This cultivar was utilized for developing a resistant hybrid. A new hybrid named CG489-71 resistant to the whitefly was identified and is being cultivated on a commercial scale.

Cotton: The genotypes of cotton viz. RS-875 (Acharya and Singh 2008), RS-2013, CSH-911, and BBR-1934 were categorized as resistant to whitefly (Acharya and Singh 2007). The negative (Ashfaq et al. 2010, Butter and Vir 1989) as well as positive (Zia et al. 2011) correlation with leaf thickness in cotton has been demonstrated. In another study, on cotton, a positive correlation between lamina thickness and hair density and the population density of whitefly was demonstrated (Khan et al. 2010, Patel 2010). The result was at variance concerning lamina thickness therefore, thorough investigations are needed to draw definite conclusions. Besides this, the leaf length, leaf lamina, hair density, plant height, leaf width were reported as positively correlated. However, the hair length was reported to be negatively correlated.

The developmental period, survival, fecundity, longevity, and sex ratio parameters of whitefly biology under 'no choice' conditions were studied by

Table 6.2. Host plant resistance in crop plants to *Bemisia tabaci*.

SNo.	Crop/resistant component	Crop variety	Country (Location)	Reference(s)
1	Blackgram/Resistant	4-5-2/L29/ UG-204	India (Punjab)	Taggar and Gill 2016
2	Cassava/Resistant	SSA, SSA7, Nase 1, Nase 2, Nase 3 (Migyera)	Colombia	Bellotti and Arias 2001, Calvert et al. 2001
3	Cassava/Resistant	ME Cu-72	Colombia	Carabali et al. 2010
4	Chilli/Non-preferrence (Antixenosis)	Ca-9, Ca-28, Ca29, Acc-5, Acc-16, Acc-18, Acc-29	India (Tamil Nadu)	Jeevananandham et al. 2018
5	Chilli/Highly Resistant	CH-27-F	India (Punjab)	Dhaliwal et al. 2015
6	Chilli/Moderately Resistant	Punjab Sindhuri, Punjab Tej	India (Punjab)	Dhaliwal et al. 2013
7	Chilli/Resistant	Surajmukhi, Japani Loungi, Pant Chilli-1, Pusa Jwala, PBC 473	India (Punjab)	Awasthi and Kumar 2008
8	Chilli/Resistant	DLS-Sel-10, WBC-Sel-5, PBC-142	India (Punjab)	Srivastava et al. 2017
9	Collard/Highly Resistant (Non-Preference)	SC Glaze, SC Landrace, Green Glaze	USA (South Carolina)	Jackson et al. 2000
10	Collard (Green)/ (Antixenosis)	VE and J genotypes	Brazil	Domingos et al. 2018
11	Cotton/Antibiosis	RS-875, RS-2017, CSH-911, BBR-1934	India (Ganganagar/ Rajasthan)	Acharya and Singh 2007
12	Cotton/Resistant	PA-183 Supriya, Kanchana LD-694	India	Jindal et al. 2007, Sidhu and Dhawan 1980
13	Cotton/Moderately Resistant	NHH-44, LK-861, RS-2013 LD694 Supriya	India	Jindal and Dhaliwal 2011
14	Cotton/Non-Preference (Antixenosis)	RS-2098, CNH911, PA-183	India	Jindal and Dhaliwal 2011
15	Cotton/Resistant	LPS-141 (Kanchan)	India	Henry et al. 1990
16	Cotton/Resistant	TH-18, TH-19, IL-88, IL-90, IL-99.R-1, R-2, R-23, R-77 (isogenic lines)	India	Kular and Butter 1995

Table 6.2 Contd. ...

...Table 6.2 Contd.

SNo.	Crop/resistant component	Crop variety	Country (Location)	Reference(s)
17	Cotton/Resistant	CIM-446	Pakistan (Faislabad)	Salman et al. 2011
18	Cotton/Resistant	Greg-25 V	Pakistan	Sayed et al. 2003
19	Cotton/Resistant	MNH-700	Pakistan	Javaid et al. 2012
20	Cotton/Resistant	La 510 ONS, Coker 413, Gumbo	Turkey	Ozgur and Sekeroglu 1986, Ozgur et al. 1988
21	Cotton/Non-Preference	*Gossypium thurberi*	USA (California)	Walker and Natwick 2006
22	Cotton/Resistant	K-68/9	Pakistan	Ullah et al. 2006
23	Cotton/Moderate Resistant	LD-694 (Highly resistant) LK-861 RS-2013 CNH-911 PA-183 Supriya	India	Jindal and Dhaliwal 2009
24	Cotton/Tolerant	NHH-44	India (Punjab)	Jindal et al. 2009
25	Cotton/Toxic and Repellent	G27, LD327, LD491, LH1556, LHH 144P	India (Punjab)	Sidhu and Dhawan 1980
26	Eggplant/Resistant	TLB	China	Hasanuzzaman et al. 2018, Hasanuzzaman et al. 2016
27	Okra/Resistant	IS-376/4/1-40x RS-2013	India	Nogia and Meghwal 2014
28	Okra/Tolerant/Resistant	Sel1-1, Sel2-2, ICI542, *Abelmoschus pungens, A. tuberculate, A. moschatus, A. manihot, A. Crinitus*	India	Sandhu et al. 1974
29	Ornamentals/Resistant/ Repellent	Marigold (African/French), Nasturtium, Calendula, Basil	India	Anonymous 2019
30	Sesame/Resistant	PR-14-2	Pakistan	Wadhero et al. 1998
31	Sesame/Resistant (antibiosis)	Fonucla UCLA-1	Venezuela	Laurentin et al. 2003
32	Soybean/Resistant	S-5	Egypt	Amro et al. 2007
33	Sunflower/Resistant	PN 2KS, SH 3322 IBD-2KS	Pakistan	Aslam and Misbah-ul-Haq 2003

Table 6.2 Contd. ...

...Table 6.2 Contd.

SNo.	Crop/resistant component	Crop variety	Country (Location)	Reference(s)
34	Squash/Non Preference	Hales Best; Amarelo Ouro	Brazil	Baldin and Beneduzz 2010
35	Tomato/Resistant	PI-134417, PI-134418 (VI), LA-1777 (IV) (Solanum habrochaites)	UK	Yu et al. 2010, Bleeker et al. 2012
36	Tomato/Resistant	FCN 3-5	Argentina	Lucatti et al. 2010
37	Tomato/Resistant	TOM687, TOM688, ZGB703, ZGB704	Brazil	Neiva et al. 2019
38	Tomato/Highly Resistant	Vio-63177	East Africa	Rakha et al. 2017
39	Tomato/Resistant	Mi-1.2	Spain	Nombela et al. 2003
40	Tomato/Resistant	Fiona and Tyking Hybrids (F-3224/F-3522)	Tanzania	Pico et al. 1997
41	Tomato/Resistant	Tiny Tim	USA	Curry and Pimentel 1971
42	Watermelon/Resistant	PI-537277 PI-346082	Israel	Coffey et al. 2015

Note: The study at serial number 3, 10 and 41 relates to *Aleurotrachelus socialis*, B-biotype-*Bemisia argentifolii* and *Trialeurodes vaporariorum* respectively.

confining the whiteflies in leaf cages. The incubation period, nymphal developmental time, total developmental time, nymphal survival, total survival, fecundity, and female longevity are affected due to antibiosis. On the other hand, the pupal period and survival, male longevity, and sex ratio did not contribute towards resistance. *Gossypium arboreum* genotype (PA183) was identified as resistant while other genotypes with moderate resistance include Supriya and LD 694 (Jindal et al. 2007). Screenhouse experiments were conducted under free-choice conditions to determine the mechanism of resistance operating in cotton against whitefly, *Bemisia tabaci* (Gennadius) (Homoptera: Aleyrodidae). Twelve cotton genotypes belonging to *Gossypium hirsutum* and *Gossypium arboreum* were assessed for oviposition preference by whitefly. To determine the basis of resistance, trichome density length, leaf lamina thickness, and compactness of vascular bundles parameters of *Gossypium hirsutum* and *Gossypium arboreum* were estimated and the correlations with the number of eggs laid were determined. The genotypes RS2098, CNH911, and PA183 were non-preferred for oviposition and exhibited an antixenosis mechanism of resistance (Jindal and Dhaliwal 2011). Greater leaf lamina thickness and more compact vascular bundles were correlated with egg-laying by whitefly.

The cotton varieties were screened for resistance against whitefly (Butter and Vir 1991). According to the study, varieties/genotypes, namely, USA-22 (sparsely hairy) had 21.5 eggs, as compared to 86 eggs on genotypes USA-13 (velvety hairy). Similarly, egg fertility and adult emergences were 57.9 and 98.3% and 96.5 and 56.1% on F414(USA140x414)414-thin and broad leaves USA-13 and f414(USA 140)141-thin broad and hairy genotypes and USA-13, respectively. The cotton varieties were screened in Punjab by applying screening techniques developed in Punjab (Kular and Butter 1996, Kular et al. 1995) and the germplasm lines (isogenic lines prepared in the background of F414), namely, TH-18, TH-19, IL88, IL90, IL99, R1, R2, R23, R77 were identified as tolerant/resistant under no-choice conditions (Kular and Butter 1995). Of these, R23 was categorized as a resistant line/variety. Under greenhouse conditions, the whitefly adults' presence (47.5 adults/3 leaves) was low on the tolerant/resistant lines F414, followed by IL88 (62.5 adults/3 leaves). On the susceptible isogenic line (R-78), it was 305 adults/3 leaves (Kular and Butter 1999). The morphological characteristics such as leaf area, leaf thickness, and hairiness were found positively correlated with the whitefly population in cotton as demonstrated by the study carried out in Punjab (Kular and Butter 1999). Another set of germplasm lines (D-53, JK-97, FBRN, IUHV-1,2-F, A102, and JK-286) was identified as resistant under field conditions in south India (Venugopal 1987). Likewise, varieties resistant to whiteflies were identified in Cukurova, Turkey and these include L a ONS, Coker 413, and Gumbo cultivars (Ozgur and Sekeroglu 1986). These cultivars showed resistance due to glabrous nature of leaf lamina, okra leaves and open canopy. Similarly, the cotton varieties were evaluated in the Nuclear Institute for Agriculture and Biology (NIAB), Faisalabad, Pakistan, for resistance against whiteflies. In this screening, cultivar CIM-446 showed comparatively greater resistance to the attack of whiteflies as revealed by the number of eggs, the lowest on the resistance (3.028 per leaf), and the highest on the susceptible cultivar (3.821 per leaf) (Salman et al. 2011). The varieties of cotton, namely, Cyto-1291 (Zia et al. 2011), NIAB-2007 (Javaid et al. 2012), and NIAB-814 were identified as resistant ones (Yousaf et al. 2015). Another study carried out in Pakistan showed Greg-25V as resistant (Sayed et al. 2003) to whiteflies on account of higher hair density (Khan et al. 1993). Also, 7-Epizingiberene is a specific sesquiterpene with toxic and repellent properties that is produced and stored in glandular trichomes in tomato (Bleeker et al. 2012). When given no choice, the whiteflies settled on a tomato, which released 7-epizingiberene and this chemical caused the death of the insect due to antibiosis (Bleeker et al. 2011).

Screenhouse experiments were conducted under free-choice conditions to determine the mechanism of resistance operating in cotton against the whitefly *Bemisia tabaci* (Gennadius) (Homoptera: Aleyrodidae). Twelve cotton genotypes belonging to *Gossypium hirsutum* and *Gossypium arboreum* were assessed for the whitefly's oviposition preference. The trichome density and length, distance from lower leaf surface to nearest vascular bundles, leaf lamina thickness, and compactness of vascular bundles were estimated for each genotype, and correlations with the number of eggs laid were determined. NHH44, LK861 (Aamaravathi), G-27, Supriya, LD327, LD-491, LPS141 (Kanchana), RS-87775, RS-2013, Suguna, Bikneri Narma, F414, H777, LH1556, LRA-5166, Ganga, LHH144 (Hybrid) (Table 6.2) resistant to Leaf Curl Virus Disease and LD694 were categorized as tolerant/moderately resistant (Puri et al. 1998, Henry et al. 1990, Sidhu and Dhawan 1980). Greater leaf lamina thickness and more compact vascular bundles were correlated with egg-laying by the whitefly (Jindal and Dhaliwal 2011).

Screenhouse experiments were conducted to know the resistance of the host plants. The field studies carried out in Imperial Valley of California reported consistently high levels of resistance in *Gossypium thurberi* against the silverleaf whitefly, *Bemisia argentifolii* (Walker and Natwick 2006). Naturally developing field infestations in plots of *Gossypium thurberi* were significantly lower than in plots of the *Gossypium hirsutum* (cultivars, DP 5415, Siokra L23, and Stoneville 474). *Gossypium thurberi* has two morphological traits (smooth-leaf and okra-leaf). Therefore, the level of resistance in this wild type genotype was significantly greater than in the cotton cultivar DP 5415 (smooth-leaf and Siokra L23). *Gossypium thurberi* and commercial cotton cultivars did not reveal antibiosis. The survival of the first instar crawler was comparatively higher on *Gossypium thurberi* than on DP 5415 in the greenhouse test. This whitefly has developed resistance to many insecticides, including insect growth regulators and systemic insecticides. Insecticide resistance in whiteflies has led to educationists developing alternative management strategies to combat the menace of this pest. Resistant varieties are often seen as lacking desirable horticultural characteristics and therefore only used when the threat of vector-borne virus disease is acute.

Watermelon: The watermelon (*Citrullus lanatus* var. *lanatus*) is an important crop produced largely in Texas, Florida, California, Georgia, Indiana, and Arizona USA (Simon et al. 2006). Watermelon is an excellent host for some pests including whiteflies in the *Bemisia* complex and spider mites. In addition to direct injury from feeding done by the *Bemisia*, there is concern about the capacity of this pest to vector diseases. Plant resistance is the most fundamental pest management tool for horticultural crops. However, as verified by the high economic losses, the current level of resistance in commercial watermelon is quite inadequate. Despite inadequate attention, research to improve the cultivated watermelon against attack by selected pests and diseases has been initiated and two accessions PI386015 and PI386018 have been identified as highly resistant to whitefly in the case of watermelon (Simmons and Levi 2002). The germplasm from wild *Citrullus* sources has improved resistance to *Bemisia* and other pests. *Trialeurodes vaporariorum* is generally the prevalent species in temperate regions while the sweet potato or silverleaf whitefly *Bemisia tabaci* predominates in warmer parts. Both species damage plants directly through sap-sucking and acting as vectors of viruses. Various biotypes (B-biotype) of the whitefly are dominant in most areas and inflict additional damage through physiological disorders such as irregular ripening of the tomato fruit. However, the Q-biotype, originally from the Mediterranean region, has been recently spotted into the USA. It is an arduous task to control these notorious creatures.

6.4 Biotechnology

The cotton crop suffered heavy losses in tonnage due to the severe attack of the key pest of cotton. However, biotechnology has come to the rescue of farmers and has solved many problems related to plant protection. The most outstanding achievement is the production of transgenic varieties of crop. First of all, Bt kinds of cotton were introduced in 2002 to manage the dreaded pest, *Helicoverpa armigera* in cotton. This technique worked well and the pest problem of bollworms were mitigated. A quantum of pesticide to the tune of 40% was reduced as a result of Bt varieties of cotton (Sanahuja et al. 2011). The chemicals recommended against bollworms were, however, found effective against sucking pests. No special separate recommendation for sucking pests was advocated. So much so, that the chemicals to tackle the sucking pests on cotton were never screened. Due to the lifting of an umbrella of pesticides during the regime of Bt cotton, the sap-sucking insects, which were on the verge of resurgence, gained momentum and there was an actual outbreak of whiteflies, which inflicted cotton failures in different regions. The farmers felt the need to have a insecticide which could help contain this new pest of cotton. There was a big change in the pest fauna: the bollworms were managed with Bt kinds of cotton while the menace of sucking pests was contained with the additional use of pesticides. In the changed scenario, the farmer had to shell out additional money to purchase potent pesticides to control the sucking pests and drastically cut down the budget for purchasing pesticides to manage the bollworms. Recent research output identified one protein with insecticidal value (Tma12) from edible fern (*Tectaria macrodonta*) which was used to create the transgenic cotton plants, and imparted resistance against the whitefly. The transgenic variety was identified and named as Tma12. This transgenic cotton will be soon released to mitigate the menace of whiteflies. The earlier experiments showed the effectiveness of Tma12 transgenic cotton against whiteflies (Shukla et al. 2016). The chemical nature of the protein was analyzed and the chemical compound was identified as lytic polysaccharide monooxygenase (Yadav et al. 2019). This discovery will go a long way to contain a devastating pest from the sucking pests that attack cotton.

Recent advances have been made in host plant resistance in important crops to avoid whitefly damage. It is an important component on which a strong IPM can be built. In cotton crop, major emphasis has been laid on identifying the important characteristics of the plant responsible for imparting resistance against whiteflies. Hitherto, the varieties grown were resistant to cotton jassid but were highly susceptible to whiteflies. Beside the information of morphological characters, biochemical compounds and nutritional contents have also been identified for use in developing resistant varieties. Amongst the morphological characteristics, thinner and glabrous leaves of cotton germplasm were identified as resistant factors and these characters contributed to the extent of 92% towards resistance as revealed by step-wise regression. Further, biochemicals compounds like tannin and phenols contribute towards host plant resistance. Other factors, like a high content of phosphorus in cotton leaves, were identified as nutrients imparting resistance to whitefly. A higher content of phosphorus in cotton leaves demonstrated resistance to whiteflies both in the laboratory as well as under field conditions. In addition to cotton, the characteristic factors imparting resistance against whiteflies in tomato, cucurbits, soybean pulses, okra, etc. were also identified. To identify resistant varieties in various crops, the screening techniques for quick identification were developed. These techniques were deployed to develop resistant varieties in crops like cassava, cotton, soybean pulses, black gram, etc. By using these techniques, these varieties can now be screened. These varieties were taken up for keeping whitefly at bay and released for general cultivation. In the case of germplasm lines, the resistant sources were utilized to develop cultivars through breeding programs.

CHAPTER 7

Natural Enemy Complex, Entomopathogenic Organisms, and Botanicals

The biocontrol of insects has been exploited to a great extent and many successful examples of pest control have been recorded. The introduced parasites, *Eretmocerus serius* for the management of *Aleurocanthus woglumi* in Cuba on citrus is always quoted as a successful example of biocontrol. After this landmark development, the work on biocontrol was initiated on a war footing. It is highly successful against insects with sedentary nature. In whiteflies, all the stages are sessile and found on the lower surface of a leaf. Moreover, the whiteflies are tough insects to check with pesticide. Therefore, the whiteflies are said to be a better candidate for the use of biocontrol agents especially those infesting vegetable crops. For the management of whitefly, the use of parasitoids is also common all over the world.

Biological control was first initiated in China as early as 304 BC, when the weaver ant, *Oecophylla smaragdina*, was used in citrus orchards to manage citrus pests. With time, another 500 species of parasitoids belonging to 23 genera and 8 families (Liu et al. 2015) were discovered and identified as potential candidates of biological control. These parasitoids were placed in two super-families, namely, Chalcidoidea (Aphelinidae, Encyrtidae, Eulophidae, Pteronilidae, and Signiphoridae) and Platygastroidea (Platygastridae) in the order Hymenoptera. Another fifty-two natural enemies (29 species of parasitoids, 8 species of pathogens, and 15 species of predators) were collected/isolated from fourteen whitefly species in Egypt. Slowly and slowly, the efforts were directed to isolate more potent parasitoids to implement on a wider scale throughout the world. However, the biocontrol work on whiteflies (*Trialeurodes vaporariorum* and *Bemisia tabaci*) was mainly confined to vegetable crops, using the *Encarsia* and *Eretmocerus* parasitoids (Wang et al. 2019). The two species of parasitoids namely *Eretmocerus mundus*, *Eretmocerus debachi* in Zheijiang, and the other *Eretmocerus mundus* and *Eretmocerus hayati* species in the Xinjiang province of China were exploited in the first instance (Zhang et al. 2015). However, the natural enemies of whitefly *Bemisia tabaci*, *Encarsia formosa* and *Encarsia sophia*, were identified as better candidates (Dai et al. 2014, He et al. 2019,

Wang et al. 2014, Tan et al. 2016). Likewise, the parasitoids *Eretmocerus warrae* has shown better performance against the host, *Trialeurodes vaporariorum* while feeding on middle-aged nymphs and adults (Hanan et al. 2015). The combination of *Encarsia formosa* and *Encarsia sophia* and predatory ladybird beetle *Harmonia axyridis* showed higher parasitism but with a reduction in adult emergence (Tan et al. 2016). In line with this, the new world records with the identification are reported in Egypt for two parasitoids, *Encarsia lutea* Masi and *Eretmocerus mundus* (Mercer) on eight species of whiteflies. Three predators (*Chilocorus bipustulatus* L., *Chrysoperla carnea* (Stephens), and *Coccinella septempunctata* L.) of a single whitefly species (*Siphoninus phillyreae* (Haliday)) were identified as predators. The field survey made in Egypt also reported local natural enemies of whiteflies as efficient parasitoids. This happens to be the first local report of six predators and three entomopathogenic fungi of whiteflies in Egypt (Abd-Rabou and Simmons 2014). The fungi are normally useful but there are situations in which fungi can have destructive effects on parasitoids sometimes. The toxic and deterrent effect of an extract of fungi (*Lecanicillium lecanii*) on the parasitoid wasp of *Bemisia tabaci* was recorded for the first time on the sweet potato (Wang et al. 2007).

The biological control contributions are discussed under the following components.

7.1 Predators

The organisms that are generalized, free-living, and requiring more prey to feed on to complete their life cycles are the natural enemies referred to as predators. These predators generally kill their host and need two or more prey organisms to complete their life cycles. The common notable predators include ladybird beetles, *Chrysoperla carnea*, praying mantis, and ant lions, big-eyed bugs, minute pirate bugs. Besides, the other notorious insect pests in polyhouse like thrips, aphids, mealybugs, etc. also require tackling so as to allow the harvest of a good crop from the polyhouse (Messelink et al. 2006). It is an important area to study and popularize biological control. The important predators of different categories are mentioned to acquaint the reader with natural enemies.

Neuroptera: The use of predator from the order Neuroptera is *Chrysoperla carnea* which has been exploited for the control of insect pests. It is a generalized predator that preys on soft bodied insects. The releases @ 50,000 adults per hectare are done to prey on the soft bodied insects including the whiteflies. These predators normally prove effective against the low population of host and therefore, the early application (60-day-old crop) of predators is beneficial. Spiders have also been named as generalized predators (Bhatt et al. 2018).

Coleoptera: In the category of general predator in the order Coleoptera are *Coccinella septempunctata* (L.) (Coleoptera: Coccinellidae), *Coccinella undecimpunctata* L. (Coleoptera: Coccinellidae). Besides these insects, important beetle, *Delphastus pusillus,* has also been identified as important predator of whiteflies, *Bemisia tabaci*. This predatory beetle has been reported as effective natural enemies of sweet potato whitefly (*Bemisia tabaci*) in screenhouse (Gloyd 1999). *Delphastus catalinae*

(Coleoptera:Coccinellidae) feeding on *Bemisia tabaci* host was studied at different temperatures and was found at its best at temperature between 22–26°C. The efficiency decreased beyond 30°C. With respect to its efficiency on different hosts, parasitization was higher (24%) on cotton followed by collards, cowpea, tomato and lowest on *Hibiscus* (Stansly and McKenzie 2007). The study further showed that it prefers eggs over nymphs thus the releases are more appropriate when eggs are more abundant. In one of the studies carried out on the prey preferences of coccinellid species, indicated more preference for puparium (6.4–7.5 daily bases), over nymphs (5.1–6.3 daily basis), and eggs. The ladybird beetle preferred *Bemisia tabaci* over *Trialeurodes vaporariorum* (Al-Zyoud and Sengonca 2004a). In a study by Hagler et al. 2004, the sweet potato whitefly, *Bemisia tabaci* (Gennadius) was fed to generalist predators to quantify the foraging behavior of the adult.

Predatory Mites: The identification of efficient predaceous mite *Amblyseius swirskii* for control of whiteflies has been considered as a landmark step in biological control. It will go a long way in managing the polyhouse insect pests. Besides insects, *Hippodamia convergens, Collops vittatus* and Phytoseiid mites (*Typhlodromips swirskii* and *Euseius scutes*) are also identified as effective predators of whiteflies (Nomikou et al. 2003). The studies carried out on two species of predatory mites (*Amblyseius swirskii* and *Euseius scutes*) demonstrated 15% more reduction in the population of whitefly in the presence of thrips as compared to single host insect (Messelink et al. 2008).

Hemiptera: The other predators from Hemiptera order include *Callistethus arcuate* and *Chilocorus* species. Recently, the sycamore whitefly, *Bemisia afer* (Priesner & Hosny) has been declared as an economic pest in Egypt and the predators viz. true bug, *Callistephus arcuate* (Hemiptera: Miridae), *Orius* sp. (Hemiptera: Anthocoridae) and *Geocoris* sp. (Hemiptera: Lygaeidae) are found associated with the species. Meanwhile, interest in the use of phytophagous mirids in the genera *Macrolophus, Dicyphus,* and *Nesiochorus* has been steadily increasing. The detail of natural enemies along with stage of host parasitized or preyed is given (Table 7.1). The true bugs, *Geocoris punctipes* (Say), and *Orius tricolor* (Say) preyed on adults while *Lygus hesperus* preyed on nymphs. The true bugs take more time than the beetles to manage whitefly species. The results demonstrated differential interaction depending upon the host, and the foraging behavior.

The molecular determination in cassava whitefly in Colombia was carried out through a pest-specific primer (Lundgren et al. 2014). Despite their effectiveness, these natural enemies could not be popularized or rated as potential agents of biocontrol. The major drawback in the popularization of predators is their slow multiplication in the laboratory. Thus, in the absence of mass production, the use of predators could not be taken up on a large scale as compared to parasitoids. The total number of generations is less and the time taken to complete a generation is also comparatively more than in the parasitoids. Above all, the transport of predators is slightly cumbersome as compared to parasitoids. On the other hand, the parasitoids are available as cards and in this form, they are easy to carry and hang on the plants. Additionally, these predators are generalized and can attack the other natural enemies of pests as hyperparasites and can cause more harm than the control of pests.

Table 7.1. Important parasites (parasitoids and predators) of whitefly.

Whitefly species	Natural enemies	Stage of host	Reference(s)
Bemisia tabaci	*Chrysoperla carnea; Chrysoperla cymbela; Chrysoperla scelestes*	Egg, Nymph	Rao et al. 1989
Bemisia tabaci	*Geocoris bicolor; Zelus* sp. *Orius* Species	Egg, Nymph	Rao et al. 1989, Dhawan 2016
Bemisia tabaci	*Menochilus sexmaculatus*	Nymph	Sangha et al. 2018
Bemisia tabaci	*Encarsia formosa*	Nymph	Zhang et al. 2020
Bemisia tabaci	*Eretmocerus* sp.	Nymph	Zhang et al. 2020
Bemisia tabaci	*Mallada boninensis*	Egg, Nymph	Sangha et al. 2018
Bemisia tabaci	*Geocoris puncticeps: Zanchius breviceps*	Nymph	Sangha et al. 2018
Bemisia tabaci	*Brumoides saturalis*	Nymph	Rao et al. 1989
Bemisia tabaci	*Serangium parcesetosum; Brumus saturalis; Encarsia lutea; Cheilomenes sexmaculatus*	Egg, Nymph, Adult	Kedar et al. 2014
Spiders	*Encarsia sophia; Encarsia lutea* Spiders	Egg (non-preferred by parasitoids) Nymph, Adult	Sangha et al. 2018
Aleurocanthus woglumi	*Eretmocerus serius*	Nymph	Begum et al. 2011
Bemisia tabaci	Parasitoid *Encarsia formosa; Encarsia sophia;* & Predator *Harmonia axyridis*	Egg Nymph	Dai et al. 2014, He et al. 2019, Wang et al. 2014
Siphoninus phillyreae	*Chilocorus bipustulates, Chrysoperla carnea, Coccinella septempunctata, Chrysoperla Zastrowi sillemi*	Nymph, Adult	Abd-Rabou and Simmons 2014, Rao et al. 1989, Sangha et al. 2018
Bemisia tabaci	*Delphastus pusillus*	Nymph	Gloyd 1999
Bemisia tabaci	*Hippodamia convergens, Collops vittatus*	Egg (preferred) Nymph, Adult	Hagler et al. 2004
Trialeurodes vaporariorum, Bemisia tabaci Trialeurodes abutilonea/ Bemisia argentifolii	*Encarsia formosa, Eretmocerus eremicus/ Eretmocerus californicus* & *Eretmocerus mundus*	Nymph, Red Eye Nymph	Ardeh 2004
Aleurodicus dispersus	*Encarsia dispersa, Encarsia guadeloupae; Encarsia haitiensis*	Nymph	Lambkin and Zalucki 2010, Mani 2010

Table 7.1 Contd. ...

...Table 7.1 Contd.

Whitefly species	Natural enemies	Stage of host	Reference(s)
Bemisia tabaci; Trialeurodes vaporariorum	*Geocoris puncticeps, Zelus* sp.	Egg, Pseudo pupa	Al-Zyoud and Sengonca 2004a, Sangha et al. 2018
Bemisia tabaci	*Encarsia formosa, Encarsia lycopersici, Encarsia porter Encarsia protransvena, Encarsia transvena, Encarsia pergandiella, Encarsi hispida, Encarsia meritoria, Eretmocerus corni, Cales noacki*	Nymph (Except the Crawler by *Encarsia*) Pseudo Pupa	Mottern and Heraty 2014, Kapadia and Puri 1990
Bemisia tabaci	*Encarsia opulenta*	Nymph, Pseudo pupa	Thompson et al. 1987
Aleurotrachelus cocois	*Eretmocerus cocois Encarsia debachi*	Nymph	Cave 2008 Rose and DeBach 1992
Aleurotuberculatus takahashi	*Eretmocerus longipes*	Nymph	Sengonca and Liu 1998
Aleurodicus dispersus/ Aleurodicus dugesii	*Idioporus affinis*	Nymph	Bellows and Meisenbacher 2000
Bemisia tabaci	*Scymnus syriacus*	Nymph, Adult	Rao et al. 1989
Bemisia tabaci Tetranychus cinnabarinus	*Amblyseius orientalis*	Nymph: eggs and Protonymphs	Rao et al. 1989, Dhawan 1999, Zhang et al. 2015
Bemisia tabaci	*Euseius soutalis*	Nymph, Adult	Rao et al. 1989
Bemisia tabaci	*Serangium parcesetosum*	Nymph, Adult	Rao et al. 1989, Sangha et al. 2018
Bemisia tabaci	*Encarsia lutea, Encarsia shafee, Encarsia transvena*	Nymph	Sundaramurthy and Chitra 1992, Dhawan 1999

7.2 Parasitoids

The natural enemies of pests continuously live with the hosts and weaken them but rarely kill them. Normally, these parasitoids require a single host for completion of their life cycle. They are obligate parasites and specialists in consuming the target host against which they have been released. They are generally called parasitoids. These possess a specialized quality of high amenability in the laboratory and can be mass-produced in larger quantities in the shortest possible period. Thus, they have a high innate capacity for an increase in numbers (rm). The whiteflies being sessile for a quite long time are thus a more ideal candidate for biological control to be practiced upon (Gerling 1992). The use of *Encarsia formosa* (Hymenoptera; Aphelinidae) was initiated in Europe in 1920s on screenhouse. Crops (tomato cucumber, eggplant, strawberry, tobacco, *Gerbera*, etc.) against whiteflies (Hoddle et al. 1998) and it successfully controlled these pests upto 1945. The ill effects of pesticides and development of resistance to pesticides in insects has triggered thinking about good qualities of biocontrol. With the result, once again this control method has gained momentum. However, the introduction of potent pestcides, the parasitoid use was almost reduced to zero level. Now the parasitoid use against whiteflies has been taken up on commercial scale. The commercial use include *Encarsia, Eretomocerus* parasitoids. These parasitoids are known to parasitize of all stages of whitefly except eggs and crawlers. These lay eggs in 2nd, 3rd and 4th instar nymphs to parasitize. The eggs of parasitoid are laid in nymphal stages (3-mature eggs in varioles). After parasitizing the host by laying eggs, the parsitized nymphs/pupae turn black with passage of time. The development of natural enemies is completed on the nymph or pseudo pupae. The host gets killed. The wasp emerges from immatures/pseudo pupae of whitefly. The host stage turns black due to the presence of *Encarsia*. After the completion of development, the adults emerge and empty holes in pupal cases become apparent and can be seen by facing pupal cases towards light. The first, effective biological control of whiteflies was with *Encarsia formosa*, initially against *Trialeurodes vaporariorum*, but also used with limited success against *Bemisia tabaci*. This was followed by *Eretmocerus eremicus*, equally effective against both whitefly species, later to be supplemented for control of *Bemisia tabaci* by *Eretmocerus mundus*. The common parasitoids of whiteflies are *Encarsia formosa, Encarsia transient, Eretmocerus mundus, Trichogramma chilonis, Trichogramma australicum, Apanteles ruficrus,* and *Bracon* species. Of these parasitoids, *Trichogramma, Eretmocerus,* and *Encarsia* are commercially available for use against tissue borers of crops and whiteflies.

It is also important to visualize the impact of varieties of crop plants on the efficacy of parasitoids. Efforts were made in this direction in Pakistan to determine the efficacy of parasitoids of cotton whitefly on commonly grown cotton cultivars such as Cyto-1291 (Zia et al. 2011) and NIAB-2007 and NIAB-814 (Yousaf et al. 2015, Javaid et al. 2012). The records of the family Aphelinidae (*Encarsia formosa* Gahan, *Encarsia lycopersici* De Santis, *Encarsia porteri* (Mercer), *Encarsia protransvena* Viggiani, *Encarsia transient* (Timberlake), *Encarsia pergandiella* Howard, *Encarsia hispida* De Santis, *Encarsia meritoria, Encarsia* species, *Eretmocerus corni* Haldeman, *Eretmocerus* species, *Cales Noacki* Howard) (Mottern and Heraty 2014), and the hyperparasitoids of the family Signiphoridae (*Signiphora aleyrodis* Ashmead, *Thysanus aleyrodis,* and *Signiphora* species) were isolated in Argentina (Viscarret et al. 2000). The species *Signiphora aleyrodis, Encarsia protransvena,* and *Encarsia transient* have been newly recorded in Argentina. Also, three more species, namely, *Encarsia opulenta, Encarsia smithi,* and *Amitus hesperidum* were identified as effective parasitoids of citrus blackfly. Among them, *Encarsia opulenta* came out to be the dominating species in Florida (Thompson et al. 1987).

The economic threshold for enduring the menace of the vector whitefly is very low and a single insect is sufficient to spread a virus infection in the whole field. However, in biocontrol strategy, some residual population is mandatory for the survival of parasitoids. The residual population is sometimes enough to spread a virus in the whole field. Thus, the biocontrol agents have neither been exploited in isolation for the management of vectors of plant pathogens nor are these effective in such a situation. The parasitoids, namely, *Encarsia formosa*, *Eretmocerus eremicus* (*Eretmocerus californicus*), and *Eretmocerus mundus* are commercially produced for the management of greenhouse whiteflies, *Trialeurodes vaporariorum*, *Bemisia tabaci* (B-biotype), and *Trialeurodes abutilonea* (*Bemisia argentifolii*). These biocontrol agents are an important component of an integrated pest management strategy to mitigate the whitefly problem in crop plants. The cards having parasitoids on the red-eye nymphs (pseudo pupae) are available for distribution and used at different places. The parasitoids are not equally effective on all host plants in screen houses due to differential levels of the whitefly population. Of the species of parasitoids, some are effective on a very low whitefly population while others may not be effective. It is therefore essential to have records of the populations of whitefly, a task for which yellow sticky traps are available. The releases of a population of whitefly are done taking into account the population.

The application of parasitoids is appropriate on the crop aged between 3- or 4 months. The thresholds of the population are already determined before the application of the bioagents. For commercial use the parasitoid was first reared on greenhouse whitefly, *Traleurodes vaporariorum* in England (1927) Now it is mass produced on tobacco for large scale production. Four methods are generally followed to release parasitoids against the whitefly, *Trialeurodes vaporariorum*, on tomato and cucumber in screen houses.

i) Direct introduction of adult *Encarsia* (Gould et al. 1975); In the first one, the ratio of two adults of whitefly/tomato plant is allowed to ensure a sufficient population of the host for the parasitoid. Once the host is established, eight parasitoids/tomato plant are released. The restricted use of this method is advocated in whitefly free pockets. This method does not require a supply of parasitoids from outside in the form of inundative releases. The breeding of parasitoids and its host continues without any restriction in the specified areas.

ii) Dribble method at sowing in anticipation of whitefly appearance at sowing (Gould et al. 1975); The second method is the 'dribble' one (1-parasitoid/plant), in which the parasitoids are released at the time of planting with the hope that the whitefly will appear as usual. The parasitoid will multiply in a sufficient number and would keep the whitefly under check.

iii) Banker plants introduction at fixed rate (Stacey 1977); In the third method both whitefly and parasitoid culture is maintained (Bank). The natural enenies and the host are kept on a plant covered with wire mesh. The mesh allows the parasitoid to escape but the host is retained in the cage. The parasitoid @ 1 banker plant/352 crop plants is introduced in the field.

iv) Inundative releases of parasitoid (Parrella et al. 1991); In the 4th method the, inundative releases are made in high number and the field is flooded with natural enemies as and when required.

The biocontrol of *Trialeurodes vaporariorum* on tomato/cucumber and *Bemisia argentifolii* on poinsettia with *Encarsia formosa* has extensively been used with success against this insect species. The plant volatile (methyljasmonate (MeJA) and methylsalicylate (MeSA)) has also been exploited, affecting the parasitism of natural enemies. In fact, the use of the plant volatile has been considered as a better agent of control of whiteflies. These volatiles help the predators and parasitoids in enhancing their searching ability of finding hosts (Yang et al. 2010) and these are now available as successful biocontrol agents in the world against whiteflies (Liu et al. 2015). Of the genera, *Amitus hesperidum* (*Aleurocanthus woglumi*), *Amitus spiniferus/Cales noacki* (*Aleurocanthus floccosus*) on citrus and *Tetraleurodes perseae* on *Persea*, and *Cales rosei* (*Aleurothrixus floccosus*) on citrus (Hart et al. 1978, Smith et al. 1964, De Bach and Rose 1976, Mottern and Heraty 2014) have been successfully established in the system while *Encarsia dispersa, Encarsia guadeloupae/Encarsia haitiensis* (*Aleurodicus dispersus*) on citrus (Kumashiro et al. 1983, Lambkin and Zalucki 2010, Mani 2010), *Encarsia hispida* (*Bemisia tabaci*), *Encarsia inaron* (*Aleurodes proletella*) on *Brassica* and *Siphoninus phillyreae* on fruit trees (Abd-Rabou 2000, Gerling et al. 2004) proved effective against whiteflies. Besides, *Dialeurodes citri* on citrus, *Encarsia noyesi* (*Aleurodicus pulvinatus/Aleurodicus dugesii*) on coconut and *Encarsia perplexa/Encarsia smithi* (*Aleurocanthus woglumi/Aleurocanthus spiniferus*), *Encarsia strenua* (*Singhiella citrifolii*) I on citrus (Hart et al. 1978, Nguyen et al. 1983, Kuwana and Ishii 1927, Bellows et al. 1990, Abd-Rabou 2000), *Encarsia tricolor* (*Aleurodes proletella*) on cabbage (Qui et al. 2007) are also considered efficient in managing whiteflies. Another genera, namely, *Entedonon ecremnus* krauteri (*Aleurodicus dugesii*) (Zolnerowich and Rose 1996) has also been identified as successful. Of the other genera, *Eretmocerus cocois* (*Aleurotrachelus atratus*) (Cave 2008), *Eretmocerus debachi/Eretmocerus furuhashii* (*Parabemisia myricae*) (Rose and DeBach 1992) did give good control of whiteflies. Also, *Eretmocerus longipes* (*Alurotuberculatus takahashi* on Jasmine) (Sengonca and Liu 1998), *Idioporus affinis* (*Aleurodicus dispersus/Aleurodicus dugesii*) (Bellows and Meisenbacher 2000) proved effective. These species of parasitoids mostly preferred nymphal stages over egg-laying or parasitization, except the *Eretmocerus* which deposited eggs between the nymph and the leaf surface (Abd-Rabou 2000, Dowell 1982, Drost et al. 1999).

In Egypt, parasitoids namely: *Encarsia inaron* (Walker), *Encarsia lutea* (Masi), *Eretmocerus* sp. and *Eretmocerus aegypticus* Evans, and Abd-Rabou (Hymenoptera: Aphelinidae) were reported. The parasitoids also responded to various semiochemicals (Birkett et al. 2003, Guerrieri 1997, Joyce et al. 1999). The kairomones are released by the host on the attack of the herbivore and these volatiles are perceived by a parasitoid. It helps the parasitoid to locate the host for survival The introduction and establishment of *Nephaspis amnicola* and *Encarsia haitiensis* have resulted in successful biological control of *Aleurodicus dispersus* in lowland and highland Honolulu (Kumashiro et al. 1983). In 1991, *Bemisia tabaci* was first noted in the southern San Joaquin Valley, infesting crops outside the greenhouses. In all, twenty-four species/strains of imported aphelinids, like *Eretmocerus*, were released in this valley: *Eretmocerus mundus, Eretmocerus hayati, Eretmocerus emeritus, Eretmocerus* sp. nr. *emiratus*, and *Encarsia sophia,* on citrus (Pickett et al. 2008). The preferences of instar of a host by *Encarsia bimaculata* and *Eretmocerus* sp. were studied on four hosts plants of the whitefly that is, collard, eggplant, cucumber, and tomato. The *Eretmocerus bimaculata* preferred the third and fourth instars of the host. The host plants did not significantly influence the instar preference of the parasites (Qiu et al. 2007). The parasitoids namely *Encarsia bimaculata* (India), *Encarsia sophia* (Thailand, Malaysia, Pakistan, Spain), *Encarsia lutea* (Israel), *Encarsia pergandiella/Encarsia hispida* (Brazil), *Eretmocerus mundus* (Spain, Israel, Italy, India, Egypt), *Eretmocerus hayati* (Pakistan), *Eretmocerus melanoscutus* (Thailand, Taiwan), *Eretmocerus staufferi* (Unknown), *Eretmocerus furuhashii* (Taiwan), *Eretmocerus emiratus* (United Arab Emirates, Ethopia), and *Amitus bennetti* (Peurto Rico) were obtained and tested in USA. Of these, the parasitoids *Eretmocerus* (*Eretmocerus mundus* in Texas, *Eretmocerus hayati* in Texas and *Eretmocerus emiratus* in California) were established in USA on *Bemisia tabaci* (Hoelmer and Goolsby 2003).

The parasitization of whitefly, *Aleurodicus dugesii* by parasitoids namely *Idioporous affinis, Encarsia noyesi,* and *Entedononecremnus krauteri* was studied. Peak parasitism was obtained in late summer with all the three parasitoids (Schoeller and Redak 2020). It was reduced by the start of spring, the variations were noted with *Idioporus affinis* only in all the three climates.

The host-plant species on which the whitefly developed affect the lifespan of *Encarsia formosa* (Van Lenteren et al. 1987). While applying this strategy, the target should be the permanent control of whiteflies. For the conservation of natural enemies during the adverse weather, suitable niches should be created to manage the host. The provision should also be made to meet the need for host for the survival of natural enemies when there is a shortage during the offseason. Excitement has been generated recently by the apparent spectacular success of the predaceous mite, *Amblyseius swirskii.* Biological control of whitefly has been commercialized using *Encarsia* and, *Eretmocerus,* the egg parasitoids. These parasitoids prefer mature stages of host for egglaying. The black scales glued on strips (each strip contains 150 pupal scales) are available and these scales @ 0.2 *Encarsia*/plant/week on spotting out the whiteflies can be tagged to the plants. The release rate can be reduced to half in case the host is not visible in the field (Hoddle et al. 1998). These parasitoids can be stored at $11.5\pm°C$ for 14 days and at $4.5\pm°C$ for 7 days to delay the emergence of adult parasitoids as revealed by the study.

7.3 Entomopathogenic Organisms

Use of Entomopathogenic fungi: The viruses, bacteria, fungi, and nematodes attacking insects are called entomopathogenic. These are useful but their use is restricted due to ultraviolet light and limited soil moisture. Only entomopathogenic fungi and bacteria are tested as entomopathogenic against whiteflies. Amongst the fungi, *Beauveria bassiana, Metarhizium anisopliae, Verticillium lecanii* have been found promising against *Bemisia tabaci* on tomato (Abdel-Raheem and Al-keridis 2017). Several others included in the list of efficient entomopathogenic fungi are *Namuryaea rileyi, Lecanicillium lecanii, Hirsutella thompsonii,* and *Paecilomyces fumosoroseus.* The use of entomopathogenic fungi is a good substitute for chemicals. These can be applied as foliar sprays, seed dressers, soil drenching, seed soaking, and injections

(Vega 2018). A study (Butt and Goettel 2000) placed the fungi *Verticillium lecanii, Aschersonia, Paecilomyces fumosoroseus, Beauveria bassiana,* and *Zoophthora* in the category of entomopathogenic fungi. These fungi, *Paecilomyces lilacinus, Lecanicillium psalliotae, Aspergilus ustus,* and *Metarhizium anisopliae* are known as promising entomopathogenic fungi (LC: 50-.38 to 26.41 × 10). A study carried out in 2000, two fungi, namely, *Beauveria bassiana* and *Paecilomyces fumosoroseus* in Florida, showed great promise in the management of *Bemisia tabaci.* Another entomopathogenic fungus (*Clonostachys rosea*) has been identified as effective against whitefly, *Bemisia tabaci* (Anwar et al. 2018). In general, the fungi are better entomopathogenics than the other microorganisms exploited as control agents. The attacked insect is easy to identify: under suitable weather conditions, the conidia attaches to the insect cuticle and then germinates, and the mycelium emerges through the skin after the growth of hyphae. Fungi can be seen on the skin and body of the insect. Additionally, the body of the host insect is filled with fluid plus the fungal growth. The skin of the host insect becomes brittle and the diseased larvae are found hanging on the branches of the tree and can be located easily (Dara 2017). Several microbial products containing entomopathogenic fungi are available in the market. The fungi, namely, *Paecilomyces fumosoroseus, Beauveria bassiana,* and *Verticillium* are effective against whiteflies (Faria and Wraight 2001, Wraight et al. 2000).

However, delayed mortality, poor ovicidal action, high cost, short shelf-life, and over-dependence on weather are limiting factors in the use of entomopathogenic fungi for successful control of whiteflies. However, these conditions can be improved upon with early application when the host population is low. The virulence, thermal requirements, and toxicogenic activity of entomopathogenic fungi were taken into account for the control of *Bemisia tabaci* and *Trialeurodes vaporariorum* (Quesada-Moraga et al. 2006). The possible role of *Verticillium lecanii, Aschersonia, Beauveria bassiana, Isaria fumosorosea* (*syn Paecilomyces fumosoroseus*), and *Zoophthora* species in the management of whiteflies *Bemisia tabaci, Singhiella cardamomi, Trialeurodes vaporariorum,* and *Aleurolobus barodensis* have been demonstrated. In China, *Paecilomyces* (*Isaria*), *Lecanicillium psalliotae, Aspergillus ustus,* and *Metarhizium anisopliae* (LC$_{50}$s of 0.36–26.44 × 10^6 spores/mL) proved to be significantly better in the control of whitefly (Dong et al. 2016). The use of the combination of both *Beauveria bassiana* and *Lecanicillium muscarium* is much superior to the use of individual entomopathogenic fungi in the management of whitefly, *Trialeurodes vaporariorum.* Another proposal involving the use of surfactant plus entomopathogenic fungi against the *Bemisia tabaci* (B-biotype) proved highly useful.

There was a problem in tackling the pesticide-resistant *Bemisia tabaci* (B-biotype) in Brazil and it has caused substantial damage. However, in this situation, the application of *Trisilo* surfactants with entomopathogenic fungi, *B. vuillemin* (a strain of CG1229) was found highly effective in checking the pesticide-resistant population of the B-biotype of the whitefly (Mascarin et al. 2014). Similarly, the use of entomopathogenic fungus, *Isaria fumosorosea* th the effectiveness of bacteria *Bacillus thuringiensis* to manage whiteflies has been demonstrated. It is being advocated to control the first instar nymph of whitefly (Al-Shayji and Shaheen

2008). The joint application of *Cordyceps javanica* entomopathogenic fungi and *Eretmocerus hayati* gave highly promising results on tomato (Qu et al. 2019). Surfactants are multifunctional and have been overlooked in formulating microbial biopesticides. Another combination of the biosurfactant rhamnolipid (RML) and conidia of two entomopathogenic fungi (*Cordyceps javanica* and *Beauveria bassiana*) against the *Bemisia tabaci* third instar nymphs induced 100% mortality within four days (Silva-Aguao et al. 2019).

Commercial Application of Entomopathogenic Fungi: Scientists have made developments in exploring effective entomopathogenic fungi for use against whiteflies and looking for better biopesticides amongst *Beauveria, Metarhizium, Paecilomyces, Verticillium,* and *Aschersonia* (Meekes 2001, Bolckmans et al. 1995). Commercial products containing entomopathogenic fungi are also available and marketed under the trade names such as Koppert (Holland-*Aschersonia aleyrodis*), Naturalis (USA-*Beauveria bassiana*)/Conidia (Columbia-*Beauveria bassiana*)/ Braconi (Columbia-*Beauveria bassiana*), Engerlingspclz (Switzerland-*Beauveria brongniarti*), Mycontrol (USA-*Beauveria bassiana*), Ostrinil/Betel (France-*Beauveria bassiana*), Bowerol (CzeckRepublic-*Beauveria bassiana*), BioPath (USA-*Metarhizium anisopliae*), BioPath (Switzerland-*Metarhizium anisopliae*), Biologic (Germany-*Metarhizium anisopliae*), Pfr (USA-*Paecilomyces fumosoroseus*) (Sterk et al. 1996) and, Mycotal/Vertalec (Netherland-*Verticillium lecanii*) (Khan et al. 2015, Masuda and Kikuchi 1993). These products are being used on a large scale throughout the globe. Many firms are engaged in production of entomopathogenic fungi for use against whiteflies.

7.3.1 Biopesticides

Use of Entomopathogenic Fungi: Many insect pests require the integration of biological and chemical control operations to contain them. Microbial pesticides/ natural pesticides containing entomopathogenic fungi/bacteria have been exploited for the management of whiteflies. For example, fungi, mainly *Beauveria bassiana, Metarhizium anisopliae, Lecanicillium lecani,* and *Isaria fumosorosea* were tested for their efficacy against the whitefly, *Aleurodicus dispersus* (Hemiptera, Aleyrodidae) on the cassava *Manihot esculenta* (Boopathi et al. 2015). Biopesticides are a good candidate for use in integrated pest management (Gelman et al. 2008). The microbial pesticides have an edge over other control tactics because of their safety to wildlife/humans/other organisms, and the environment, along with many more benefits already discussed (Usta 2013). Better plant and root development were also noticed with microbial pesticides. However, the reduced efficacy with UV rays, species-specific mode, and nonavailability of standard formulations, and rapid degradation in storage are the limiting factors of the microbial pesticides. Thus, while biopesticides can be a good tactic to be a part of IPM, biochemicals and microbial entomopathogenic fungi should be roped in.

Use of Bacteria: The toxin-producing bacteria, *Chromobacterium subtsugae* (Martin et al. 2007) and *Photorhabdus luminescens* (Blackburn et al. 2005) were identified as entomopathogenic organisms against insects. A bacterial toxin of high molecular weight protein complex a (Tca) of *Photorhabdus luminescens*, has been found toxic to neonates of sweet potato whitefly, *Bemisia tabaci* (B-biotype), and caused 50% mortality of nymphs. The toxin produced by these bacteria is a chemical compound that acts as an insecticide to control whiteflies and can be included for better management of whiteflies. The addition of arabinose, mannose, ribose, and xylose to the artificial diet causes mortality of whitefly nymphs and adults. The high molecular weight protein complex (Tca) when fed to whiteflies orally caused significant mortality of the target insect. With the advancement of science, another bacterium, *Macrolophus caliginosus*, has been exploited to manage the whitefly *Bemisia tabaci* (Alomar et al. 2006). At another location, the effectiveness of bacteria *Bacillus thuringiensis* against whiteflies has been demonstrated by Al-Shayji and Shaheen (2008). It is considered as highly effective against the first instar nymph of the whitefly.

7.4 Botanicals

In the earlier days before the introduction of organic pesticides, man was dependent on plant extracts for solving ailments of plants. Later on, the pesticides of plant origin were developed, and which provided some relief from the farmers' problems. Products with insecticidal value were developed from plants. These included nicotine sulfate, rotenone, pyrethrins, etc. Margosa-O (Azadirachtin-based) from plants and other insecticides such as calcium sulfate (Miles 1927) were the primitive insecticides. The first insecticide registered by the environmental protection agency of the USA was Margosan-O. Now more than 2,121 plant species with insecticidal values have been explored and identified. Besides the insecticide-yielding plants (*Ageratum* and *Tagetes,* neem, etc.), several plants with antifeedant, attractant, and repellent properties have been discovered (Purohit and Vyas 2004). Neem products (Neemark 0.4%/Neemark 0.5%) are now available in the market on account of their efficacy against *Bemisia tabaci* on cotton (Phadke et al. 1988). Margosan-O was tested for controlling *Bemisia tabaci* on cotton (*Gossypium hirsutum*) in Phoenix, Arizona (Table 7.2). Biorationals were found effective, causing a significant reduction (60%) in the population of nymphs of whiteflies (Flint and Parks 1989). In another instance, cotton leaves sprayed with neem product containing azadirachtin (@138.7 ml a.i./ha) caused significant reduction in the population of whiteflies (Mordue and Nisbet 2000). The lower concentration was, however, ineffective. The use of neem oil @0.5% and neem seed extract @5% caused a reduction in the whitefly population to the extent of 56.4 and 43.7%, respectively (Nimbalkar et al. 1994). In a subsequent study, the efficiency of neem oil (2%) against eggs (74.4%) and nymphs (100%) of whitefly after 24 hrs was demonstrated (Roychoudhury and Jain 1993). The application of neem oil for the management of the initial infestation of whitefly has been advocated in insecticide resistance management (IRM) strategies (Kranthi and Russell 2009). Neem oil @2.5 liter/ha during the early part of the season is suggested for IRM strategies focused on sucking pests. These products have not been popularized because of their slow action. It is one of the biggest constraints in the adoption of botanicals, that is, their lack of knockdown against the whitefly. It is necessary to use repeated applications due to low toxicity and rapid biodegradability. The second constraint is the unavailability of standard and economically viable

Table 7.2. Use of biorational remedial measures against the whitefly.

SNo.	Species	Pesticide	Category	Reference(s)
1	*Bemisia tabaci*	Azatin; Neem azal	Azadirachtin	Greer 2000
2	*Bemisia tabaci*	Naturalis-O; *Clonostachys rosea*	Entomopathogenic (Fungi)	Greer 2000, Anwar et al. 2018
3	*Bemisia tabaci*	Hot pepper wax; Botanigard	Plant-based	Greer 2000
4	*Bemisia tabaci*	Neemguard, RD-9, Repelin Neemolin	Plant-derived oil	Puri et al. 1991
5	*Bemisia tabaci*	Garlic Gard	Garlic-based	Greer 2000
6	*Bemisia tabaci*	M-pede; Safer	Soap/surfactant	Greer 2000
7	*Bemisia tabaci*	Fish oil resin soap, Nirma	Soap/surfactant	Natarajan et al. 1991, Puri et al. 1991
8	*Bemisia tabaci*	Trilogy 90EC; Triact 90EC	Neem oil	Greer 2000
9	*Bemisia tabaci*	Rakshak Gold, Econeem, Nimbecidine, Achook	Neem-based/Botanical	Natarajan and Sundaramurthy 1990
10	*Bemisia tabaci*	Neem oil	Azadirachtin	Mann et al. 2001
11	*Bemisia tabaci*	PFR-97	Entomopathogenic	Greer 2000
12	*Bemisia tabaci*	Golden Soybean natural spray oil	Soybean oil	Greer 2000
13	*Bemisia argentifolii*	Predator	*Delphastus catalinae; Nephaspis oculatus*	Liu and Stansly 1999
14	*Bemisia tabaci*	Predator	*Amblyseius swirskii*	Stansly and Natwick 2010
15	*Trialeurodes vaporariorum*	Entomopathogenic (Fungi)	*Beauveria bassiana* and *Isaria foumier*	Mascarin et al. 2013, Quesada-Moraga et al. 2006
16	*Aleurodicus dispersus*	Entomopathogenic (Fungi)	*Beauveria bassiana* and *Isaria fumosoroseus*	Boopathi et al. 2015
17	*Bemisia tabaci, Singhiella cardamom, Trialeurodes vaporariorum,* and *Aleurolobus barodensis*	Entomopathogenic (Fungi)	*Paecilomyces fumosoroseus, Beauveria bassiana,* and *Verticillium*	Faria and Wraight 2001, Wraight et al. 2000
18	*Bemisia tabaci, Singhiella cardamom, Trialeurodes vaporariorum,* and *Aleurolobus barodensis*	Entomopathogenic (Fungi)	*Paecilomyces, Lecanicillium psalliotae, Aspergillus ustus, Metarhizium anisopliae*	Quesada-Moraga et al. 2006

Table 7.2 Contd. ...

...Table 7.2 Contd.

SNo.	Species	Pesticide	Category	Reference(s)
19	*Trialeurodes vaporariorum*	Entomopathogenic (Fungi)	*Beauveria bassiana* and *Lecanicillium muscarium* in combination	Mascarin et al. 2013
20	*Bemisia tabaci* (B-biotype)	Entomopathogenic plus *Trisilo* surfactant	*Beauveria bassiana* (Vuillemin Strain)	Mascarin et al. 2014
21	*Bemisia tabaci*	Entomopathogenic (Fungi)	*Clonostachys rosea*	Prada et al. 2008, Anwar et al. 2018
22	*Trialeurodes vaporariorum & Aleurolobus barodensis*	Entomopathogenic (Fungi)	*Hirsutella thompsonii, Lacanicillium lecanii, Lacanicillium longisporum, Isaria fumosorosea, Paecilomyces lilacinus & Lecanicillium muscarium*	Fargues et al. 2005
23	*Bemisia tabaci*	Entomopathogenic (Mixture)	*Cordyceps javanica & Beauveria bassiana* (Mixture)	Silva-Aguao et al. 2019
24	*Bemisia tabaci*	Entomopathogenic (Bacteria)	*Photorhabdus luminescens,* and *Cromobactehum substage, Macrolophus caliginosus*	Alomar et al. 2006
25	*Aleurodicus dispersus*	Entomopathogenic (Fungi)	*Beauveria bassiana, Metarhizium anisopliae, Lecanicillium lecanii & Isaria fumosorosea*	Boopathi et al. 2015
26	*Bemisia tabaci*	Entomopathogenic (Bacteria)	*Bacillus thuringiensis*	Al-Shayji and Shaheen 2008
27	*Bemisia tabaci*	Entomopathogenic plus parasitoid	*Cordyceps javanica & Eretmoceus hayati* (Mixture of Fungus & Parasitoid)	Qu et al. 2019a
28	*Bemisia tabaci*	Predator	*Chrysoperla carnea*	Zia et al. 2008
29	*Trialeurodes vaporariorum*	Entomopathogenic (Mixure)	*Beauveria bassiana & Lecanicillium muscarium*	Malekan et al. 2013

formulations of neem. These are also rapidly inactivated by UV light, temperature, and leaf surface pH conditions. However, these neem products can play a significant role in the management of whitefly in the organic production of vegetables. The extract from tomato containing 2-tridecadone showed promise against whitefly.

Mass Production and Commercial Use: The procedure for mass culturing of whitefly on tobacco plants was suggested (Lynch and Johnson 1989). In this technique the two cultures of tobacco plants infested with whitefly and the other one of parasitoids and whitefly are maintained. The parasitization of healthy 3rd and 4th instar pseudo nymphs is done by introducing the parasitoid. The black parasitized pseudo pupae are obtained on host leaves. The eggs carrying host plants from oviposition culture are taken to place in the chamber containing cultures of parasitoid. The parasitized black pseudo pupae are now available and the strips of parasitized pupae can be prepared to use or transport to other areas. The parasitoid when not in use can be stored in the cold storage. The effect of weather parameters was studied on biological characteristics. It was found that parasitoids can be kept in cold storage from 4.5±°C and 11.5±°C for 14 d and 7 d, respectively without any ill-effects (Lopez and Botto 2004). The parasitoid pupae stored at the above temperature can delay the emergence of parasitoids or can be transported to long distances for use. Similarly the predators *Chrysoperla carnea* was successfully reared on eggs, brewer's yeast, honey, sucrose and ascorbate for 10 generations. It is also important to provide artificial diet to adults during the period of nonavailability of host. Swap of cotton dipped in honey solution (10%) can be hanged in the rearing cage as food for adults (Thompson 1999). Adults are also fed on aphids. The other method is to rear adults of predators on plants enclosed in three types of transparent cages (Sattar and Abro 2011). In this technique, a pair of male and female is kept in cages viz; Perspex cage, transparent glass, wooden cage. Of these, glass cages are better to culture the predator. After preparing the egg cards (eggs glued on card) the predator, *Chrysoperla carnea* is ready for use or transport to far off places (Zia et al. 2008). These cards are hanged on to plants. On hatching, the larvae would start feeding on nymphs of whitefly alongwith the adults of predator. The study carried out in Pakistan proved the efficiency of green lace wing in controlling whitefly using egg cards (@30,000 eggs/acre). Multiple releases were made at an interval of 15-days (Zia et al. 2008).

This chapter provides an overview of the natural enemy complex of whiteflies. The recent advances in the use of biopesticides against the whiteflies have been reviewed. Many recent developments have taken place in the field of natural enemies and botanicals. A large number of predators has been identified but these could not equate with the abundance of parasitoids. Much headway has been made in using parasitoids against whiteflies. Based on the efficacy of this category of natural enemies, a number of firms have come forward to mass produce natural enemies for use against this notorious pest. Though the parasitoids are available there is a need to further refine the technology for mass production and to apply under field conditions. Latest technology of mass rearing of parasitoids and predators is suggested. There is also a need to develop techniques for the mass production of predators. It is important to lay more stress on the conservation of natural enemies. Several entomopathogenic fungi have been identified. Earlier, *Beauveria bassiana, Metarhizium anisopliae* were commonly used on whiteflies. *Aschersonia, Paecilomyces,* etc. have also been added now to the list of efficient entomopathogenic fungi. However, there need is to develop a methodology to reduce their dependence on weather. Likewise, recent advances have been made in the application of botanicals but still, the products with quick knockdown are lacking. The existing products in this category act slowly on the host and thus could not be popularized. New products (biorationals) have been introduced and are likely to stay as a vital part of the integrated pest management of whiteflies. These biorationals include entomopathogenic fungi, plant extracts, soaps and detergents, oils, natural enemies and have been found highly effective against whiteflies. The exact biorational products used against whiteflies have been presented in the table. Besides, there are many more products from biorational category which are available in the market for use against whiteflies throughout the globe.

CHAPTER 8

Insect Behavior Alteration Approaches

◇◇

It is a unique tactic in which the behavior of the target species is altered to the disadvantage of insects. The major benefit of this tactic is its compatibility with all other methods applicable for use against insect pests. Above all, this technique doesn't create pesticide-resistance problems in insect pests. The chemicals involved are such that the insect hardly possesses the ability to develop resistance. In this technique, the behavior of the whiteflies is manipulated and utilised in three ways viz:

  i) Mulches,
 ii) Semiochemicals, and
iii) Changes in the pest physiology

With the altered behavior, the whiteflies are either attracted to the host or repelled away from the host. The whiteflies are lured to the source killed with an insecticide. The other way is to repel the whiteflies away from the target crop by artificially treating the crop to be protected. Inspite of its benefits over other methods of pest management, it has not been tried or exploited on a large scale. Thus, these approaches can be utilized to control whiteflies. The tactic is still in the infancy stage. It is expected that soon it will prove to be one of the best tactics of the integrated whitefly management system. This kind of control measure is always long-lasting and free from side effects. The information available on the use of these approaches involved in alteration of the behavior of whiteflies is discussed.

8.1 Mulches

Mulch is a popular material used to cover the field entirely to conserve moisture in the deficient fields and had nothing to do with the control of pests in the earlier days. The covers used were mainly of straws, polythene sheets, or aluminum sheets. At many places, scientists have described this aspect under the cultural practices. The other item responsible for modifying the behavior of whiteflies is through the UV rays. It is a known fact that the flight behavior of whiteflies is greatly modified by the ultraviolet rays of the environment. The whiteflies are attracted to the UV light of the intensity of 320–400 nm through optical barriers, which thereby alters the behavior of this insect. This aspect of behavior was exploited for the management of whiteflies on tomato in greenhouses. It was later on found that reflection of mulches reduces the insect settling. Hitherto, farmers had been ignorant about the repellent effect of mulches on the settling of whiteflies in the field. From then, mulches (either living mulches or plastic/synthetic mulches) have been used in pest management as the pest's reflection (from the mulch) disrupts the settling of sucking pests like whiteflies and aphids (Summers et al. 2005, 2004a, 2004b). It has been demonstrated that various colored mulches (straw, aluminum, or plastic colored) not only conserve moisture but are instrumental in successfully reducing the density of insects also. The use of mulches has shown promise against vector whiteflies in tomatoes (Csiznszky et al. 1999, Csiznszky et al. 1995). Further, it has been confirmed that both reflective plastic and wheat straw mulches repel silver leaf whiteflies in cantaloupe. The silver leaf whitefly population (adults and nymphs) has been greatly reduced with the use of plastic/wheat straw mulches. With the application of this technology, cantaloupe crop was protected against whitefly vector. In Fresno (California) alone, around there has been a 13% reduction in area of cantaloupes due to the emergence of serious pests (Jetter et al. 2001). So much so, the California Melon Research Advisory Board (2003) took immediate steps and listed it as a priority measure to manage whitefly. At the UC Kearney Research and Extension Center (KREC), the use of only reflective plastic mulch and straw mulch against silver leaf whitefly in cantaloupes was advocated. The chemical control of vectors enhances the incidence of non-persistent stylet borne/gut borne viral diseases (Jetter et al. 2001, Gibson et al. 1989, Ferro et al. 1980). The use of insecticides against vector is known to increase the activity of vector. As a result the whitefly moves from plant to plant very rapidly; the probing of the host plant is also fast, particularly in case of the spread of stylet-borne viruses. The development of pesticide resistance is a major factor that has restricted their use in crops suffering from viruses. In line with this, the development of resistance to imidacloprid and other insecticides among whiteflies has been reported and it is of major concern (CMRAB 2003, Elbert and Nauen 2000, Prabhaker et al. 1998). However, as the whiteflies acting as vectors are repelled by UV wavelengths, the outcome is the delay in symptoms of plant viruses by 3–6 weeks (Summers et al. 2004, Summers and Stapleton 2002).

In a study conducted by Frank et al. (2005), synthetic mulches (reflective and white) and live mulches (live mulches of buckwheat, *Fagopyrum esculentum,* white clover, *Trifolium repens),* were tested against whiteflies. Out of these mulches, the live buckwheat and two synthetic and reflective are commonly used and found promising against whitefly vectors of plant viruses. The incidence of whitefly (*Bemisia tabaci*)-borne viruses viz. Tomato Mottle Virus in Florida and Tomato Yellow Leaf Curl Virus in Jordon was drastically reduced with the use of aluminum or silver reflective mulches (Csizinszky et al. 1995, Suwwan et al. 1988). The living mulches are responsible for delaying the whiteflies on bare ground as well (Hilje and Stansly 2008). The masking of tomato target plants with established living mulches (*Arachis pintoi, Drymaria cordata, Coriandrum sativum*) and aluminum foils proved highly beneficial in the management of *Bemisia tabaci* in Costa Rica (Greer and Dole 2003, Csizinszky et al. 1999).

Additionally, UV-absorbing plastic films for excluding whiteflies have also been put to use in recent years (Antonious and Snyder 2008). Other scientifically based findings on the use of UV-protecting plastics reported significantly better control of the whitefly, and the Tomato Yellow Leaf Curl Virus (Monci et al. 2019). Plastic/muslin mulches generally used to exclude the whiteflies are quite popular in Israel and Egypt. The protective cover is used to enclose the screen house (Berlinger et al. 1983). Further, the study demonstrated that the UV-deficient greenhouse affects whitefly attraction and flight behavior (Antignus et al. 2001). On the other hand, the UV-absorbing plastic sheets, when used as a cover, protected the crops against viral diseases vectored by whiteflies (*Bemisia argentifolii*) (Ben-Yakir et al. 2012, Costa and Robb 1999, Antignus et al. 1996). In another study the use of plastic films/UV radiations showed a significant reduction in virus incidence in tomato from 96–100% to 6–10% in greenhouses (Kumar and Poehling 2006). The use of reflective mulch against *Bemisia tabaci* in watermelon (*Citrullus lanatus* var. *lanatus*) (protective and field conditions) was found highly effective (Simmons et al. 2010). Polyethylene mulch with a reflective silver stripe and yellow summer squash, *Cucurbita pepo* L. trap crop were tested alone and in combination to reduce densities of *Bemisia argentifolii*; the silver reflective mulch reduced pest population through the reduced fecundity (Smith et al. 2000). However, the repellent-metalized mulches offered limited protection compared with genetic tolerance to the virus (Smith et al. 2019). The additional use of companion crops (basil) and covers as a part of integrated management against *Bemisia tabaci* was tried in tomato in Kenya and found highly promising (Mutisya et al. 2016).

8.2 Semiochemicals

The volatile secondary metabolites between plants and other organisms have long been seen in crop protection (Pickett and Khan 2016). The ecofriendly novel approaches based on a modification of the behavior of whitefly are likely candidates in IPM in the future (Darshanee et al. 2017, Darshanee et al. 2016). These are the chemicals released by one species of an organism which elicit a response in the receiving individuals of the same species or of another species. Semiochemicals were first identified in transgenic hexaploid wheat when it was used in defence against the attack by an aphid species called (E) Beta-Farnesene. It was perceived by the aphid, *Myzus persicae,* and used to protect themselves from predators (Bruce et al. 2015). Subsequently, a large number of chemicals in the form of semiochemicals were identified. Normally, these are categorized into two parts;

i) Allelochemicals

ii) Pheromones.

Of these, the former category is defined as the chemicals released by an organism which elicit a response in the receiving individuals of a different species; the latter category called pheromones include chemicals released by an organism to the outside environment which elicit a response in the receiving individuals of the same species. Allelochemicals are further divided into allomones (where the benefit goes to the emitter of chemicals), kairomones (the benefit is harvested by the receiver of a different species), apneumones (chemicals emitted by the dead material elicit a response in the individuals of a different species), and synomones (the benefit goes to both the emitter as well as the receiver). Of these four categories, the allomones and kairomones have been exploited for managing insect pests. Pheromones have been utilized for monitoring the population of various insect pests, particularly, the lepidopterous (moths) and coleopterous (bark beetles) pests. Pheromones have also been tried and advocated for mass trapping, male-confusing techniques, and mating purposes in insects. The use of these chemicals is limited but there is a scope for their use against insects preying on vegetables.

As has been mentioned in earlier chapters, the whiteflies, namely, *Bemisia tabaci* and *Trialeurodes vaoporariorm*, are serious pests of crops, especially vegetable crops. Hitherto, the management of whiteflies was mainly tackled through the use of pesticides to which these insects have developed resistance at a very faster rate. To replace this component of pesticide use or to reduce the dependence on pesticides, behavior-modifying chemicals such as semiochemicals can be called on to fill this gap. Though the use of semiochemicals is still in the infancy stage and has not been tested on the field scale yet, there is every possibility that these compounds emitted by plant species will be utilized soon and a sound system of plant pest management would be in place. This tactic is comparatively more reliable and efficient for managing insect problems as one of the components of IPM. Gas chromatography (GC) with electroantennography (GC-EAG) and a detector (flame ionization detector) is an essential tool to identify semiochemicals within a blend of complex compounds. This noble approach is still in the experimental stage. However, the basic/preliminary studies to evaluate the use of semiochemicals in pest management have indicated great success. The pest control achieved through this approach will be of permanent nature.

It is a fact that the semiochemicals can replace the pesticides-based strategy through the exploitation of natural volatile signal process with the manipulation of insect behavior. For example, the semiochemical is released by the organism (plant) but received by the individuals of other species from the parasitoid category (*Encarsia formosa*—a parasitoid of whitefly) and the advantage goes to the receiver as the compound enhances the searching ability of the parasitoid. The changes in whitefly behavior are brought about with volatile chemicals or blends of such chemicals (Darshanee et al. 2017, Cao et al. 2008). Whiteflies are known for having stronger olfactory senses than any other pest species. It is unique characteristic of this pest. The olfactory organs are present on its antennae and can be exploited by semiochemicals. The emitted volatiles are eventually characterized by electrophysiology or olfactometry through antennal apparatus. To this end, scientists have studied the ultrastructure of antennal sensilla in three whitefly species *Bemisia tabaci, Trialeurodes vaporariorum,* and *Aleyrodes proletella* (Zhang et al. 2015, Mellor and Henderson 1995). The volatile binding and chemosensory proteins were detected in *Bemisia tabaci* during the transcript analysis (Wang et al. 2017). The chemosensory proteins of *Bemisia tabaci* have been shown to bind volatiles (Liu et al. 2016). The whitefly olfactory system being highly developed is able to isolate the stereoisomers of volatile organic compounds (VOCs) (Bleeker et al. 2011). The semiochemicals emitted by tomato plants are perceived by either the whiteflies or the pest's natural enemies with detection by their antennae. In this context, several terpenoids have been exploited for the control of whiteflies as kairomones are perceived by the natural enemies of whitefly in tomato (Bleeker et al. 2009, Degenhardt et al. 2003). In one method, paper cards saturated with semiochemicals (Sesquiterpenes Zingiberene and Curcumene and the monoterpenes p-Cymene, alpha terpene, and alpha phellandrene) are attached to the branches of the tomato. On hanging these cards, the emission of chemicals leads to the trigger of a strong repellent of whiteflies. When semiochemicals were studied using T and Y-type olfactometer involving the insect antennae, two types of effects were recorded. These effects are either attractant ((E)-Linalool, Phenols, (E)-Caryophyllene, (R)-Limonene, 2-Ethyl-1-Hexanol, 3-Hexen-1-ol, 1,8Cineole) or repellent (R-limonene, Geranyl nitrile, myrcene E-Ocinene, Citral, Alpha pinene) (Schlaeger et al. 2018). These semiochemicals have been identified using *Trialeurodes vaporariorum* and *Bemisia tabaci* as more agriculturally important than the other known species.

The wild types lines containing a high content of Acyl sugars and zingiberene in tomato have been identified as highly resistant to the B-biotype and *Bemisia argentifolii* of *Bemisia tabaci* (Neiva et al. 2019). As a result, the reduction in female fecundity and the heavy mortality in nymphs population was conspicuous in an experiment. It was discovered to be a type of antibiosis resistance. The resistant varieties with a high content of the above said chemical have been identified. Another aspect of behavior modification through semiochemicals is being tapped to utilize in the control of whiteflies. This comprises of planting react to the attack of this herbivore and emitting volatile compounds that are perceived by insect fauna. These volatile compounds are emitted into the atmosphere and perceived by other organisms. Secondary plant metabolites containing small lipophilic molecules (SLMs) are promising breakthroughs in crop resistance to pests. It is known that the wild ancestors of the pyrethroids, neonicotinoids, and butenolides had an abundance of SLMs as compared to present-day cultivars (Birkett and Pickett 2014). The semiochemicals are still in the experimental stage. The volatile terpenoids emitted by wild tomato plants (*Solanum pennellii, Solanum habrochaites,* and *Solanum peruvianum*) are perceived by whiteflies and these are being exploited for whitefly management. An indirect but important approach to control whiteflies is to identify plant-mediated semiochemicals that attract natural enemies. For example, (Z)-3-Hexen-1ol, (E)-4,8-dimethyl-1,3,7 the nonatriene (DMNT), and 3-octagon are emitted in higher quantities by the bean plant, *Phaseolus vulgaris*, when attacked by *Trialeurodes vaporariorum* (Birkett et al. 2003). The synthetic form of these VOCs, in presentation individually or in a blend, enhanced the attraction of parasitoid, *Encarsia formosa*, towards the *Arabidopsis thaliana* plants which emit myrcene and had been hitherto attacked by *Bemisia tabaci* (Tu and Qin 2017). With the application of synthetic myrcene semiochemical to *Arabidopsis thaliana* plants, they attracted more *Encarsia formosa* parasitoids which reduced the whitefly population (Zhang et al. 2013).

Semiochemicals are an important component of push-pull technology which combines repellents and attractants in the same cropping system. The pest is deterred from the crop plant (push) and lured to a more attractive source (pull) at the same time (Cook et al. 2007). This strategy is especially suited for the control of greenhouse pests, such as *Trialeurodes vaporariorum*, for example, because of the confined area (Cook et al. 2007). Semiochemicals can be incorporated in the cropping systems through cultural practices, such as intercropping by planting semiochemical-emitting plants near the crop. The intercropping of tomato (*Solanum lycopersicon*) with coriander (*Coriandrum sativum*) helped the males and females in recognition of volatiles (kairomone effect), indicating the role of semiochemicals in selection. The tomato intercropping with coriander caused an odor-masking effect on tomato volatiles, thereby interference in the host selection of *Bemisia tabaci* (Togni et al. 2010).

The co-occurrence of *Trialeurodes vaporariorum* (Westwood and *Bemisia tabaci* (Gennadius)) on tomato in a screenhouse in Japan was studied via a Y-tube olfactometer. The distribution of whiteflies (*Trialeurodes vaporariorum* and *Bemisia tabaci*) adults within the canopy of the plant were found confined to upper and middle leaves, respectively. A similar trend was noted concerning oviposition. The selection of leaflets was due to the different volatile compounds emitted by the tomato leaflets (Tsueda et al. 2014). Using the same apparatus, a Y-tube olfactometer was used to investigate the repellent effects of celery on Q-biotype of *Bemisia tabaci*, when it was planted along with cucumber (*Cucumis sativus*). The effectiveness of the repellent qualities of the whitefly's less-preferred vegetables viz. celery, asparagus lettuce, Malabar spinach, and edible amaranth in the suppression of two biotypes of the sweet potato whitefly, *Bemisia tabaci* (Gennadius) (Hemiptera: Aleyrodidae) on cucumber, *Cucumis sativus* L. (Cucurbitaceae) was assessed. Intercropping of celery/Malabar spinach with cucumber significantly reduced whitefly numbers on cucumber. Y-tube olfactometer behavioral assays revealed strong repellency felt by the whiteflies. Of the two whitefly biotypes, the biotype B showed the greatest repellency when faced with asparagus lettuce extract, while celery and Malabar spinach caused immense repulsion in the case of Q-biotype. In both the situations, two major volatiles, D-limonene from celery and geranyl nitrile from Malabar spinach, were identified (Zhao et al. 2014). The Western European celery varieties (Juventus and Ventura) and the Chinese celery variety (Jinnan) showed strong repellency against the whitefly. The repellency could be due to the presence of D-Limonene, β-myrcene, and (E)-β-ocimene compounds, D-Llimonene (Europe), and β-myrcene (Chinese) which were experienced at different locations (Tu and Qin 2017). The reduction in the adult population of *Bemisia tabaci* on tomato was recorded with the planting of coriander (*Coriandrum sativum*), Greek basil (*Ocimum minimum*) plants or citronella grass (*Cymbopogon* species), nearby. These plants were used either as intercropping or live mulch (Carvalho et al. 2017). The economically viable possibility for field application of active semiochemicals is the deployment of synthesized VOCs via sprays or slow-release dispensers (Cook et al. 2007). The studies in this direction at field level focusing on whitefly as the target pest are lacking. However, the settling of *Bemisia tabaci* on tomato plants was significantly reduced in a greenhouse experiment. The bottles containing a mixture of (R)-limonene (1%), citral (slow-release agent), and olive oil (antioxidant) (in the ratio of 63:7:30), were used as a 'push' treatment and yellow sticky traps as the 'pull' treatment (Du et al. 2016, Sacchetti et al. 2015). The glandular trichomes are present on the tip of the hair and are a rich source of oils/fragrance and can be utilized to strengthen this kind of technology (Glas et al. 2012). The volatile compound emitted by the host plant as a result of a whitefly attack is released into the environment and perceived not only by whiteflies but the neighboring plants of the host species as well (Zhang et al. 2019).The studies were conducted in China to identify the volatiles in tomato and eggplant those can trigger a response in the behavior of the *Trialeurodes vaporariorum* (Darshanee et al. 2017). The volatile identified was ((Z)-3-hexen-1-ol) and terpenoids alfa-pinene, (E)-beta-caryophyllene, alfa humulene, azolene showed preference to whitefly adults both males and females. A volatile compound called salicylic acid (SA) is produced by the tomato in response to phloem-feeding whiteflies (Shi et al. 2016). A recent study conducted on tomatoes demonstrated the application of salicylic acid analog which strengthened the defense mechanism against the whitefly and Tomato Yellow Leaf curl Virus disease (Monci et al. 2019). *Arabidopsis thaliana* plants triggered the emission of SA on initiation of an attack by the whitefly as well as a gene-encoding chemical (ocimene/myrcene synthase) which is perceived by the parasitoid *Encarsia formosa* under greenhouse conditions (Jhang et al. 2013). Likewise, whitefly feeding on tomatoes/*Arabidopsis*, activate Jasmonic acid (JA)/Ethylene (ET) in addition to SA, which guard the plants against pathogens (Puthoff et al. 2010, Zarate et al. 2007, Kempema et al. 2007). As a result of feeding of whitefly, SA and JA are emitted in *Arabidopsis thaliana* that influenced the development of whiteflies. The mutant that activated the production of SA and suppressed the production of JA that enhanced the development of nymphs but in mutant that increased the production of JA and decreased SA content slow down the development of nymphs. With the application methyl jasmonate, the drastic reduction in the development of nymphs in latter case was recorded. This aspect can be exploited for management of whiteflies, through SA and JA defense mechanism. Thus the response of plants to whitefly feeding suggests that JA/ET and novel defense pathways are induced (Zhu-Salzman et al. 2004, Walling 2008, Zarate et al. 2007, Walling 2000). This compound is then also perceived by the plants grown in the vicinity of the host. The adjoining plants act as guards against the host in the adjoining fields. Another study carried out on tomatoes demonstrated the reduction in the production of volatiles in an environment with a higher content of nitrogen, thus leading to them holding a low attraction of whiteflies (Islam et al. 2017). The greenhouse whitefly, *Trialeurodes vaporariorum* (Westwood), reared on tomato *Lycopersicon esculentum* (Miller) treated with nitrogen at 308 ppm N in Spain reduced the thickness of the leaf, increased egg survival, size of pupal exuviae, and female tibial length whereas, the females developed on fertilized with 84 ppm N showed the lowest oviposition frequency. *Trialeurodes vaporariorum* populations raised on high nitrogen levels demonstrated a higher intrinsic rate of increase (Jauset et al. 2000). Both the species of whiteflies got affected when there was an increased dose of nitrogen in the tomato crop.

8.3 Tissue Culture

Tissue culture is also being exploited in the management of whiteflies but this technology is still in the development stage in this regard. Studies were conducted to control the greenhouse whitefly through the disruption of osmoregulation. The greenhouse whitefly *Trialeurodes vaporariorum* was allowed feeding access to transgenic plants of tobacco (Raza et al. 2016). The genetics of double-stranded RNA indicated that both Aquaporin (AQP) and Sucrose Gene Alpha-Glucosidase (AGLU) are involved in the disruption of osmoregulation as assessed through the PCR. The osmoregulation disruption caused 70% mortality of greenhouse whitefly after six days. An earlier study indicated somaclonal variation as a tool to develop resistant varieties of *Torenia fournieri* plants (Hadi and Bidgen 1996). The callus culture of *Trialeurodes vaporariorum* was initiated on modified Muashige and Skoogsalt medium (MS) with 2.26 uM 2,4 Dichlorophenoxyacetic MS medium amended with 2.46 uM Indolebutyric acid and 8.48 uM benzylatedenine to study whitefly regeneration. It significantly reduced oviposition and adult emergence of greenhouse whitefly *in vitro*. These chemicals are either repellents or attractants (Schlaeger et al. 2018). The semiochemicals like limonene and caryophyllene are preferred by whiteflies while myrcene, ocimene, etc. are repellent semiochemicals. The volatiles are emitted as a result of attack of whiteflies in crop plants and are known to alter the behavior of herbivore.

It is a unique tactic and can be exploited for the management of whitefly in integrated control programs as it is free from ill effects of other tactics. Behaviour modifying approaches discussed in the write-up include mulches and semiochemicals. The disruption of osmoregulation has also been included in this chapter. Hither to the use of mulches was confined to conservation of moisture. But now a days, this tactic is more important as control tactic as it is compatible with other control methods and mitigates the pesticide resistance problems. The mulches are mainly of straws, polythene sheets, or aluminum sheets which are tested and recommended/advocated against pests. The other approach responsible to modify the behavior is the UV rays as the whiteflies have attraction to UV light of 320–400 nm intensity. Semiochemicals in the form of allelochemics and pheromones-have been discussed in relation to whitefly. The paper cards impregnated with semiochemicals (Sesquiterpenes Zingiberene and Curcumene and the monoterpenes p-Cymene, alpha terpene, and alpha phellandrene) to repel whiteflies are commercially available for use. The Western European celery varieties (Juventus and Ventura) and the Chinese celery variety (Jinnan) showing strong repellency against the whitefly are available for cultivation as resistant varieties. The role of disruption of osmoregulation has been discussed.

CHAPTER 9

Mechanical and Physical Measures

◇◇◇

The mechanical and physical measures were the most common control practices earlier. In those days, the chemicals of plant origin were in use. Soon, toxic chemicals made from carbamates, organochlorines, organophosphates, etc. came into the market. With the easy availability of potent chemicals and quick relief from insect pests, the use of soft chemicals or non-chemical measures was almost wiped out of the scene. The exploration of more and more potent chemicals continued until the side effects of these chemicals became apparent. Once the side effects of these toxic chemicals became known to the general public, scientific pursuits for finding substitutes for toxic chemicals gained momentum. This resulted in research efforts targeting the development of age-old physical measures. To this end, plant protection machines were developed. Simultaneously, the cultivation of crops in the screen houses became popular and this organic cultivation of precious crops under protective cover started increasing at a faster rate. So much so that the demand for organic produce has also tremendously increased. With the popularization of organic produce, the need was felt to harvest organic produce without chemicals. The production of machines is the outcome of the protective cultivation of crops.

9.1 Mechanical Measures

Mechanical measures are now in use to control insect pests. In this component, the crops/plants are raised under protective places where the vector population does not reach. Several mechanical measures are used to mitigate insect pest problems. These measures are as under:

- Barbed mesh screenhouses can be put to use to avoid insects or vectors. Screens of brass with mesh size of 0.213 mm^2 or less are known to deny whiteflies (Bethke et al. 1994). Thus, the crops escape the invasion of insects. All the crops cannot be grown in a protective cover, so only the precious crops can be protected with this technique and that too over a small area. The important point to be kept in mind is the low threshold of the insect vector; therefore this measure cannot be put to use in such a situation. However, it is highly suitable for the control of nonvectors.

- The control tactics should be selected as per the requirement of control operations. In this context, the use of homemade organic products and the use of spray of water under pressure are effective tactics to dislodge the soft-bodied fragile insect pests (Greer 2000).

- Another mechanical technique is to manage the whitefly by sucking and removing insects from smaller areas. The vacuum machines are now available to reduce the pressure of pests. Also, the whiteflies with massive populations can be sucked by the vacuum machines (Weintraub and Horowiitz 2001). Nowadays, tractor-propelled vacuum for mass catching of insects is in use.

- The protective cover (Berlinger et al. 1983), mulches (Simmons et al. 2010), oils (Butter and Rataul 1973), and surfactants (Greer 2000) are in use for reducing whiteflies' damage in glasshouses or other protected/open field cultivation of crops. The Sunspray Ultra-fine Spray Oil™ (mineral oil) is still being used on tomatoes grown in Greenhouses to control greenhouse whitefly (Williams and Pat 1995). It is better than M-Pede™ insecticidal soap being advocated against whiteflies. The whiteflies remain hidden on the underside of a leaf and evade the exposure to pesticides, but the lower leaf surface contributes towards the heavy buildup of the population. In case the threshold is low, or the whitefly is acting as a vector of viruses, this technique will not work.

- The hidden population of whitefly consisting of nymphs and pupae can easily be reduced by plucking the highly infested leaves of cotton and destroying the debris of cotton carrying the immatures of whiteflies. Moreover, the whiteflies acting as a vector of viruses can acquire the virus from the leftover foliage and can inoculate viruses in some crops, particularly the tomatoes. By plucking the leaves, the further spread of the virus can be checked.

- Whitefly-borne viruses appear in a crop suddenly as these whiteflies are transported by air. The mulching can provide great help in repelling the aerial vectors. These mulches can be of aluminum or UV light (Antonious and Snyder 2008). The floating row cover for an optimum period in tomato has been found useful in minimizing the incidence of the Tomato Leaf Curl Virus via reducing whitefly population through the modification of behavior (Al-Shihi et al. 2016). The use of shiny metallic mulch in the field is quite useful for repelling the whiteflies and therefore, it is exploited as a tool to check whitefly menace in many places. The transparent polyethylene mulch in cantaloupe (*Cucumis melo* L.) in a tropical region reduced whitefly populations and virus incidence as compared to control without mulch (Orozco-Santos et al. 1999).

- Vacuuming is an easy and highly useful practice on a small-scale collection of insects in a protected crop. With the success of hand-operated machines, the electrostatic sweepers were developed and are in operation. Recently, the electrostatic insect sweepers developed in Japan were tried against whiteflies. The results were highly satisfactory (Jakkawa et al. 2015). These are now being utilized for large scale protection of crops. Greenhouse tomatoes have been menaced by *Bemisia tabaci* and the application of an electric field screen and a portable electrostatic insect sweeper with eight insulated conductor wires (ICWs) were arranged at constant intervals along a polyvinylchloride (PVC) pipe and covered with a cylindrical stainless net (Takikawa et al. 2015).

9.2 Physical Measures

The whiteflies are quite manageable in the case of small-scale control using physical measures. These measures are deployed to check the whitefly infestation. The screenhouses should be such as to deny entry to the population of whiteflies or these protective plants can be covered with plastic films to check the entry of whiteflies. For that purpose, the mesh size for the construction of the screen house is 0.19 mm/230 × 900 um or less to exclude the whiteflies. As mentioned earlier, brass screenhouses with a mesh size 0.213 mm^2 or smaller can also be used to exclude whiteflies (Bethke et al. 1994). In case the mesh size is bigger than the prescribed one, the entire house can be covered with muslin/nylon nets/plastic film (Tanaka et al. 2008, Kakutani et al. 2012, Bethke and Paine 1991, Berlinger et al. 1983). These covers can be electrically charged (Takikawa et al. 2016, Takikawa et al. 2015). This strategy is extremely useful to protect the crops from the attack of whiteflies/viral diseases. Floating row covers are also put to use to get rid of whiteflies (Mutisya et al. 2016, Gogo et al. 2014, Ben-Yakir et al. 2012, Qureshi et al. 2007).

9.2.1 Release of Natural Enemies

The natural enemies are available in two categories viz. predators and parasitoids (Liu et al. 2015). Among the predators, the large-scale use of *Chrysoperla carnea* is common. The predators are slightly difficult to mass rear in the laboratory, so they are not exploited to that extent. The other category of natural enemies is the parasitoids. These are mass-reared and have been popular in the control of a large number of insect species. Amongst parasitoids, the *Trichogramma* species is quite common. The trio cards containing the parasitoids in eggs are hanged on branches. The adults on emergence from the pseudo pupal stage immediately start parasitizing the host eggs. Initially, these have been exploited for the management of tissue borers of crops. But now it has been extended to many other insect pests such as moths and butterflies. It includes tomato, horn worm, *Helicoverpa armigera,* codling moth, cutworm, armyworm, cabbage looper, fruit moths, diamond back moth, sugarcane borers, tent butterfly, etc. There are more than 10 species of egg parasitoid *Trichogramma chilonis, Trichogramma, Tricogramma australicum, japonicum, Trichogramma pintoi, Trichogramma oleae, Trichogramma cacoeciae, Trichogramma bourarachae,* etc. For the management of whiteflies, the parasitoids of *Encarsia* species and *Eretmocerus* species are available in the form of cards in the market and these parasitoids are mentioned elsewhere in this book along with hosts insect. These cards are becoming extremely popular. There are preconditions for the successful utilization of these natural enemies. The major condition is to ensure the host population. For more information, please refer to Chapter 7 in this book.

9.2.2 Pest Monitoring/Surveillance

Traps containing the sex pheromones are generally used by insects for finding mates, enhancing foraging activity, and selection of habitat. These are also used to monitor population for initiating sprays against pest species. These are also lured to the source and killed. Beside, these traps are placed to lure the pest to trap out the population. To make this strategy successful, the trap must trap out 95% population of pest species. The other category of traps are yellow sticky traps exploited to trap out the population of whiteflies. The study carried out on *Aleurocanthus spiniferus* in China garden demonstrated the attraction of this tea garden towards yellow color. The whiteflies are attracted towards yellow color as these whiteflies have a preference for yellow followed by, pink, red, white, and others (Wang et al. 2015). These yellow sticky traps are used for applying control measures on whiteflies (McHugh 1991). These traps are available in the market to monitor the population of whiteflies in the different agroecosystems. To estimate the population, generally, 1–4 yellow sticky traps per 300 square meter areas are enough. Traps are utilized in two ways: for monitoring the population of host insects and for trapping larger populations of host insects from the small areas. These traps target adult whiteflies and need to be replaced at an interval of one week. The traps are generally available in 6 cm × 15 cm size and can be prepared at home. The cardboard of the specified size is taken and painted yellow with petroleum jelly or motor Mobil oil and the ready traps can be placed just above the crop canopy. The entry of whiteflies is generally through doors meant for the entry of a worker, so a trap can be placed near there. The yellow sticky traps should be about 10 cm above the crop canopy. In case the population of whitefly trapped is 5 per trap/week, the crop must be sprayed with pesticides so that it may not reach economic injury level to inflict economic loss.

9.2.3 Blasting with Water

Blasting the whitefly with water with heavy pressure can kill it. It should be done once a week and carried on continuously for three weeks. This approach can work well on a limited scale to dislodge the soft-bodied weak fliers.

9.3 Use of Botanicals

As many as 2,121 plant species have been reported to possess pest control properties. Of these, 1,005 plants have insecticidal, antifeedant, attractant, and repellent properties (Purohit and Vyas 2004). The botanicals are substances derived from plants and are known to contain antifeedants, attractants, and repellents (Deletre et al. 2016). Phadke et al. (1988) observed the efficacy of neem products against *Bemisia tabaci* on cotton and observed that Neemark (0.4%) was as effective as endosulfan in controlling whitefly, while Neemark (0.5%) was more effective than fenvalerate. Flint and Parks (1989) tested Margosan-O for control of *Bemisia tabaci* on cotton (*Gossypium hirsutum*) at Phoenix in Arizona, USA. There was a significant reduction (60%) in the population of immature stages of whiteflies on cotton leaves after applications of aqueous sprays containing azadirachtin (138.7 ml a.i./ha) but at a lower concentration, it was ineffective. In another study, it was demonstrated that the effectiveness of neem oil (0.5%) and neem seed extract (5%) on whitefly reduced the population to the extent of 56.4% and 43.7%, respectively (Nimbalkar et al. 1994). Further, the neem oil (2%) had a good success rate on the eggs and nymphs of whitefly after 24 hrs with 74.4 and 100% mortality, respectively (Roychoudhury and Jain 1993). Application of neem oil for the management of the initial infestation of whitefly has been advocated in insecticide resistance management (IRM) strategies (Kranthi and Russell 2009). Neem oil @2.5 liter/ha during the early part of the season is suggested for sucking pests under IRM strategies for Bt cotton and validated when used in farmers' fields during 2008 and 2009. One of the biggest constraints in the adoption of botanicals is the non-acceptance of the product by farmers as neem formulations do not give a knockdown effect. Field trials with neem formulations in cotton have not shown satisfactory control with seed kernel extracts; it is necessary to have repeated applications. The second constraint is the availability of standard and economically viable formulation of neem as it is rapidly inactivated by UV light and changes in temperature, leaf pH, etc. However, these products are useful in the organic production of vegetables.

This chapter provides an overview of the various physical and general measures used against whiteflies. The physical measures were the main weapon to mitigate the insect pest problems in the earlier period when the organic pesticides were not developed. These measures have been exploited against the dreaded pest, the locust, as well as others. There is no substitute for such measures until today. The ill-effects of pesticides such as resistance in insect pests, resurgence of secondary pests, residues problems, etc. has become evident over time. Once again, these measures have gained momentum. Machines have also been introduced to tackle insect pests. Of these machines, the vacuum machines have been manufactured and used to manage soft-bodied insects, mainly the sucking insect pests. It is highly successful in glasshouses. With the success of vacuums, electrostatic sweepers have also been developed to manage whiteflies in open fields of crops. Besides these, other measures like blasting the whiteflies with water and removal of leaves carrying the large population of immatures of whiteflies are important to reduce the population of this pest. The use of yellow sticky traps has been in use since time immemorial and there is no substitute for these measures so far. The yellow sticky traps are used for monitoring and surveillance. It is the best option today to trap the population of whiteflies. It was initially meant for monitoring sucking pests in the glasshouse but now it has been exploited for monitoring whiteflies in an open field. The chapter further mentions physical measures like the use of mulches and the cultivation of crops in isolation/green house. These measures are very effective against whiteflies. The nonchemical methods such as the release of parasitoids also fall under this category. To popularize these non-chemical measures, there is a need to further refine and develop these methods.

CHAPTER 10

Cultural Measures

◇◇

The extensive application of pesticides has eroded the use of cultural measures and practical wisdom to check the menace of insect pests, including whiteflies. However, these measures are not only effective against whiteflies but are highly compatible with the environment and there is no fear of developing resistance (Hilje et al. 2001). Cultural control practices are aimed at avoiding or preventing whitefly infestations, eliminating sources of whiteflies, and keeping whiteflies out of growing areas. They are important components in tackling the menace of insect pests as a part of the integrated pest management program (Hilje et al. 2001). These practices are more relevant to prevent the population build-up (Abd-Rabou and Simmons 2012). These practices include crop rotation, selection of variety, planting/harvesting dates, trap crop, irrigation, mowing, hoeing, etc. for managing pest problems in crops.

10.1 Altering Sowing/Harvesting Time

It is instrumental in checking the menace of insects. The alteration in sowing dates of cotton from January and July to August and September increased oviposition from 1.6–6.3 eggs/cm^2 to 10.6–24.7 eggs/cm^2 in the case of whitefly of cotton (Dhawan and Simwat 1999). This is important for vectors of plant pathogens. By changing the sowing or harvesting the dates, the susceptible plants sometimes escape the severe attack of plant viruses vectored through whiteflies and this category of resistance is sometimes called ecological resistance. The plants are susceptible but behave like the resistant ones. A late sown crop of cotton invites heavy infestation of whiteflies in Punjab, India (Singh and Butter 1997); the early planting of tomato in spring or late in fall in Cyprus helps the crop avoid contracting the Tomato Yellow Leaf Curl Virus, which is vectored by whiteflies (Ioannou 1987). Similarly, the change in the planting time of tomato in Egypt reduced the spread of the Tomato Yellow Leaf Curl Virus (El-Gendi et al. 1997). Thus, it can be seen that the early sown crop always suffers less from an attack by whiteflies.

10.2 Crop Geometry

The plant population in cotton has been demonstrated as a measure to prevent the attack of sucking pests, particularly whiteflies. It has been seen that closer spacing of cotton favors whitefly buildup as compared to a wider spacing. The wider spacing of the cotton crop @ 38.5 cm had a lower whitefly population as compared to closer spaced crop under unsprayed conditions (Arif et al. 2006, Butter et al. 1992b). This is because the pesticide application in closer-spaced cotton crop is not thorough, and thus the buildup of the pest continues.

10.3 Crop Thinning and Roguing

Crop thinning is important to maintain the plant population at the recommended level. The recently introduced Cotton Leaf Curl Virus is transmissible through *Bemisia tabaci* in Punjab. In case the Leaf Curl Virus appears early in the season, the whitefly vector injects the virus in the plants, which then display symptoms in about twenty days depending upon the season. The early removal of diseased plants would go a long way in delaying the appearance of the Leaf Curl Virus in cotton. This process of the uprooting of diseased plants can be made feasible by increasing the seed rate at sowing. This practice has been introduced in Punjab and is working well (Butter 2018). Another study conducted by Ioannou (1987) for three years involving removal of old and overwintering plants of tomato in March and April in southern coastal areas of Cyprus completely checked the Tomato Yellow Leaf Curl Virus. On discontinuing the practice in the subsequent year, the incidence of the infected plants was 15% recorded in 1983 and rose to 40% in 1984. This practice of roguing is also helpful in delaying the appearance of Tomato Yellow Leaf Curl virus in tomatoes. The tomato crop transplanted early in spring or late in the autumn season suffers less in Cyprus.

10.4 Crop Sanitation/Crop Free Period

Crop sanitation involves the removal of crop debris containing virus concentration to prevent it from being picked up and spread in the crop raised in the vicinity of the harvested field. During the period of the outbreak, it is always desirable to avoid the cultivation of whitefly host crops on yearly bases from mid-June to mid-August in the summer season to break the cycle of whitefly carryover. During the period of the outbreak of Tomato Yellow Leaf Curl Virus, a crop-free period from June to July was observed in the Dominican Republic (Hilje et al. 2001, Gilbertson et al. 2007) along with prohibiting the cultivation of crops like brinjal, beans, chili, cucurbits and cotton in that period (Salati et al. 2002). The removal of overwintering tomato drastically reduced the incidence of Tomato Yellow Leaf Curl Virus in the Dominican Republic in spring (Ioannou 1987). The early planting of the crop in spring/late in fall is highly useful in tackling whiteflies. Likewise, a month-long free period observed in the desert areas of Israel reduced the whiteflies (Ucko 1998). The same model was emulated in Florida, USA, where it was found highly beneficial in reducing Tomato Yellow Leaf Curl Virus in the tomatoes there.

10.5 Isolation of Crop

Crops grown under a protective cover or in isolation do not suffer from the attack of insect pests in nature (Berlinger et al. 1983). The key crop can be separated by keeping buffer areas (either keeping the land fallow or by planting the nonhost crops). If this strategy is followed, the virus infection in a crop can be managed. In the case of whiteflies, this procedure can prove highly beneficial to reduce the damage inflicted by them. The valuable crops can be grown under a protected cover, to keep the whiteflies away from the host. This kind of isolation is extremely useful for harvesting a healthy crop. The concept of raising crops in screenhouses has emerged from the concept of crop isolation, particularly regarding the management of vector-borne diseases.

10.6 Use of Fertilizers and Irrigations

The fertilization of crops has a profound effect on the population of whiteflies. Under greenhouse conditions, the population of B-biotype (*Bemisia tabaci*) was significantly higher on poinsettia fertilized with heterogeneous fertilizer as compared to unfertilized plots of poinsettia (Bentz et al. 1995). Furthermore, egg-laying and adult emergence in *Trialeurodes vaporariorum* were also higher in heavily fertilized tomato crops with nitrogeneous fertilizers (Park et al. 2009, Jauset et al. 2000, Dhawan and Simwat 1999, Bentz and Larew 1992). The nitrogen nutrient in cotton leaf is generally higher in a crop which is heavily fertilized with nitrogeneous during the active growth. With the higher dose of nitrogen fertilizer, the crop attains the rank growth and the leaves become succulent and the succulency enhances its susceptibility to whitefly (Rote and Puri 1992).

The scenario viz. insect pest changed with respect to cotton with the introduction of the Cotton Leaf Curl Virus disease in the region. It was a devastating malady, causing cotton crop failures. Many efforts were directed to contain this viral disease and major emphasis was laid on checking this malady with the use of crop varieties resistant to the whitefly. A cotton hybrid (Fateh) resistant to the virus was developed in Punjab, India and tolerant varieties for general cultivation by farmers were introduced (Kular and Butter 1995). Studies carried out in India indicated that in cotton highly fertilized with phosphorus, there was a corresponding reduction in the population of whitefly (Butter et al. 1992). The population of whitefly adults was 240.7/30 leaves and nymphs 82.5/30 cm^2 in a crop of cotton heavily fertilized with phosphorus (60 kg p$_2$O$_5$/ha) as against 493.6 adults/30 leaves and 137.3 nymphs/30 cm^2 in soil without phosphorus. The highly fertilized (nitrogen, phosphorus, potash) cotton crop (@20–60–60 kg/ha) is highly favorable for the build-up of whiteflies (Natarajan 1986). With the increase in the dose of fertilizer to 200–100–100 kg/ha of NPK, a further increase in the population was evident in the study carried out in India (Rote and Puri 1992). In another study in this area conducted using NPK at 50–25–25 kg/ha, respectively, and double the dose of NPK to cotton, the build-up of the population of whitefly was apparent (Prohit and Despande 1991). However the study carried out on cotton in Punjab, India demonstrated the role of phosphorus in imparting resistance to whitefly both under field and screenhouse conditions (Butter et al. 1992). The study further showed non-preference of cotton to whiteflies for feeding, as revealed by low population density of nymphs and adults. The egg laying by whitefly was also drastically reduced with higher dose of phosphorus. This was understood from the feedback from the cotton farmers who used to apply phosphorus to cotton during the active crop season despite no recommendation. However, further study is required to determine the exact mechanism of resistance.

The impact of irrigation water was studied through different techniques of irrigation viz. flooding, furrow irrigation, and sprinkler. Drip irrigation proved detrimental to whiteflies and as a result, the incidence of viral diseases was reduced considerably (Abd-Rabou and Simmons 2012a). It is also important to work out the effect of the frequency of irrigation on preferred crops. The intensity of irrigation generally exerts influence on whitefly build-up. The frequently irrigated crop generally attains the rank growth and it enhances the succulency of crop to insect species. However, this aspect needs to be investigated. During the monsoon, the rainfall is negatively correlated with the whitefly population (Thriveni 2019). The weather after the termination of monsoon becomes highly favorable for the build-up of cotton whitefly and the insect inflicts losses in cotton in north India.

10.7 Barrier Crop/Eradication of Overwintering Hosts

The barrier crops planted around the field help protect the main crop. To be successful, the barriers plantations should be taller than the host crop. This will prevent the arrival of whiteflies from outside. Also, these crops will create congenial conditions for a suitable niche conducive for the survival of natural enemy complex and build-up of natural enemies. The practice of planting sorghum crop as a barrier crop around tomato has been extensively implemented in Brazil. It is known to significantly reduce the density of whiteflies as well as encourage the parasitoids. The whiteflies being poor fliers cannot fly beyond 2 meters height, thus the barrier crops sown are generally taller ones. Sorghum, maize, and *Sorghum halepense* (Baru) are planted as barrier crops around the tomato. It was demonstrated in the Jordan valley in Israel that weeds act as reservoirs of the Tomato Yellow Leaf Curl Virus and the whiteflies pick up the virus in June and July and spread the virus. The planting of sorghum or maize around cotton crop fields helps in managing the whiteflies (Reddy and Rao 1989). Sweet pepper is a reserviour host of *Bemisia tabaci* and it should be kept in mind white planting main crops nearby (Kil et al. 2014). The weed species (*Cynanchum acutum*) acting as over-wintering hosts are removed and the spread of the Tomato Leaf Curl Virus is checked in these areas (Cohen et al. 1988).

10.8 Elimination of Weeds

The whiteflies are categorized as polyphagous pests due to a very wide host range of crop plants. Most of the weed flora also harbor many dreaded diseases of viral origin. It is, therefore, more important to be cautious in the management of whiteflies which are vectors of viral diseases. The unchecked growth of weed hosts of whitefly around the cotton field and overwintering hosts namely, *Lantana* sp., *Solanum* sp., *Euphorbia* sp., *Datura* sp., *Hibiscus* species are helpful in carrying over of whiteflies (Dhawan 1990). Therefore, their elimination is mandatory to check the whitefly/viral diseases. The elimination of overwintering hosts of the Tomato Yellow Leaf Curl Virus transmitted by the whitefly and early planting of tomato in Cyprus reduced the incidence the virus (Ioannou 1987). *Amaranthus retroflexus* has also been identified as a host of whitefly (Smith et al. 2015). In addition to the removal of alternate hosts, keeping a month-long crop-free period in the infested area is considered a better option to eliminate the virus incidence in tomatoes.

10.9 Use of Mixed Crops/Inter-Crops/Trap Crops

There are many instances wherein it has been shown that the plantation of crops either as border rows or as a mixed crop in the main crop protects the host crop against the attack of whiteflies. The planting of border rows of coriander and fenugreek (both non-hosts of whiteflies) is quite useful in protecting the main crop. This is particularly useful for vector whiteflies of non-persistent stylet-borne viruses. As the vectors of stylet-borne viruses could get entrapped in the border crops, these vector species would not be able to transmit the virus and the spread of the virus will get checked. The same would apply to non-host crops grown as mixed crops or as an intercrop. The stylet-borne viruses are checked within seconds to minutes if kept on non-hosts/fasting. The trap crop is generally more preferred by the key pest and is planted with the main crop; for example, the cucumbers are more preferred by whiteflies than the tomatoes. With the planting of cucumber along with tomato, the whiteflies concentrate on cucumber and will not devour the main crop tomato, thus saving the main crop. The population buildup of whiteflies can easily be controlled with pesticides on the preferred crop. It is also important to take note of the fact that cucumber is not a host of the Tomato Leaf Curl Virus. This is a common practice in Sudan and many other Middle Eastern countries (Ioannou 1987). The intercropping of tomato with cucumber/squash/corn/capsicum (El-Serwiy et al. 1987)/french bean/ celery is useful for reducing the whitefly population as well as the spread of viral diseases in Jordon (for example, the Tomato Yellow Leaf Curl Virus) and many other places (Verma et al. 2013, Abd-Rabou and Simmons 2012, Pei-Xiang et al. 2011, Schuster 2004, Youssef et al. 2001, Al-Musa 1982). Castor, marigold *Nicotiana rustica*, wild brinjal (*Solanum khassianum*) in the ratio of 1:10 protects the target crop (Sundaramurthy and Chitra 1992, Puri et al. 1994, Dhawan and Simwat 1999). The brinjal and cucurbits are good plants to act as preferred hosts while maize is a good barrier crop. These trap crops attract the whiteflies and later on can be managed with a limited use of pesticides (Choi et al. 2016, Choi et al. 2015). There were studies carried out on species of whiteflies, namely *Aleurotrachelus socialis* and *Tetraleurodes variabilis*, regarding the effect of intercrops on the development of the insect (Gold et al. 1990). The result indicated heavy egg-laying by these whiteflies on cassava intercropped with cowpea as compared to cassava alone. The intercropping of target crops with okra, moong, and pigeon pea are also useful options.

10.10 Border Plantation/Wind Breaks/Modification of Habitat

To recap, the planting of border rows with tall-growing trees and hedges to protect the field is quite a useful practice in checking the extent of damage done by insect vectors. The tall-growing trees and hedges will prevent the incoming whitefly population transported on air currents and would save the key crop. The hedges and other tall-growing plantation would go a long way to provide suitable niches for the survival of parasitoids and enhance the parasitization. Among the crops, sorghum and maize are good barrier crops and can be given priority as barrier plantation. It is desirable to provide floral plants arranging nectar as food to the natural enemies to enhance their efficiency (Jervis and Heimpel 1998). Besides, in the event of an insufficient host population, necessary steps can be taken to arrange a host population to overcome the shortage.

This chapter overviews how cultural measures are highly important as preventive technology to check the buildup of whitefly in the crop agroecosystem. This aspect is specifically important in the situations where the whiteflies are acting as vectors of plant pathogens. Measures pertaining to barrier crops, trap crops and modification of habitat, management of trap crops, mixed crops, border crops, use of irrigation and fertilizers, elimination of weed hosts and overwintering hosts, modification of harvesting and sowing dates, crop sanitation, crop free period, crop cultivation in isolation, etc. were discussed. These measures were neglected in attempts to control whitefly pests due to the easy availability of pesticides. Besides, these measures did not have a knockdown effect and were therefore disregarded as significant in a control strategy. Their resurgence and growth in popularity are developments that had taken place to control whiteflies as vectors of plant viruses rather than as an insect pest. Of these measures, most of the practices are highly useful for preventing the buildup of whiteflies. Protective cultivation in glasshouses is becoming popular due to the concept of organic farming. Such measures are likely to be in the forefront of IPM strategy viz. whiteflies in the years to come.

CHAPTER 11
Non-Chemical Measures

◇◇◇

Whiteflies are soft-bodied insects with poor flight activity that use air currents to spread and migrate to far-off places. However, within the field, the adults fly away from one plant to another adjoining plant by making whirling movements. To control these weak fliers, non-chemical methods are considered sufficient to harvest precious high-value crops grown under a protective cover. In the era of organic farming, farmers are showing great interest in these non-chemical measures. At present, 179 countries (650,000 organic producers) have taken up the cultivation of organic farming. The total area under organic produce is 7,20,000 hectares (Farag-El-Shafie 2019). Organic farming is the method of cultivating crops without the use of heavy-duty chemical pesticides and chemical fertilizers. However, mild pesticides with high safety posing no serious threat to the environment and natural enemies are used. These pesticides are rapidly degradable and safe, and include biopesticides, inorganics, botanicals, microbials, oils, and soaps (Table 7.2). The organic farming principles are highly matched with integrated pest management (IPM) for prevention, avoidance, monitoring, and control of pests. The ill-effects of toxic chemicals (pesticides and fertilizers) are being mitigated and in most situations, their use has been stopped in crop production due to the growing popularity of organic farming.

Hitherto, farmers had resorted to the use of unwanted mixtures of pesticides without label claim on a crop and the repeated application of these pesticides with the same mode of action. Many new molecules of pesticides had been added to the list of recommended chemicals with erroneous considerations. The use of neonicotinoids has been on the verge of withdrawal from European countries on account of the danger to the beekeeping industry. There are many more countries dependent on the beekeeping industry for honey production that are likely to raise an alarm against other such pesticides as well. Safety measures to avoid direct hitting of foraging bees would be required to strictly enforce the regulation. Ways and means are needed immediately to develop strategies to ensure the safety of the beekeeping industry. Failing on this front, governments are in the process of saying goodbye to this category of pesticides. Above all, green chemicals with sufficiently long residues are available. Stringent rules are the need of the hour to ensure the proper use of chemicals. The identification of these potent chemicals requires time and money, and thus these expensive molecules must be used judiciously so that these chemicals remain effective for a longer period. It takes years to identify a potent molecule and it costs millions for the pesticide industry. It is, therefore, essential to stick to their judicious use.

The adoption of non-chemical measures is essential to minimize the toxic effects of chemicals. It can be done on a small scale to protect high-value crops. Non-chemical measures need to be included in IPM as one of the tactics. A few non-chemical measures have been identified to mitigate the menace of whitefly. Whiteflies have developed resistance to most organic pesticides by 1980. Therefore, the need to minimize the use of toxic chemicals and to include non-chemical measures in pest control has stimulated the scientific community to lay more stress on the development of these measures. Beyond this, efforts have been made to increase farmers' dependence on biorationals. These biorationals have been advocated at many places in the world (Greer 2000) and include parasitoids (Prarasitoids like *Encarsia formosa, Eretmocerus mundus*), entomopathogenic fungi (*Beauveria bassiana* Naturalis-O/Botanigard; *Paecilomyces fumosoroseus*-PFR-97), insecticidal soaps/surfactant (M-pede/safer), Azadirachtin/Garlic-based products (Neem Azad, Azatin, Garlic gard), oils (Hot pepper wax), and oils (golden Natural spray oil-soybean oil) (Greer 2000). However, care must be taken to ensure that the whitefly population will be checked in an open field or field covered with a covering of polythene or a polyhouse. The whiteflies population can be assessed with yellow sticky traps—a widely accepted technology. These technologies can be utilized in two ways viz. for surveillance of whiteflies and mass trapping to eliminate the most population. The whiteflies are known to exude honeydew on which sooty mold develops. The presence of ants on plants to feed on honeydew is an indication of the attack of whiteflies; on disturbing the leaves of a host plant, the whitefly adults will flutter and soon settle down on the same plant or the adjoining one. The leaves, particularly the topmost leaves, are sticky to touch in the morning hours and indicate the presence of whiteflies in a crop. The recognition of the presence of whiteflies in the crop in such a situation becomes easy and thus, subsequent control measures can be taken up.

The organic production of vegetables is picking up at a faster rate in protective enclosures. This kind of cultivation is likely to face more pest problems. Protective cultivation has almost doubled since 1990 in the USA (Weintraub et al. 2017).

The successful raising of crops in greenhouses can be done with the use of measures which have been elucidated in the following paragraphs:

- For the management of whitefly, crop cultivation is done in polyhouses so as exclude the population of sucking pests. For the exclusion of the target insect, the screen houses should have a mesh size between 0.46–0.462; 0.266 × 0.818; 0.230 × 0.900 to prevent the entry of insects. The screenhouses with 50 mesh size screen to exclude the whiteflies are available in the market and can be put to use.

- Encouraging the release of natural enemies, predators (*Delphastus catalinae/ Delphastus swirskii/Macrolophus pygmaeus/Macrolophus caliginosus/ Amblyseius cucumeris/Amblyseius fallacis/Amblyseius swirskii*), parasitoids (*Encarsia formosa/Eretmocerus mundus/Eretmocerus eremicus*), and parasitoids (*Galendromus occidentalis/Macrolophus pygmaeus/Neoseiulus swirskii/Neoseiulus californicus/Phytoseiulus ferbimitis/Phytoseiulus persimilis*

Stethorus puncticillum) is important (Weintraub 2009, Weintraub and Berlinger 2009). The effective parasitoids and predators have already been identified and their availability in the market has already been ensured. These natural enemies will take care of thrips, aphids, mites, and mealy bugs, including whiteflies by feeding on immature stages. These facilities should be improved upon via the improvement of ventilation, selective and soft pesticides, spray application techniques, etc.

- The application of semiochemicals, both allelochemicals (kairomones/ allomones) and pheromones (monitoring/mass trapping/male confusion technique/attraction) should not be ignored (Islam et al. 2017).

- The cultural measures mentioned in Chapter 10 which have been ignored should be given priority in pest management programs (Hilje et al. 2001).

- The mass trapping of adult whiteflies on yellow sticky traps and monitoring of population for applying control tactics to keep the pest under control is essential (McHugh 1991).

- The regular vacuuming of whitefly adults on a limited scale on high-value crops is a useful proposal to control whiteflies. Electrostatic sweepers are available and must be put to use (Weintraub and Horowitz 2001).

- The use of mulches is highly useful and must be included as a tactic of an integrated strategy. Mulching with reflective aluminum foils/cultivating the target crop in established mulches in the field is a good approach to repel the whiteflies and protect the crops. This is extremely applicable in high-value crops raised under protective covers (Vincent et al. 2003).

- The washing of plants with simple water using a strong pressure hose can destroy the major population of adult whiteflies. It is highly useful in small areas and that too in polyhouse crops (Weintraub and Horowitz 2001).

- Potting plants of parsley (*Petroselinum crispum*), basil (*Ocimum basil*), dill seed (*Anethum graveolans*), and marigold in and around the main crop is a better proposal to repel the whiteflies.

- To spray the plants with a homemade solution of detergent and lemon juice in an equal ratio (one tablespoon of washing detergent plus one spoon of lemon juice) to protect the crop from whiteflies is another proposal (Greer 2000).

- The periodic release of predators (ladybird beetle and green lacewings) that feed on immatures on crops grown under protective cover is a useful practice. The produce realized under such protection is better and free from pesticide-residue problems (Liu et al. 2015).

- The use of commercially available formulations of pathogenic organisms/ parasitoids, like the wasps hailing from genera *Encarsia* spp., *Eretmocerus* spp., and pathogenic fungi can be exploited (Boopathi et al. 2015). Ready-made packages of the parasitoids are available as parasitoids cards in the market.

- Neem oil/Horticultural oils/Mineral oils are effective against whiteflies and must be exploited for managing the whiteflies. The clogging of spiracles causes suffocation and kills the target insect through this process (Butter and Rataul 1973).

- Soaps are also effective against whiteflies (Greer 2000) and must be included in an integrated pest management system as one of the tactics. These soaps and detergents are known to kill the whiteflies through the penetration in cuticle and disrupt cell membrane, leading to leakage of cells, halt respiration and dehydration and death of insect. Besides, these soaps also interfere in cell metabolism and hormonal balance. Above all these products are responsible for blocking of spiracles and disruption of respiration.

- It is important to make use of yellow sticky traps to monitor and kill the insects by mass trapping (McHugh 1991). These traps in the screenhouse can be quite useful to trap the population of whiteflies in screenhouses.

- Ready-made entomopathogenic fungi (*Beauveria bassiana/Metarhizium anisopliae*) are available for use on fields in covered cultivation of crops (Boopathi et al. 2015).

- The physical removal of plant parts/debris infested with whitefly immatures (nymphs and pupae) at regular interval in the screenhouse is helpful to check the population of whitefly. The damaged leaves on a plant carrying should be shredded and destroyed. It will go a long way in the elimination of a large chunk of the population of whiteflies.

- Under field conditions, the cultivation of cotton in a citrus orchard in south-western districts of Punjab in India should be discouraged. If cotton is to be planted in citrus orchards, *Gossypium arboreum* species of cotton should be chosen as this is resistant to the whitefly-borne Cotton Leaf Curl Virus. This practice can only be followed in areas where this viral disease is prevalent.

The introduction of potent pesticides and their availability at door step eroded the use of these age old practices. The limited use of these techniques is due to their slow action and lack of knockdown effect.

The chapter provides an overview of bio-rational approaches available with the farmer. These biorational include mulches to repel whitefly, regular mass trapping adults on sticky traps, use of detergents (soaps), neem oils, entomopathogenic fungi, and use of water spray to dislodge the adults of whiteflies adults. Such physical measures have gained momentum nowadays. Homemade remedies such as a mixture of detergent and lemon can be a good proposal in obtaining good control of insects, especially the whiteflies. The use of plants acting as repellent is an even better proposal to repel the whiteflies. The plants could be of basil, dill seed, marigold, and parsley. Besides these, the use of parasitoid/predators in controlling whiteflies is beneficial.

CHAPTER 12

Weather Parameters and Forecasting Models

Whiteflies are always present in the cotton agroecosystem but do not attain the pest status till the month of July. The whitefly build-up starts with the onset of monsoon season and assumes alarming proportions as the monsoon recedes by the end of August. It is important to note that during the year, these tiny flies continue to increase in number and the peak population is witnessed in the months of September/October.

12.1 Weather Parameters and Whitefly

The global warming of weather has a profound effect on the flora and fauna of the earth. The whiteflies are no exception to it (Aregbesola et al. 2019). The impact of global warming viz. relating to the spread and multiplication of whiteflies was examined critically. The collective impact of various parameters has been known on these tiny creatures. However, among the parameters of weather, the temperature has been identified as contributing the most. A significant influence of minimum temperature and afternoon relative humidity showed a negative impact on the build-up of whiteflies in a study carried out by Dhawan and Simwat (1998). Prolonged drought and high temperatures also add to the build-up of whiteflies (Rao 1987).

More than 1,556 species of whiteflies have been identified hitherto but two to four species have been taken to describe typical whitefly. The species of whitefly, *Bemisia tabaci, Trialeurodes abutilonea, Trialeurodes vaporariorum,* and *Tetraleurodes perseae* were considered to describe to study the effect of temperature on important parameters of biology with special reference to fecundity, oviposition, and survival with overall development (Cui et al. 2008). The temperature being more important than other abiotic factors, the information was gathered on this parameter to pave the way for sound analysis. The experiment conducted on whiteflies in which the pest was exposed to temperature between 37 to 45°C for one hour. All males of the species *Trialeurodes vaporariorum* (> 39°C) and *Bemisia tabaci* (> 41°C) died while the fecundity of females was drastically reduced. The males were more sensitive than females in both species. The pronounced effect of temperature on oviposition, nymphal survival and reproduction was recorded on whitefly, *Bemisia tabaci* (Curnutte et al. 2014). According to the study, the reproductive rate declined by 36.4% at 33°C, the optimum temperature was 28–33°C. The temperature of 27/28°C was the most favourable to whitefly development (Singh and Butter 1985). Further the temperature was negatively correlated with adult body size, and length of males and females. Among all other factors, the favorable weather remained the critical component in the occurrence of outbreaks. The cumulative use of pesticides against whiteflies developing on hairy varieties of cotton was the dominating factor that contributed the most to their buildup (Singh and Butter 1997). Of the weather parameters, the temperature has a profound effect on poikilothermic animals (Table 12.1). As whiteflies are poikilothermic animals, their body temperature does not remain constant, therefore, the whiteflies are bound to be influenced by the changing weather. Considering the importance of weather parameters, the prediction models of whitefly have been made available in cotton (Kaur et al. 2009). The incidence of whitefly was observed throughout the crop season and the maximum population of whitefly adults was recorded on 34th (2014) and 31st SMW (2015) (standard meteorological week). The cultivar of cotton, Western Niroga 151 BGII recorded 45.95 adults/leaf in 2014 and 49.70 adults/leaf in 2015. The result showed significant negative correlation between population of whitefly and temperature and significant positive correlation between whitefly population with morning and evening relative humidity. The weather parameters with the whitefly population were studied in India. It includes the maximum and minimum temperature, rainfall, sunshine, relative humidity both in the evening and morning. The developmental thresholds from eggs to pupa were estimated using linear regression (Lower = 10°C and Upper = 32°C). The day length and the maximum and the minimum daily temperature were considered to calculate the hourly changes in temperature during the day. Finally, the effective degree-hour was calculated using hourly temperature data. In the second part, the output of effective degree-hour (of the first part) was used to calculate the transit time of all stages.

Earlier chapters in this book have briefly mentioned the relationship between the whitefly buildup and humidity. In fact, the whitefly showed a significant but positive correlation with minimum temperature in the morning, and relative humidity in the morning and evening. A study conducted by Jha and Kumar (2017) showed the correlations between the population of whitefly and maximum temperature (–0.481), and population and minimum temperature (–0.483). The study further reported positive correlation between whitefly population and relative humidity (Morning 0.514 & Evening 0.483), and whitefly population and sunshine (–0.641**)/wind speed (0.007). Whitefly population and all these abiotic factors (temperature, relative humidity, sunshine and wind speed) contributed to the extent of 55.70% as revealed by the step-wise regression (R2). The negative correlation between the whitefly population (*Bemisia tabaci*) and weather parameters (temperature, relative humidity, and precipitation) was obtained in a study carried out on cotton genotypes with nectaries and without nectaries cottons in Pakistan (Umar et al. 2003). The population of whiteflies was maximum on 34th (56.48 adults/leaf) and 31st SMW (135.72 adults/leaf) during 2015 and 2014, respectively. The significant but negative correlation between the population of whitefly *Bemisia tabaci* and maximum temperature –0.562* but positive significant correlation with relative

Table 12.1. Impact of weather parameters on the whitefly population.

Max. Temp	Min. Temp	RH (M)	RH (E)	Rainfall	Wind velocity/ Evaporation	Sunshine	Source
–0.419	–0.297	–0.152	–0.167	–0.456	–0.718**	0.002	Bhatt et al. 2018
0.880**	0.634**	–.767**	–.695**	---------	0.579*/0.847**	---------	Burade et al. 2019
–0.481	–0.483	0.514	0.483	--------	0.007	–0.641	Jh and Kumar 2017
–0.562*	0.234	0.671**	0.559*	–0.272	0.067	–0.146	Janu and Dahiya 2017
–0.593*	0.290	0.627**	0.639**	–0.205	-----------	---------	Rolania et al. 2018
0.250	0.158	------	--------	------	---------	-----------	Meena et al. 2013
0,036	0.454*	0.128	0.389	–0.194	-------------	---------	Shera et al. 2013
–0.371	–0.795*	–0.368	–0/839*	–0.797*	-------------	–0.873*	Dhawan and Simwat 1999
0.340	0.240	–0.380	–0.130	–0.020	0.410	0.420	Kaur et al. 2009
0.574* (Mean)	--------	0.716** (Mean)	--------	0.291	-------------	---------	Arif et al. 2006
0.691*	0.584*	0.460*	0.262	0.217	-----------	---------	Nemade et al. 2018

(* and **) denotes Significance (P = 0.05/5%) and (P = 0.01/1%), respectively.
--------- denotes information not available.

humidity *0.671 (morning)/0.559* (evening) (Janu and Dahiya 2017) was obtained. However, a negative nonsignificant correlation between the population of *Bemisia tabaci* and maximum/minimum temperature, relative humidity (morning/evening), and rainfall was apparent (Patra et al. 2016). Though the population of whiteflies was comparatively less during the years of the study. A highly significant negative correlation with wind speed (–0.718**) was obtained (Bhatt et al. 2018). Multiple factors such as temperature, rainfall, and relative humidity were also worked out and these three parameters affected the whitefly's development up to the extent of 66.4%. The results were in complete conformity with the earlier study (Janu and Dahiya 2017). The higher density of cotton plants was favored by the large population of whiteflies (Arif et al. 2006). The density of cotton was higher in closer spacings (Butter et al. 1992) as compared to widely planted cotton. The data presented in Table 12.1 showed that correlation of whiteflies population on okra with weather parameters, Tmax (r = 0.880**), Tmin (r = 0.634**), EVP (r = 0.847**), WV (r = 0.579*) was positive and significant. Relative humidity RH-I (r = –0.767**) and RH-II (r = –0.695**) were negatively significant (Burade et al. 2019). Another experiment reported the significant positive correlation of temperature (maximum-0.691*/minimum-0.584*) and morning relative humidity (0.460*) with population of whitefly (Nemade et al. 2018). The collective contribution of temperature,relative humidity and rainfall demonstrated up to 77% contribution based on step-wise regression (R2). However, a significant negative correlation between the population of whitefly and maximum temperature/relative humidity (morning and evening) was obtained in a study carried out in Haryana, India (Rolania et al. 2018). The data also showed a negative correlation between the population of whitefly and rainfall. Another study carried out on weather parameters and population of whitefly showed negative correlation between population of whitefly and temperature (both maximum and minimum), evening relative humidity and rainfall and positive correlation with morning relative humidity and sunshine (Sharma and Kumar 2014). The population of whitefly started building up from the 20th SMW reached peak during the 42nd SMW and declined by 52nd SMW. Accordingly, another study reported that temperature, relative humidity, and evaporation had definite relationships with the whitefly population. Of these, temperature (maximum-34–36°C)/minimum-24–26°C) and evaporation were positively correlated with the buildup of the whitefly population. While relative humidity (morning: 80–90%; evening: (30–40%)) and the buildup of the whitefly population were negatively correlated. A study carried out in south India indicated the collective influence of weather parameters. Similar studies were carried out in central India and have shown the effect of temperature 28–36°C and 62–92% relative humidity coupled with scanty rainfall from August to January (Jayaswal 1989).

Though the data have been generated on the impact of weather on population of whiteflies yet the results are conflicting. The internal analysis of experiments indicated that in most cases the experiments lack drawbacks in the lay out plan. The data were collected from other experiments sown with different objectives but the additional information was gathered on whitefly build up. The conflicting reports concerning the correlations between the whitefly population and above weather parameters in different regions indicate the need for further investigations by laying down the separate planned experiments to draw definite conclusions.

12.2 Forecasting Models

To predict the population build-up of whitefly in the cotton crop ecosystem in Punjab, India, a systematic study was carried out during 2005 and 2006 on Bt-cotton (cv Ankur 651/RCH 34) of the variety *Gossypium hirsutum* with three dates of sowings (Kaur et al. 2009). An increase in maximum (above 32°C)/minimum temperature (above 25°C) and sunshine (above 9 hr) (r = 0.12) with reduced relative humidity led to the buildup of *Bemisia tabaci* population wef. 22nd Standard Meterogical Week (SMW). The increase in temperature and sunshine hours caused reduction in the relative humidity (RH). The correlation coefficient between the population of whitefly and maximum (r = 0.34)/minimum (r = 0.24) temperature was evident. Further, the negative correlation between whitefly population and pan evaporation/relative humidity (r = –0.38/–0.13) were noted. The study made by Kaur et al. (2009) endorsed the earlier work (Umar et al. 2003, Singh and Butter 1985, Khan and Ulla 1994, Panickar and Patel 2001). In all, two peaks of the population (23rd to 25th SMW/32nd to 36th SMW) 2005/2006 were recorded. The investigation illustrated the whitefly activity from the 33rd SMW to the 43rd SMW with the peak population during 37th SMW. The population of whitefly in Punjab started building up in July with the on set of monsoon and the two peaks in August and September were reported (Kaur et al. 2009). The findings of this study are very much in line with the early results done on cotton (Ullah et al. 2006, Ozgur et al. 1990, Luo et al. 1989). In another study, the population exhibited a positive significant correlation with maximum temperature and a positive non-significant correlation with minimum temperature, negative non-significant correlation with the relative humidity, and negative significant correlation with the wind speed and rainfall (Thriveni 2019).

Outbreaks occur only when the average relative humidity exceeds 60%. A temperature of 45° to 46°C kills the adult whiteflies, but below this limit, both the speed of development and the rate of reproduction are positively correlated with temperature. Ovipositional activity reaches its maximum between 33°C and 37°C (Trehan 1944).

Many studies have also been initiated to examine the influence of weather parameters on the efficiency of natural enemies of whiteflies. It is also clear that the natural enemies cannot escape the change of climate. One such study conducted in this direction by Das et al. 2017 showed the influence of weather (temperature) on the colonization of two predators (*Geocoris* and Assasins bug) in Bangladesh. The result of the study demonstrated favourable impact on predators.

The development period, survival rate, longevity, and fecundity of two whiteflies, *Bemisia tabaci* B-biotype and *Trialeurodes vaporariorum* (Homoptera: Aleyrodidae) was compared at 15°C–24°C temperature. The egg development of *Bemisia* was better than that of *Trialeurodes* at 15°C, 18°C, and 24°C. A significantly slow rate (and consequently) longer pseudo-pupal development and lower survival rate were found in *Bemisia tabaci* B-biotype at 15°C compared with those at 18°C, 21°C and 24°C. Significantly, higher fecundity was found in *Bemisia tabaci* (B-biotype) at 24°C as compared to 15°C–21°C of *Trialeurodes vaporariorum*. The fecundity of *Trialeurodes vaporariorum* was significantly lower at 24°C than at any other temperature. Further, 2nd instar larval development was found in *Bemisia tabaci* and *Trialeurodes vaporariorum* at 15°C as compared with that at 18°C, 21°C, and 24°C. However, significantly shorter 3rd instar larval development was found in *Trialeurodes vaporariorum* as compared to that of *Bemisia tabaci* at 15°C, 18°C, and 24°C. The adaptive divergence of tolerance to relatively low temperature may be an important factor that results in the interspecific differentiation between the seasonal dynamics of these two whiteflies in China (Xie et al. 2011).

This chapter recounts the studies that were carried out to assess the role of weather parameters such as temperature (maximum and minimum), relative humidity (morning and evening), rainfall, wind speed, and sunshine. The influence of weather was examined in relation to species such as *Bemisia tabaci, Trialeurodes vaporariorum,* and *Trialeurodes ricini*. The negative correlation between maximum temperature (–0.562*) and population of whiteflies, and the positive correlation between minimum temperature (0.671**) and population, and positive correlation between population and relative humidity was obtained; additionally, the negative correlation (–0.641**) between the population and sunshine was recorded. The changing weather, particularly the short and mild winter with overall warm climate added to the build up of whiteflies. The changed climate was helpful in carrying over of higher population of whiteflies, which led to the outbreak of whiteflies. However, it is clear that further systematic studies are still required to precisely predict further whitefly outbreak.

CHAPTER 13
Vector of Plant Viruses

◇◇

Arthropods are well-known vectors of plant pathogens and viruses, the whiteflies also fall in this category (Butter and Rataul 1977). The whitefly transmits DNA (*Begomoviruses*) and RNA plant viruses (*Crinivirus, Ipomovirus,* and *Torradovirus*). The interaction of these viruses containing both DNA and RNA is reviewed and discussed with special reference to genome structure, epidemiology, and control of whitefly vectors.

13.1 Plant Virus Transmission

The most significant viruses belong to the *Begomovirus, Crinivirus, Closterovirus, Torradovirus, Ipomovirus,* and *Carlavirus* genera (Fig. 13.1). The important viruses hailing from these genera are the Cucumber Vein Yellowing Virus (CVYV), Cucurbit Yellow Stunting Virus (CYSDV), Tomato Chlorosis Virus (ToCV), and Tomato Torrado Virus (TTV) and are associated with whitefly outbreaks (Navas-Castillo et al.

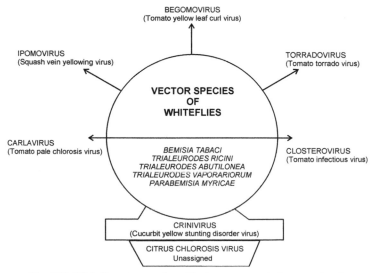

Fig. 13.1 Whitefly vector species with virus genera and viruses vectored.

2014). These viruses are from the categories of stylet borne non-persistent (Tomato Torrado Virus), foregut-borne non-persistent (Tomato Infectious Virus, Cucurbit Yellowing Stunting Disorder Virus, Sweet potato Mild Mottle Virus, and Tomato Pale Chlorosis Virus), and Circulative-persistent viruses (Tomato Yellow Leaf Curl Virus). The viruses belonging to these genera are transmissible through *Bemisia tabaci*, a devastating species of whitefly. The newly discovered Geminiviridae DNA virus (Citrus Chlorotic Dwarf Virus) in Turkey has not been assigned a genus (Loconsole et al. 2012).

Unlike the aphid-borne viruses, stylet-borne viruses are acquired and inoculated from the epidermis/mesophyll/parenchymatous cells within minutes and retained at the acrostyle in the stylet. The acrostyle is a point in the stylet where the salivary and food canals meet (acrostyle). The viruses retained at this point get detached with the salivary secretions and inoculated in the plant through the stylets. However, the foregut-borne viruses are attached to the inner linings of the esophagus in the alimentary canal. The helper component helps the virus to attach itself there, thereby preventing the further movement of the virus. With the release of the insect's saliva, the helper component gets into the plant via the stylet. The movement of the virus does not take place beyond the esophagus. The stylet-borne viruses are ineffective when the vector is feeding on non-host plants or has fasted for hours. These two types of plant viruses: transmitted by whiteflies are not retained in the body of vector and there is no transovarial transmission of plant viruses. The third category of viruses is the persistent circulative type (for example, Tomato Yellow Leaf Curl Virus) in which the virus circulates in the body of vector but does not multiply there. The virus is picked up through stylets while sucking the sap and the gets into the esophagus, crossover the filter chamber and enters into mid gut and finally delivers in to the haemolymph. It later enters into the stylet and passed on into the phloem with salivary secretions from salivary glands via stylets (Czosnek et al. 2017). The circulative viruses are those which are known to circulate in the insect vector starting with entry into stylet followed by -esophagus-alimentary canal-blood-accessary salivary glands -stylet- and back to the plant through inoculation. After acquisition access, the virus was detected in head (10 minutes), midgut (40 minutes), haemolymph (90 minutes) and salivary glands (5.5 hours) with PCR and it was regarded as circulative virus transmissible through whiteflies (Ghanim et al. 2010a). The well-worked virus among the whitefly-borne viruses is Tomato Yellow Leaf Curl Virus.

The whitefly-borne viruses are retained in the body of vector from days to possibly throughout life of vector. Further, the whitefly-borne viruses are not seed transmissible except for the Tomato Yellow Leaf Curl Virus which has recently been reported as seed-borne (Kil et al. 2016). The identified species of whiteflies are beyond 1,556 at this moment but of these species, only five (*Bemisia tabaci, Trialeurodes abutilonea, Trialeurodes vaporariorum, Trialeurodes ricini, Parabemisia myricae*) are designated as vectors of plant viruses. The two latter species are yet to be confirmed as vectors.

These viruses are ingested when the pest sucks excessive sap. The excess sap in the diet of whiteflies makes the diet poorer in nutrition. To enrich the food, special cells (Gottlieb et al. 2008) are present in the insect which harbor symbiotic bacteria called endosymbionts. As mentioned in an earlier chapter, endosymbionts are of two categories: primary and secondary. The primary symbionts contain carotenoid-synthesizing genes (Santos-Garcia et al. 2012, Sloan and Moran 2012). The primary bacteria, namely, *Portiera aleyrodidrum* is found in all the whitefly populations. However, the secondary endosymbionts are not present in all populations of whiteflies. The colonizing bacteria are almost similar in closely related species of whiteflies. The secondary endosymbionts present in whiteflies generally have a lot of diversity. These are *Rickettsia, Arsenophonus, Hamiltonella, Cardinium* and *Fritschea, Wolbachia, Orentia-like organism*. The endosymbionts are variable in intensity due to changes amongst species, biotypes, or races. Of these, *Hamiltonella* is present in 33% of the total whitefly population (Chiel et al. 2007) of the B-biotype, while *Wolbachia* and *Arsenophonus* inhabit 87% of the population of the Q-biotype of whiteflies. The biotypes of whiteflies (B-biotype and Q-biotype), have also been known to synthesize a GroEL protein called chaperone through the endosymbionts present in these two biotypes GroEL protein also ensures the safe transport of viruses in the blood of whitefly (Morin et al. 2000). Heat Shock Protein 70 (HSP70) (Czosnek and Ghanim 2012, Gotz et al. 2012) has also been identified as a helper component in the blood of the body of the vector. Similar to this, a protein (63 KDa), synthesized by endosymbionts in the whitefly ensured the movement of Tomato Leaf Curl Virus from cell to cell in the vector whitefly (Morin et al. 1999, Douglas 1998).

13.2 Coat Protein (CP)

To understand the mechanism of virus transmission, it is essential to understand the role of such determinants as the coat protein. The coat protein is known to play a role in the acquisition and retention of the virus on the chitin inner linings of the gut wall where foregut-borne viruses are retained (Ng and Zhou 2015, Ng and Falk 2006).

To this end, the molecular interaction between *Begomoviruses* and an important vector, *Bemisia tabaci* was studied. The mechanisms and proteins encoded by the insect vector and its bacterial symbionts have been elucidated (Rosen et al. 2015). The role of coat protein has been demonstrated through the gene replacement carried out in African Cassava Mosaic Virus (whitefly-borne virus) and the Beet Curly Top Virus (leafhopper-borne virus). By exchanging the protein coat of these two viruses and inoculating the tobacco plants with two new mutants (whitefly- and leafhopper-borne) symptoms exactly similar to those of the African cassava mosaic were produced and both the mutants were transmissible through whiteflies. These experiments highlighted the role of coat protein in the transmission of the Cassava Mosaic Virus. Earlier, it was considered that transmission of the virus is through the nucleic acid only and the protein coat does not play any role in transmission. However, in a study conducted by Briddon et al. (1990), the role of coat protein in the transmission whitefly-borne viruses has been demonstrated.

13.3 Virus Diseases

The whiteflies are known to transmit viruses belonging to *Begomovirus* (Tomato yellow Leaf Curl Virus), *Torradovirus* (Tomato Torrado Virus), *Closterovirus*

(Tomato Infectious Virus), *Ipomovirus* (Squash Vein Yellowing Virus), *Carlavirus* (Tomato Pale Chlorosis Virus), and *Crinivirus* (Cucurbit Yellow Stunting Disorder Virus). Among these viruses the Citrus Chlorotic Dwarf Virus has not been assigned a genus so far (Butter 2018, EPPO 1996). The virus diseases belonging to different genera transmissible through whiteflies (Brown et al. 1990) are briefly narrated ahead (Fig. 13.1; Table 13.1). The Tomato Yellow Leaf Curl Virus disease (*Begomovirus*): was the first virus to be recorded, in Israel in 1930. With ssDNA, this virus had monopartite (old world) and bipartite (New world) genomes (Brown et al. 2012, Seal et al. 2006). It was recorded on tomato, potato, tobacco beans, and pepper. The diseased plants can be recognized by their stunted growth, curling of leaves, puckering and thickening of leaf lamina with vein thickening and chlorosis of leaves. The out growth on veins (Cup like structures) on the lower leaf surface is present in advance stage of disease and these are called leaf enations (Plate 13.1). The Tomato Yellow Leaf Curl Virus is the only one from the whitefly-borne viruses mentioned that was classified as a seed-borne virus (Kil et al. 2016). It belongs to the category of circulative viruses, which are retained in the body of vector for life but do not multiply there. It is a Begomovirus, with circulative ssDNA, encapsulated genome and exists in two isolates (Sardana and Israel isolates). Although it is transmissible through the biotypes (Q and B), the two differ in transmission efficiency (Pan et al. 2012). The differential transmission is due to the endosymbionts responsible for the safe transport of the virus into the blood and the production of the chaperone. Coat

Table 13.1. Virus genera transmissible through whiteflies (Modified from Butter 2018).

S No.	Mechanism	Virus genera	Virus	Vector species	Source
1	Non-Persistent stylet-borne	*Closterovirus*	Tomato Chlorosis Virus	*Trialeurodes vaporariorum*	Wisler et al. 1998
2	Non-Persistent foregut-borne	*Torrado Virus*	Tomato Torrado Virus	*Bemisia tabaci; Trialeurodes vaporariorum; Trialeurodes abutilonea*	Verbeek et al. 2014
3	Persistent Circulative	*Begomovirus*	Tomato Yellow Leaf Curl Virus	*Bemisia tabaci; Trialeurodes ricini?*	Brown et al. 2012, Seal et al. 2006
4	Non-persistent foregut-borne	*Crinivirus*	Cucurbit Yellow Stunting Disorder Virus	*Bemisia tabaci*	Qiao et al. 2011
5	Non-persistent foregut-borne	*Ipomovirus*	Sweet potato Mild Mottle Virus	*Bemisia tabaci*	Colinet et al. 1996
6	Non-persistent foregut-borne	*Carlavirus*	Tomato Pale Chlorosis Virus	*Bemisia tabaci; Trialeurodes vaporariorum*	Antignus and Cohen 1987
7	Transmission	Unassigned	Citrus Chlorotic Dwarf Virus	*Parabemisia myricae*	Loconsole et al. 2012

Plate 13.1. Cotton Leaf Curl Virus disease.

protein is responsible for the acquisition of viruses (Ng and Zhou 2015, Ng and Falk 2006). The Tomato Leaf Curl Virus (ToLCV) disease is managed through both chitosan and *Pseudomonas* sp. Chitosan, which has enhanced biocontrol efficacy of *Pseudomonas* sp. against ToLCV (Mishra et al. 2014).

Cucurbit Yellow Stunting Disorder Virus disease (*Crinivirus*): It was first recorded in 1982 in the United Arab Emirates and is transmissible through *Trialeurodes abutilonea* and *Trialeurodes vaporariorum* (Qiao et al. 2011). Detailed information on this virus is lacking but it is known that it is whitefly-borne.

Tomato Chlorosis Virus disease (*Closterovirus*): It is a synonym of Potato Chlorosis Virus and was first recorded in 1996 in Florida. It is a flexuous, rod-shaped virus and is transmissible through *Trialeurodes vaporariorum*, and *Trialeurodes abutilonea*. It is characterized by yellowing of leaves with interveinal chlorosis along with thickened veins.

Tomato Torrado Virus disease (*Torradovirus*): It is also called a chocolate virus. This is a spherical virus, transmissible through *Trialeurodes vaporariorum*, and *Bemisia tabaci* whiteflies. The acquisition and inoculation of the virus are two hours each but the optimum time for acquisition and inoculation is 6 and 8 hours, respectively. The retention of this virus by the vector is for 8 hours (Verbeek et al. 2014). This virus is of persistent type.

Tomato Pale Chlorosis Virus disease (*Carlavirus*): It is a strain of Cowpea Mild Mottle Virus (Antignus and Cohen 1987) purified from tobacco, *Nicotiana glutinosa* and was first recorded in Israel. Enzyme–linked immunosorbent assay and an immunoelectron microscopy decoration test demonstrated serological relationship with Cowpea Mild Mottle Virus. The disease is characterized by severe mosaic pattern being formed or necrosis of leaves with yellow spots appearing. There are also symptoms of malformation of the leaves, stem crinkling, and bud necrosis associated with this malady. It is transmissible through whiteflies, *Bemisia* spp., and *Trialeurodes* spp. with a nonpersistent relationship. Being a Carlavirus, it has unique relationship with whiteflies in that Carlaviruses are actually transmissible through aphids but this virus is an exception, and also transmissible through the whitefly vector. It is a single-stranded RNA virus of molecular weight of 2.5×10^6 with a genome size of eight bands that possess total five open reading frames of protein with a virus genome of 2,500 nucleotides. Mostly, it is the legumes that are the hosts of whiteflies.

Sweet Potato Mild Mottle Virus disease (*Ipomovirus*): It is a filamentous, non-enveloped, positive ssRNA virus belonging to the *Ipomovirus* genus. It is prevalent in East Africa and Philippines and can be recognized from the mild mottling of leaves along with marks patches, spots and streaks spread in an irregular pattern. It has a wide host range which includes 14 plant families containing 45 plant hosts of this virus. It is transmissible through *Bemisia tabaci.*

Citrus Chlorotic Dwarf Virus disease (*Geminiviridae*): It is an ssDNA virus containing 5 open reading frames and large genome (3.64 vs 2.5–3.0 kb) that affects twelve varieties of citrus (Loconsole et al. 2012). Further, it is not transmissible through mechanical means. With an increase in the inoculation period from 24 to 48 hours, the transmission rate shoots up from 75 to 100 percent. Although it is considered a whitefly-transmissible virus bayberry (*Parabemesia myricae*), this fact needs confirmation.

Cotton Leaf Curl Virus (*Geminivirus*) disease: This is a begomovirus around 18–20 mm in diameter and 30nm long circular ssDNA. It is caused by the Cotton Leaf Curl Virus (CLCuV) and is a monopartite Geminivirus (*Geminivirus*), vectored by the whitefly in a persistent, circulative manner. It causes vein thickening, and the leaf enations are conspicuous on the lower leaf surface of cotton (Plate 13.1). The CLCuV damages the *Gossypium hirsutum* while *Gossypium arboreum* is resistant to this virus as the cuticular wax acts as a physical barrier (Khan et al. 2015a).

Cassava Mosaic Virus disease. The Indian Cassava Mosaic Virus (ICMV) and Sri Lankan Cassava Mosaic Virus (SLCMV) are responsible for the spread of this disease in India (Karthikeyan et al. 2016). The partial dimer clones of the persistent and non-persistent isolates of SLCMV and the re-emerged isolate of ICMV were infective in *Nicotiana benthamiana* upon inoculation. Studies on pseudo-recombination between SLCMV and ICMV in *Nicotiana benthamiana* provided evidence for trans-replication of ICMV DNA B by SLCMV DNA A. The management of whitefly tropical whitefly integrated pest management project was launched in 1996 to take care of 6 aspects such as (I) whitefly vector and viruses in cassava and sweet potato in sub Saharan Africa (II) whitefly borne viruses in mixed cropping system of Mexico, Central America and the Caribbean (III) virus problems of eastern and southern Africa for mixed cropping system (IV) South Eastern region of Asia for mixed cropping (V) *Trialeurodes vaporariorium* as pest in mixed cropping in Andean high lands (VI) and whitefly as pest of South America (Morales 2007). In this project all tactics will be tested to develop suitable control strategy. In this project efforts are made to combine preventive measures, cultural measures, biological controls mechanical/physical approaches along with chemical control to keep whitefly and ciruses under check.

The whitefly is more important from the point of vectors than an actual pest as it is known to transmit several *Geminivirus/Begomoviruses* in nature. More than 1,556 species of the whitefly species have been identified but it is fortunate that all species are not vectors of plant viruses. The chapter has provided an overview of style-borne, foregut-borne, and persistent-circulative viruses transmissible through whiteflies. The role of the protein coat in the transmission of whitefly-borne viruses has been discussed. The virus-vector transmission has been discussed taking into consideration the virus genus and the viruses associated with that genus. There has been a brief discussion of the viruses within different genera such as Tomato Torrado Virus (*Torradovirus*), Tomato Yellow Leaf Curl Virus (*Geminivirus*), Cucurbit Yellow Stunting Disorder Virus (*Crinivirus*) Sweet Potato Mild Mottle Virus (*Ipomovirus*), Tomato Chlorosis Virus (*Closterovirus*), Tomato Pale Chlorosis Virus (*Carlavirus*), and the newly identified virus Citrus Chlorotic Dwarf Virus (*Geminiviridae*) and their ability to spread through vectors.

CHAPTER 14
Management Strategies

◇◇◇

To formulate a successful whitefly-management strategy, one should be well versed with the pest fauna of crop, the available techniques and weather conditions. As the whiteflies are poikilothermic animals, these are affected the most by global warming. The influence of climate change is thus more pronounced on whiteflies. There has been a rise in temperature (5–6°F) during the last four decades due to global warming. The ideal conditions are available during the 26th SMW which include maximum (35–40°C) and minimum temperature (25–30°C), and sunshine hours (4–8 hours) (Maharshi et al. 2017). The population records are taken using yellow sticky traps during the monitoring and surveillance. The tactics used depend on the population threshold. The application of pesticides already poses a serious threat to mankind through insecticide residues and pest-resistance problems (Castle et al. 2010, Horowitz et al. 2018). The excessive use of chemicals has overshadowed the inclusion of all the slow-acting tactics. The chemicals which were in common use like organochlorines (endosulfan), organophosphates (triazophos, monocrotophos, etc.), carbamates (Carbaryl), and synthetic pyrethroids (deltamethrin, cypermethrin, permethrin, fenvalerate, etc.) failed miserably to keep the pest problems under check (Palumbo et al. 2001). Despite their failure, these continued to be used in crop protection and applied in different forms, either by increasing the dosage or by mixing two or more pesticides. The pest problems got further accentuated, and the crop raising entered into a highly uneconomical zone. The social problems in the farming community become apparent with great unrest at many places. The introduction of synthetic pyrethroids in 1983 in India gave some relief but faded away soon. These chemicals created another problem of the appearance of sucking pests like mealy bugs, whiteflies, and aphids. The crop failures became apparent throughout the globe. These pest populations reached a level almost touching the category of key pests.

Although the majority of the population of whitefly remained unexposed to pesticides because the immature stages were confined to the lower leaf surfaces and that too on the lower canopy leaves that evaded the pesticide targeting, the wide host range and the rapid migration via air currents led to spread of whiteflies to far-off places. The pesticide failure crunch led to the discovery of more potent non-neurotoxic pesticides in mid-nineties namely insect growth regulators (IGR)

(Horowitz and Ishaaya 1996), buprofezin, pyriproxyfen, and nuvaluron (Castle et al. 2010, Ishaaya et al. 2003, Horowitz 2003) which were safe for the pest's natural enemies (De Cock and Degheele 1998). Of these, the pyriproxyfen-a juvenile mimic category is responsible for disturbing the flow of hormones, thereby affecting the embryogenesis (Ishaaya and Horowitz 1995, Ishaaya et al. 1994, Koehler and Patterson 1991). Imidacloprid (neonicotinoid) was first commercial product patented by Bayer Crop Science in 1985. By 2011 thiamethoxam (Syngenta), clothianidin (Sumitomo Chemical/Bayer Crop Science), acetamiprid (Nippon Soda), thiacloprid (Bayer Crop Science), dinotefuran (Mitsui Chemical), nitenpyran (Sumitomo) were introduced (Tomizawa and Casida 2005). After the introduction of these insecticides more potent pesticides (neonicotinoids) with translaminar properties entered the market (Horowitz et al. 2018). Imidacloprid and thiamethoxam were the first pesticides in this category (Luo et al. 2010, Roditakis et al. 2011, Nauen and Denholm 2005). Efforts were made to find out the ways and means to tackle this newly emerged problem. To be successful, timely surveillance was stressed by researchers. Emphasis was also laid on the preventive measures needed before the sowing of the crop to contain the multiplication of whiteflies. The cultural measures include the use of resistant varieties, elimination of most preferred hosts, crop-free period, and alteration of sowing/harvesting dates which create a disadvantage for the whiteflies. The other measures such as protective cultivation of the crop, use of an adequate dose of fertilizers, maintaining proper spacings while sowing, the use of desirable seed rate, and the provision of trap crops are also useful to mitigate the whitefly menace. Besides the cultural measures, the application of reflective devices (altering the behavior of the whitefly) is a desirable option either to repel/or attract the whiteflies. The separate measures to check the spread of plant viruses through the management of whiteflies vector should also not be overlooked. While it is easier to contain whiteflies, it is a far more cumbersome proposal to contain the whitefly as a vector.

14.1 Pest Monitoring and Surveillance

For tackling the whitefly menace, the timely monitoring and surveillance of the crops is mandatory. For this purpose, the use of yellow sticky traps in both under-protective and protective conditions is advocated. The use of yellow sticky traps is the best approach to monitor the whitefly population at this moment. To be successful, placing the yellow sticky traps @1–4 traps/1000 sq. ft. areas have been suggested (McHugh 1991). They have been used to both trap the population within small areas and for monitoring and surveillance. It is an important tool to monitor the population and to do surveillance as the farmers can then decide the tactics to be used or trap out the population. Initially, for a heavy adult population, it is more appropriate to trap the population. The adult population of the whitefly can also be trapped with electrostatic sweepers in the greenhouse/protected crop. Once egg-laying has taken place, it is important to take a strategic decision to select and apply the tactic. The use of insect growth regulators is most appropriate to treat the eggs and thus check the whitefly population. After the hatching of eggs and with the availability of a mixed population, it is appropriate to depend on the judicious use of neem products/

biopesticides and soap/detergents/natural enemies based on the threshold. Further increase in the population of whiteflies should be taken care of with the application of insect growth regulators like buprofezin. The single application of growth regulators is enough to tackle whiteflies in a protective crop. The other alternative is to tackle the whitefly build-up through the use of innovative technology related to the application of electrostatic sweepers for sucking the adults. The crop should be thoroughly monitored/examined throughout the crop season. Early surveillance is the key to detect whiteflies, select tactics, and check the low population. The control of whiteflies at this stage would mitigate the migration of pests by way of nipping the population at the bud stage. As shown by yellow sticky traps, the level of population is lower than or near the threshold level, the insecticide identified should be from the botanicals, mineral oils, horticultural oils, and neem oils or predators/parasitoids. The use of yellow sticky traps to attract and mulches to repel the population is still common in Israel and can be put to use elsewhere as well (Avidov 1956). The use of yellow sticky traps is also the best option in the country like Israel.

14.2 Breeding/Miscellaneous Strategies

Breeding Strategy:

- To achieve long-lasting and permanent control of such pests, it is important to develop resistant varieties of crop plants to have a strong foundation. Regarding measures to tackle the menace of this pest, the important practices (Butter and Dhawan 2001) are elucidated in the proceeding paragraphs. However, to start with, it is desirable to depend on transgenics. Already some progress has been made and a Tma 12 (protein isolated from edible fern, *Tectaria macrodonta* introduced) transgenic variety has been developed in India (Shukla et al. 2016). It is expected that this innovative research will be seen in action on the farmer's field soon.

- To start with, the breeding strategy needs to be explored. The crops with available resistant germplasm/varieties should be selected and introduced to lay the sound foundation of integrated whitefly management. The problem of cotton bollworms was mitigated with the development of transgenic crop and their subsequent introduction. The problems of sucking pests, particularly the whitefly, should be tackled through the development of resistant varieties. For this purpose, the jassid resistant hairy/pubescent varieties need to be replaced immediately as the whitefly has preference towards such pubescent varieties. The total and sudden replacement of hairy varieties in cotton could jeopardize the control strategy. It is thus advised that this should be done in a phased manner. The continuation of a recommendation of hairy varieties would be an open invitation to the cotton leafhopper with the result that it would regain its earlier position to continue as a key pest of cotton. Taking into consideration the bases of resistance, the cotton varieties resistant to both jassid and whitefly should be the target. Morphological (thinness of leaf lamina and density of trichomes, angle of insertion of hair), biochemicals (total phenols and tannins including gossypol), and nutritional bases of resistance (higher content of phosphorus) are identified in cotton and resistant varieties, which can be developed with the incorporation of desired traits (Butter and Vir 1989, Butter et al. 1992, Butter et al. 1992a). In other crops, the bases of resistance should be identified along the pattern of cotton. A recent study carried out in Pakistan demonstrated the susceptibility of Bt cottons to whiteflies as compared to conventional varieties of cotton (Atta et al. 2015). The population of adults of whitefly/leaf on transgenic and conventional cotton cultivars was 2.38 and 2.36 adults, respectively. Therefore, while planning whitefly management, this point should be well taken care of.

- The varieties of cotton with yellow-green foliage are highly preferred by a key pest, *Helicoverpa armigera*, as well as the whitefly throughout the world and should be avoided; the varieties with purple/red foliage should be the prime focus (Husain and Trehan 1942).

- The cotton varieties with okra/semi-okra leaves/narrow leaves permitting light and air penetration should be the target (Butter and Vir 1989). It would not only check the whitefly menace but would also pave the way for early maturity of the crop. The cotton growers would thus, also be free from many other problems encountered with a prolonged crop season.

- Emphasis should be on the development of early maturing, narrow leaves, and sympodial plant type cotton varieties. The whitefly resistant/tolerant cultivars should be a priority. Besides, such varieties are needed to sow, as per recommended agronomic practices, staggered sowing over a long period should be discouraged. The late sowing/late matiring varieties are always more prone to whiteflies.

- The leaf sap rich in gossypol, tannins, and phosphorus content should be given weightage in the breeding of cotton varieties. In addition cotton varieties with more than 6 pH of the sap should be avoided. The whiteflies prefer to feed on older leaves late in the season as the pH level of older leaves (7.25 pH) is higher than the younger leaves (6.0 pH) (Berlinger et al. 1983a).

- The glossy foliage of host varieties are also known to repel whiteflies as these are non preferred.

Miscellaneous Strategy

- Besides the breeding strategy, the liberal use of cultural measures (Hilje et al. 2001) and biocontrol agents (Liu et al. 2015) should be encouraged. The management of irrigation, drip irrigation, and fertilization (phosphorus) should be taken into account (Butter et al. 1992a). Besides the use of intercrops (squash, corn, french bean, capsicum) (Schuster 2004, Youssef et al. 2001), barrier crops (maize), and trap crops (eggplant) should be introduced. Melon is a good trap crop for the management of whitefly. Avoid the cultivation of lettuce and cucurbits around a cotton field. It is also fruitful to make use of an eggplant as a trap plant in screenhouse along with systemic insecticides (Choi et al. 2016). The experiment was conducted in Korea with eggplant as trap plant using systemic insecticide (dinotefuran/cyantraniliprole/pyridaben/clothianidin) in which mortality up to 50% was recorded. Amongst the systemic insecticides, dinotefuran was the most promising and this caused 88.4% mortality of adults of *Bemisia tabaci*.

- The inclusion of newer approaches like semiochemicals should be given weightage in order to prepare a long-lasting proposal for managing the whitefly menace (Islam et al. 2017).
- The biorational (plant pesticides, entomopathogenic fungi, use of natural enemies, homemade remedies, and physical measures) measures should be priority areas. It is important because of the increased areas coming under organic farming. The organic farming has already covered 7,20,000 hectares. Besides, screenhouse cultivation of precious crops is gaining importance nowadays; thus, the mechanical and physical measures are likely to play a major role in future (Weintraub and Horowitz 2001).
- The general practice for raising a good crop involves clean cultivation without crop residues while elimination of weed hosts with off-season weed flora acting as reservoirs of the pest should be followed (Dhawan and Simwat 1999).
- The cultivation of the preferred hosts of whitefly (brinjal, cucurbits, tomato, potato, cauliflower, crucifers) in the cotton belt during/and in the offseason should be discouraged (Butter and Dhawan 2001).
- The use of biological control agents (predators, parasitoids, and entomopathogens) (Al-Shihi et al. 2016, Naranjo 2001) is a must to achieve long-lasting and permanent control over the pest.
- The homemade remedies/water spray is effective in the beginning, should be exploited, and is free from ill effects. It can serve as a supplement in areas with deficient irrigation facilities (Greer 2000).
- The selective use of chemicals safe to natural enemies (De Cock and Degheele 1998) as and when required based on the economic threshold level (Sukhija et al. 1986) is essential to save the crop from the devastating attack of pest. IGR should be preferred over others and the farmers should start the protection strategy with such noble chemicals (Horowitz and Ishaaya 1996).
- The chemical control with selective insecticides safe to natural enemies and applied based on the economic threshold is appropriate and safe for the environment (Liu et al. 2015).
- Always avoid the repeated application of single insecticide and absolutely avoid the use of mixtures of insecticides. Tank mixing can sometimes be useful.
- Thorough coverage is essential for this pest as the insect is generally found hidden on the lower surface of leaves. There is a need to develop better pesticide application equipment to target these hidden whiteflies. The effect of spray technology has been confirmed through a study carried out in India. The population of whitefly adults (5.9/leaf) and nymphs (8.4/cm^2 area on the leaf) in a cotton field were sprayed with an Akela sprayer, where the remaining population was comparatively higher as compared to cotton sprayed with mist blower where the population was 3.6 adults/leaf and 1.33 nymphs/cm^2. Thus, the mist blower was found more efficient on account of thorough coverage as compared to the other kind of sprayer (Anonymous 1989).

- The biorational materials like soap and detergents should be included in the control operations of whitefly. The oils are known to cause suffocation and block the spiracles to cause mortality in whiteflies (Metcalf and Metcalf 1993). The detergents are responsible for breaking the surface tension and thereby efficiently managing the insect pest. The oils and soaps are of immense value. These are also known for their safety. The mineral oils have already been demonstrated as effective agents against whitefly on tomato (Butter and Rataul 1973).

- As a part of an integrated pest management (IPM) strategy, the liberal use of polypropylene sheets (Agryl) and UV-absorbing screens (BioNet) and films are useful (Doukas and Payne 2007, Raviv and Antignus 2004, Antignus et al. 1998). Emphasis is laid on including tactics like biocontrol (natural enemies/ biopesticides), mechanical and physical measures, and cultural practices. The tactics to be used will depend on the situation, that is, whether the crop is raised under a protected cover or is grown in the open field, or if the tactic is meant to manage the pest or the vector. It is also important to have a complete knowledge of vector or a category of virus as the tactics applicable to controlling different categories of viruses are different. For example, if it is a stylet-borne non-persistent virus or it is of the gut-borne category, the choice of pesticide is different concerning the persistent type of plant viruses. The tactics are different for the same crop raised in the field or in screen houses. The management strategies are also different for protected situations and under open field conditions (Fig. 14.1).

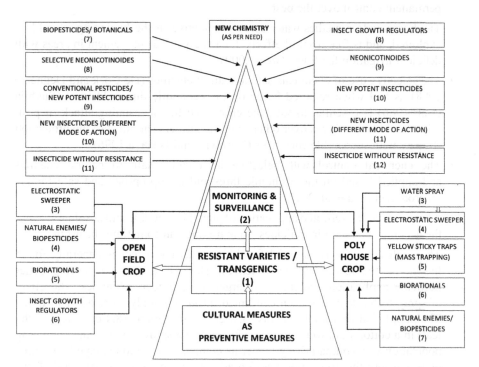

Fig. 14.1 Suggested integrated whitefly management model.

14.3 Selection of Insecticides

The discovery of DDT was earlier considered the Golden Era in pesticide history. After DDT, organophosphates and carbamates and synthetic pyrethroids were introduced. The Golden Era got diminished due to the ill effects Neonicotinoids, insect growth regulators and many other pesticides with a unique mode of action and selectivity towards natural enemies were introduced (Casida and Quistad 1998). After this, introduction of transgenic has eroded the dominance of organic pesticides.

14.3.1 Neonicotinoids

The insecticides with systemic and translaminar action, pronounced residual activity and a unique mode of action called neonicotinoids were in the market. Imidacloprid was the first new molecule followed by acetamiprid (1995), thiamethoxam (1998), thiacloprid/clothianidin (2000), and dinotefuran (2002) (Elbert et al. 2008). These chemicals were target specific, safe to non-target organisms, and least disturbing to the environment. The high target specificity made these insecticides globally acceptable in integrated pest management strategies and pest resistance management programs (Jeschke et al. 2010). Thiamethoxam has been declared harmful to honey bees, thus its continuous use is questionable in cotton (Henry et al. 2012).

14.3.2 Entomopathogenic Fungi

Fungi: For the management of whiteflies, entomopathogenic fungi like *Beauveria bassiana, Metarhizium anisopliae, Verticillium lecanii* (Abdel-Raheem and Al-keridis 2017), *Nomuryaea rileyi, Lecanicillium lecanii, Hirsutella thompsonii, Paecilomyces fumosoroseus* can be applied through seed dressers, soil drenches, seed soaking, and injections (Vega 2018). A study advocated *Verticillium lecanii, Aschersonia, Paecilomyces fumosoroseus* (Osborne and Landa 1992), *Beauveria bassiana* (Boopathi et al. 2015) as efficient entomopathogenic fungi against whitefly. *Paecilomyces lilacinus, Lecanicillium psalliotae, Aspergilus ustus,* and *Metarhizium anisopliae* are also promising entomopathogenic fungi. Already, *Beauveria bassiana* and *Paecilomyces fumosoroseus* are in use in Florida against *Bemisia tabaci* (Wraight et al. 2000). *Clonostachys rosea* has also been identified as effective against the whitefly, *Bemisia tabaci* (Anwar et al. 2018). In all, *Beauveria, Metarhizium, Paecilomyces, Verticillium, Aschersonia* (Meekes 2001, Bolckmans et al. 1995) have been advocated. Commercial products are also available (Mycotal contains *Verticillium lecanii*, Botanigard contains *Beauveria bassiana*, and PreFeRal contains *Paecilomyces fumosoroseus* (Stansly and Natwick 2010)).

14.3.3 Botanicals

It is important to include botanicals along with the other pesticides in integrated pest management. The pesticides of inorganic materials—preferably, the products of plant materials—such as nicotine sulfate, rotenone, pyrethrins, Margosan O (Azadirachtin-based) should be included (Miles 1927). More than 2,121 plant species with insecticidal value have been identified besides the insecticide-yielding plants

(*Ageratum* and *Tagetes,* neem, etc.). In addition to botanical pesticides, numerous plants with antifeedant, attractant, and repellent properties have also been identified (Purohit and Vyas 2004). Of these, neem products (Neemark 0.4%/Neemark 0.5%) is effective against *Bemisia tabaci* on cotton and should be considered (Phadke et al. 1988). Margosan-O is another neem product that contains *Bemisia tabaci* in cotton cultivation, as discovered in Phoenix, Arizona, and is a good inclusion in the pest's management (Mordue and Nisbet 2000, Flint and Parks 1989). Another substitute can be neem oil @ 0.5% and neem seed extract @ 5% to contain whiteflies (Nimbalkar et al. 1994). Neem oil (2%) is effective against eggs and nymphs of the whitefly (Roychoudhury and Jain 1993) and can be included in resistance management strategies (Kranthi and Russell 2009). Neem oil @2.5 liter/ha during the early part of the season is suggested for sucking pests under integrated resistance management (IRM) strategies. These products are slow-acting with low toxicity, lack of knockdown, rapid biodegradability, repeated applications, inactivation by UV light, and dependence on the weather, which limit their use.

14.4 Suggested Integrated Whitefly Management Model (SIWMM)

A sustainable integrated pest management (IPM) strategy was followed after the development of resistance in insects to pesticides 1980s onward. The major components in IPM are pest identification, regular monitoring, determination of the economic threshold, the selection of suitable tactics, and an assessment of the system. To achieve results, it is important to use tactics for prevention, avoidance, monitoring, and identification and have these tactics ready to use as the situation warrants it. The sequence of tactics is intelligently drawn as per the need in the given agroecosystem. Before that, it is essential to monitor the population of eggs, nymphs, and adults using appropriate monitoring techniques and suitable sampling procedures. The insecticide resistance management strategy is suggested taking into consideration the key pest fauna of cotton in the Indian/Asian region (pink bollworm-*Pectinophora gossypiella, Helicoverpa armigera, Earias species, Bemisia tabaci*), Australian region (Whitefly –*Bemisia argentifolii, Helicoverpa species Mite infestation-Tetranychus urticae*), in Arizona, USA (Lygus bug-*Lygus hesperus, Pectinophora gossypiella* (now eliminated), *Helicoverpa armigera,* whitefly B-biotype), and in Israel (*Bemisia tabaci, Helicoverpa armigera, Pectinophora gossypiella*) (Horowitz et al. 2018). The repeated use of insecticides and pesticides which create pesticide resistance in the target pest should be avoided. The selection of the pesticides rests with the plant protection expert depending upon the pest species present in the crop.

- The management of whitefly is required in both protected cultivation and unprotected ones (Fig. 14.1; Table 14.1). The control of whitefly in the open field is slightly different from the control operations followed for crops raised under the protected covers. The model of whitefly management in a protective cover and open field cultivation should be followed separately. To prevent the build-up of whiteflies it is essential to go for agricultural practices and these are quite effective against whiteflies. The first and foremost duty of the farmer is

Table 14.1. Suggested integrated whitefly management model.

Activity No.	Polyhouse crop (protected crop) with all preventive measures (mainly the cultural measures)	Open field crop (unprotected crop) with all preventive measures (mainly the cultural measures)
1	Resistant varieties/Transgenics	Resistant varieties/Transgenics
2	Mass Trapping & Monitoring/ Surveillance	Monitoring/surveillance for application of tactic (based on economic threshold)
3	Water spray under pressure (Mechanically dislodges adults)	Use of electrostatic sweeper
4	Electrostatic sweeper (Elimination of winged adults of whitefly)	Biopesticides (Neem-based/Plant origin)/ Natural enemies (*Encarsia/Eretmocerus*)
5	Yellow sticky traps (Mass Trapping)	Biorationals (Neem oil, Soybean oil, Neem seed kernel extract/Neemguard/Nimbicidin/Godrej Achook)/Soap, Surfactant (Fish oil resin soap)
6	Biorationals (Neem oil, Soybean oil, Neem seed kernel extract/Neemguard/ Nimbicidin/Godrej Achook)/Soap, Surfactant (Fish oil resin soap)	Insect growth regulators (Buprofezin/Pyriproxyfen/Lufenuron)
7	Natural predators *Chrysoperla carnea*/Biopesticides (Entomopathogenic fungi)/Any commercial product-Koppert/Naturalis/ Conidia/Braconi/Engerlingspolz/ Mycontrol/Ostrinil/Betel/Bowerol/ BioPath/Biologic/Pfr/Mycotal/Vertalec/ Botanigard and PreFeRal	Biopesticides Entomopathogenic fungi (PreFeRal)/Botanigard/Mycotal and botanicals (Neem oil, Soybean oil, Neem seed kernel extract/Neemguard/Nimbicidin/Godrej Achook
8	Insect growth regulators (Buprofezin/ Pyriproxyfen/Lufenuron)	Selective neonicotinoids (Imidacloprid/Acetamiprid/Thiamethoxam, etc.)
9	Neonicotinoids (Imidacloprid/Thiamethoxam/ Acetamiprid, etc.)	Conventional insecticides (Triazophos/ Acephate, etc.)/new potent insecticides from Diamides group (Chlorantraniliprole/ Spiromesifen/Spirotetramat)
10	New potent insecticides Diamides (Flubendamide/Chlorantraniliprole)	New insecticides (different mode of action) (Pymetrozine), Pyridine Corboxmide (Flonicamid)
11	New insecticides (different mode of action) (Pymetrozine/Spiromesifen/ Spiroteramat), Pyridine Corboxmide (Flonicamid-Lygus bug)	Insecticide without resistance to whitefly (different pesticide at different locations)
12	Insecticide with zero resistance to whitefly/safety for natural enemies/no repeat	Need-based application (new pesticide without repeat)

* The sequence of activities (tentative, need-based use, and dependent on availability of product/new molecule) followed to manage whiteflies. The preference should be given to alternate use of pesticides of two different groups.

to look for the situations in which the management tactics are to be used. It is always desirable to select the field to avoid the whitefly attack. In this context is advisable to avoid the cultivation of cotton in citrus orchards due the appearance of virus vectored by whitefly. If the cultivation of cotton is to be done in such situations, look for plantation of desi cotton (*Gossypium arboreum*) which is resistant to Geminivirus. The cultivation of inter crops such as okra and moong should be avoided.

- To lay the strong foundation, the emphasis should be laid on resistant varieties/ genetically modified plants. Headway has been made to develop resistant varieties in some crops (cotton, soybeans crops, pulses, tomato, and cassava). The varieties resistant to whiteflies in other crops such as watermelon, muskmelon, cucurbits, and brinjal, have also been developed. The transgenic variety of cotton (Tma12) (Shukla et al. 2016) has been developed and is likely to find a place soon in India. It is a gene isolated from edible fern and introduced in cotton to produce protein in cotton leaves, which toxic to whiteflies. This tactic serves as a foundation to raise strong IPM. Still, efforts are needed to identify resistant varieties in other important crops.

- The exclusion of whiteflies is the best approach. It is important to construct the greenhouse as per specification using a mesh of size 0.19 mm to 0.82 mm or the entire structure can be covered with the plastic (0.213 mm²) as per recommendation (Bethke et al. 1994). To make a screenhouse, screens of 50 mesh screen size (50 holes/per square inch) are available in the market and can be utilized for excluding the whiteflies. Besides this screenhouses covered with UV-absorbed polyethylene film also reduce the population of whitefly to the extent of 50%. A screenhouse can be covered with a plastic film that permits a higher level of UV light (Price 1999, Raviv and Antignus 1984). Care should be taken to permit normal light inside the protective structure. Before sowing, it must be ensured that the nursery of vegetable crops should be free from viral infection. In all, five species of whitefly are known to act as a vector of plant viruses.

- Hitherto successful cultural measures have been eroded due to the ease in the use of pesticides. An ideal tactic of IPM, cultural measures must find a place in the strategy to keep the whitefly population under check. Several measures are clubbed together in cultural tactics. The first useful practice is to keep the area free from the crop at least for two months from mid-June to mid-August in the traditional crop cultivation areas. Furthermore, it is a polyphagous pest and its host range includes a large number of crops and weed flora. The whitefly migration takes place in two distinct phases throughout the year: from cotton to other hosts and back to cotton via spring through potential hosts. It migrates to alternate hosts (*Brassica* spp., *Solanum nigrum, Solanum melongena, Solanum incarnium, Glycine max, Cucums* sp., *Mentha viridis*, several weeds) in winter and from these hosts to spring hosts (*Euphorbia, Lantana camera, Ipomoea batatus, Ricinus communus, cardofana, Abutilon grandifolium, Duranta* sp., *Convolvulus arvensis, Brassica* spp.) to cotton, cucurbits (melon) and the shoe flower *Hibiscus esculentus* (Rafiq et al. 2008, Husain and Trehan 1933). This kind of migration from the main crop to carry-over hosts also exists in California, USA (Johnson et al. 1982), Israel (Melamed-Madjar et al. 1982, Melamed-Madjar et al. 1979), Thailand (Nachapong and Mabbett 1979, Mabbett et al. 1980), and Iran (Habibi 1975). If the areas are infested with weed hosts that are suitable as hosts for carrying over and the survival of whitefly, it is essential to clean the vicinity of weed hosts (*Euphorbia, Lantana*, etc.). Eggplant (Brinjal), melon, and squash are preferred by this whitefly and thus should be avoided. Similarly, flowering plants such as *Petunia Hibiscus, Zinnia,* etc. are helpful for the successful carry-over of the pest. Another cultural control measure is the use of trap crops. The crop should have intercrop/trap crop (brinjal and cucurbits as trap crop in tomatoes) growing in a mixed crop pattern to prevent the spread of vector-borne viral diseases through whitefly vector (Choi et al. 2016, Choi et al. 2015). It is further necessary to make use of balanced fertilizers. A higher dose of phosphorus can reduce the whiteflies in cotton (Butter et al. 1992a). The crop geometry should also be taken into account.

- As soon as the egg-laying is spotted in the crop, it can be checked with the application of an ovicidal chemical Insect Growth Regulators (IGR). The IGR (like buprofezin, pyriproxyfen, etc.) can be sprayed to control the eggs as a first-round measure as recommended in southwestern USA (Dennehy et al. 2005). Pyriproxyfen is also advocated early in the season. After the applications of IGRs, the next spray should be done after a week as the hatching takes place within seven days after the spray of IGR; in this round, the IGR should not be used and the next spray should be done with any of the neonicotinoid pesticides (Ishaaya et al. 2003). The neonicotinoids that are safe for mammals, natural enemies, and other useful fauna should be selected. Besides, care should be taken to discard the neonicotinoid to which the whitefly has developed resistance. It is better to use neonicotinoids with different modes of action. Another alternative is to make use of the biopesticides which are safe and highly compatible with the other tactics of the IPM.

- Another essential component is biological control. Already potent parasitoids, predators, and entomopathogens have been identified (Gerling et al. 2001). Of these, *Encarsia formosa* and *Eretmocerus eremicus* are liberally used in screen houses/glasshouses to manage whiteflies. Besides, predators like *Coccinella septempunctata, Delphastus catalinae, Nephaspis ovulates, Chrysoperla carnea* are suitable to use as well (Liu et al. 2015). *Amblyseius swirskii* mite has been put to use to manage whiteflies in protective crops (Stansly and Natwick 2010). The combination of biocontrol and IGR can be applied collectively in the beginning. The *Encarsia formosa* in combination with IGR is better than the individual tactic in managing the whitefly (Gill 2000). Similarly the use of *Eretmocerus eremicus* plus IGR is better than the individual tactic in the control of whitefly (Hoddle et al. 2001). The use of these two tactics should not be applied in case the population is < 1 nymph/10 cuttings as in the case of poinsettias. Besides, the predator *Delphastus pusillus* is effective as it feeds on eggs. The larvae of this predator normally consume 1000 eggs/larvae per day. As indicated earlier innovative releases of the beetle can be a suitable option.

As long as the population is low in the crop, the biocontrol agents can be made use of as these are effective when there is a low density of the pest. If the virus is a stylet-borne virus, the biocontrol agents should not be used to prevent the whitefly. It would rather enhance the spread of such plant viruses. The threshold of the vector is low as compared to other pests. Even single whitefly can spread the virus in the whole field. The persistent plant viruses can be checked through the application of biocontrol agents like *Encarsia, Eretmocerus* sp. Neem-based products, mineral oils, and pesticides with ovicidal activity should be selected (Butter and Rataul 1973). These neem products are comparatively safe for the whitefly's natural enemies.

- The whitefly had developed resistance to organic pesticides by 1980; thus, the use of biorational, is helpful in such situations. The biorationals include oils, surfactants, insecticidal soaps, neem products, entomopathogenic fungi products, and parasitoids and predators (Greer 2000).

- Commercial formulations based on fungi are available for use. The important formulations containing entomopathogenic fungi are *Beauveria bassiana, Isaria (Paecilomyces), Aschersonia, Verticillium, Metarhizium anisopliae.* They can be used at this stage. In case the need arises, the collective use of these two tactics is a good option (Bolckmans et al. 1995). Specialized parasitoids like *Encarsia formosa* and predatory mites are quite beneficial to check the growing whitefly population. To start with, the trap out strategy with yellow sticky traps can prove highly beneficial, particularly in the protective cultivation of crops.

- In small areas, the battery-operated vacuum pump developed in Japan (Electrostatic insect sweeper) can be put into use to suck the large chunk of whitefly adults (Jakkawa et al. 2015, Weintraub and Horowitz 2001). In case the need arises, the insect sweeper can be put to use. It is eco-friendly and it is the need of the hour.

- The use of aluminum-silver-painted silver polyethylene mulch is desirable to protect the crop. The use of sawdust mulch is prevailing till today in Israel (Avidov 1956). The use of sawdust in cucumber and aluminum and blue polyethylene mulch in zucchini reduced the whitefly vector population to the extent of 50% and also delayed the appearance of the viral disease in these crops. The use of mulch is a good tool to repel whiteflies to protect the crop.

- Biopesticides can be put to small-scale use in the greenhouse to check whiteflies using microbial pesticides. These biopesticides normally contain entomopathogenic fungi. These fungi are either *Beauveria bassiana* or *Paecilomyces fumosoroseus.* These biopesticides can be used in combination with NeemAzal, or NeeAzad, or Latin. The use of such chemicals is an extremely good option as these products are safe for natural enemies based on the surveillance report.

- The management strategy should be supplemented with the placement of whitefly repelling plants such as basil, calendula, etc. In the event of non-deployment of these measures, there is every possibility of a more whitefly build-up.

- Blasting with water under high pressure is good practice to dislodge the adult flies and get rid of them. This practice can be done once in a week and be continued for three weeks. In case the population of whiteflies still persists, it is advisable to check the use of commercial cards which maybe carrying parasitized red-eye nymphs of whiteflies. After parasitization, the color of the pupae 4th instar changes to black or amber-brown. The pupae parasitized by *Encarsia formosa* are black in appearance and can be utilized for assessment of biocontrol agents. To contain the whitefly menace the inclusion of Surf/or other detergents can be sprayed. Ensure the presence of a sufficient population of parasitoids before resorting to these tactics. In case the population of the parasitoid is low, supplement the population with inundative releases. In the absence of parasitoids, resort to home remedies to contain the whiteflies.

- In case an increasing population of whiteflies is revealed through surveillance, resort to the use of insect growth regulators (IGR) (Precision/Apploud) when there is a low host population. The action of IGR on the eggs will further prevent the build-up of the population of whiteflies. The IGR has juvenile action interfering in chitin formation which is required for the formation of cuticles; it also inhibits the molting process. If the population of whitefly increases after hatching of eggs in the next generation, select one chemical from the neonicotinoids group instead of insect growth regulator. This would help to contain the whitefly in a protective crop. But the use of neonicotinoids should not be done for more than two rounds during the crop season. It is better to select the neonicotinoids based on different modes of action. The monitoring of chemicals should be done regularly for the development of resistance. In case the resistance is recorded against a chemical(s), it (they) should be withdrawn from the field.

- The use of new chemicals to which the whitefly has not developed resistance should be given weightage. Care should be taken to use them judiciously to prolong their life; that is, to prevent the creation of resistance in the whiteflies through overuse. The new chemicals with a different mode of action include pyriproxyfen (Juvenile hormone mimic), buprofezin (chitin inhibitor), spiromesifen and spirotetramat (reduction in lipid synthesis) (Nauen et al. 2008), cyantraniliprole (calcium depletion), dinotefuran, diafenthiuron, flubendiamide, flonicamid, etc. and are potent pesticides against whitefly. Under this category, the diafenthiuron-thiourea derivative is safe for the whitefly's natural enemies, pollinators and mammals and affects only the mitochondrial ATP system in whiteflies (Streibert et al. 1988, Ruder et al. 1991). The predesign before the application of IGR helps in lowering the dose of rational chemicals. Studies were conducted in line with this concept on the reduced dose of chlorpyriphos with a predesign of lufenuron and the technique followed proved highly safe for the whitefly's natural enemies in the cotton crop ecosystem (Aulakh and Butter 2009). The other pesticide, pymetrozine (of the pyridine group) is known to affect the salivary system of whiteflies. The disturbance in the salivary system means the enhancement of hunger followed by the death of insects due to hunger (Kayser et al. 1994). Another group of pesticides affecting the lipid biosynthesis belonging to Spiro cyclic type, which are also safe for natural enemies, predators, and parasitoids bumble bees, was introduced and belong to the group containing spiromesifen and spirotetramat (Bielza et al. 2009, Nauen and Denholm 2005). Cyantraniliprole (Cyazypyr) was identified for managing whiteflies through the depletion of calcium reserves and it is safe for mammals and other beneficial organisms (Sattelle et al. 2008). Another new molecule, dinotefuran, was tested with many others against two invasive species (MEAMI and MED) of whiteflies and it was found more toxic than for controlling whiteflies (Qu et al. 2017). It significantly reduced the fecundity and survival of these two invasive species. These new chemicals cannot be ignored. These products are based on differential modes of action. Thus, the pesticide tactic is bound to stay in any management program. It is therefore essential to select a safe molecule with a label claim, use IGR for the first spray, maintain right intervals (that is, < 7 days for growth regulators and biorationals and > 7 for organic insecticides), restricted to 1–2 sprays of an insecticide, avoid mixtures of insecticides, prefer alternative use of different groups, and withdraw the chemical to which the whitefly has developed resistance. Always follow the judicious application of pesticides based on economic thresholds. The plant protection technology suitable for the sessile stages of the whiteflies on the lower leaf surfaces should be selected and this is slightly different from the other routine spray equipment. The spray should be directed towards the lower canopy of the plants and that too on the lower leaf surface for a thorough coverage of the plant.

• The entry of whiteflies in the glasshouse is generally through the negligence of workers; it is thus important to keep double doors to exclude whiteflies. This condition normally forces the need for yellow sticky traps in the protective crop as well. Fumigation used to be done in the earlier days with tetrachloride ethane (Parker 1928).

Based on experience, it was stressed that organic remedies (organic insecticidal soap, vacuuming, blast off with water, and sticky traps), host-plant resistance, biocontrol, cultural measures, physical tactics, and a judicious use of potent chemicals should be applied to achieve long-lasting results. The control operations would be highly compatible with the environment as suggested (Horowitz et al. 2011).

A sound management strategy is essential to control the difficult problem of whiteflies. Simultaneously, monitoring and surveillance from the beginning are the most important. The whitefly has developed resistance to most insecticides and resistance-management strategy is an important step to take care of whiteflies.

This chapter highlights the various strategies suggested based on information generated using yellow sticky traps. A tentative sequence is suggested by plant protection experts. To lay a strong foundation, the first step is to look for the breeding methodology to develop resistant varieties. To have a strong foundation, resistant varieties should be put in place. At this moment, highly susceptible varieties of host plants are being grown that need to be replaced. After this step, it is useful to follow all preventive measures to tackle this problem by introducing cultural practices. These practices are applicable before the sowing of the main crop to prevent the build-up of whiteflies. These practices hitherto remained neglected on account of potent pesticides. After the utilization of cultural measures, other tactics should be identified and kept ready for use. The plant protection expert should be intelligent enough to pick the strategy and implement it in the correct sequence. It will go a long way to mitigate the problems of whiteflies in the agroecosystem and polyhouses.

References

Abdel-Baky, N.F. (2006). Impact of temperatures and plant species on the biological features of the castor bean whitefly *Trialeurodes ricini*. Fourth International *Bemisia* Workshop International Whitefly Genomics Workshop December 3–8, 2006, Duck Key, Florida, USA.

Abdel-Raheem, M.A. and Al-Keirdis, L.A. (2017). Virulence of three Entomopathogenic fungi against *Bemisia tabaci* (Gennadius) (Hemiptera: Aleyrodidae) in tomato crop. Journal of Entomology 14: 155–159.

Abd-Rabou, S. (2000). Role of *Encarsia inaron* (Walker) (Hymenoptera: Aphelinidae) in biological control of some whitefly species (Homoptera: Aleyrodidae) in Egypt. Shashpa 7: 187–188.

Abd-Rabou, S. and Simmons, A.M. (2012). Some cultural strategies to help manage *Bemisia tabaci* (Hemiptera: Aleyrodidae) and whitefly transmitted viruses oaps, oils, and detergents on the sweet potato whitefly. Florida Entomologist 76: 162–167.

Abd-Rabou, S. and Simmons, A.M. (2012a). Effect of three irrigation methods on incidences of *Bemisia tabaci* (Hemiptera: Aleyrodidae) and some whitefly transmitted viruses in four vegetable crops. Trends in Entomology 8: 21–26.

Abd-Rabou, S. and Simmons, A.M. (2014). Survey of natural enemies of whiteflies (Hemiptera: Aleyrodidae) in Egypt with new local and World records. Entomological 38–56.

Abdullah, N.M.M., Singh, J. and Sohal, B.S. (2006). Behavioural hormoligosis in oviposition preference of *Bemisia tabaci* on cotton. Pesticide Biochemistry and Physiology 84: 10–16.

Acharya, V.S. and Singh, A.P. (2007). Screening of cotton genotypes against whiteflies, *Bemisia tabaci* Genn. Journal of Cotton Research and Development 21: 111–114.

Acharya, V.S. and Singh, A.P. (2008). Biochemical bases of resistance in cotton to the whitefly, *Bemisia tabaci* Genn. Journal of Cotton Research and Development 22: 195–199.

Agarwal, R.A. (1969). Morphological characteristics of sugarcane and insect resistance. Entomologic Experimentalism ET Applicata 12: 767–776.

Aharoni, A., Giri, A.P., Verstappen, F.W., Berea, C.M., Sevenier, R., Sun, Z., Jongsma, M.A., Schwab, W. and Bouwmeester, H.J. (2004). Gain and loss of fruit flavor compounds produced by wild and cultivated strawberry species. Plant and Cell 16: 3110–3131.

Aheer, G.M. (1999). Morphological factors affecting resistance in genotypes of cotton against some sucking insect pests. Pakistan Entomologist 21: 43–46.

Ahmad, K.A., Dwivedi, H.S. and Dwivedi, P. (2015). Pesticide scenario of India with particular reference to Madhya Pradesh: A review. New York Science Journal 8: 69–76.

Ahmad, M., Arif, M.I. and Ahmad, Z. (1999). Insecticide resistance in *Helicoverpa armigera* and *Bemisia tabaci*, its mechanisms and management in Pakistan. pp. 143–150. Proceedings of the Regional Consultation on Insecticide Resistance Management in Cotton, June 28–July 1, 1999, Multan, Pakistan.

Ahmad, M., Arif, M.I. and Ahmad, Z. (2000). The resistance of cotton whitefly, *Bemisia tabaci* to cypermethrin, alphacypermethrin, and zeta cypermethrin in Pakistan. pp. 1015–1017. Proceedings of the Beltwide Cotton Conference, National Cotton Council, January 5–9, 2000, Memphis, TN USA.

Ahmad, M., Arif, M.I., Ahmad, Z. and Denholm, M.C. (2002). Cotton whitefly (*Bemisia tabaci*) resistance to organophosphates and pyrethroid insecticides in Pakistan. Pest Management Science 50: 203–208.

Ahmad, M., Arif, M.I. and Ahmad, Z. (2001). Reversion of susceptibility to methamidophos in the Pakistani populations of cotton whitefly, *Bemisia tabaci*. pp. 874–876. Proceedings of the Beltwide Cotton Conference, National Cotton Council, January 9–13, 2001, Memphis, TN USA.

Ahmad, M., Arif, M.I. and Naveed, M. (2010). Dynamics of resistance to organophosphates and carbamate insecticides in the cotton whitefly *Bemisia tabaci* (Hemiptera: Aleyrodidae) from Pakistan. Journal of Pesticide Science 83: 409–420.

Ahmed, A.H.M., El-High, E.A. and Bashir, N.H.H. (1987). Insecticide resistance in cotton whitefly (*Bemisia tabaci* Genn.) in the Sudan Gezira. Tropical Pest Management 33(1): 67–72, 103, 107.

Ahmed, M.Z., Ren, S.X., Xue, X., Li, X.X., Jin, G.H. and Qiu, B.L. (2010). Prevalence of endosymbionts in *Bemisia tabaci* populations and they're *in vivo* sensitivity to antibiotics. Current Microbiology 61: 322–328.

Ahmed, M.Z., Ren, S.X., Mandour, N.S., Greeff, J.M. and Qiu, B.L. (2010a). Prevalence of *Wolbachia* supergroups A and B in *Bemisia tabaci* (Hemiptera: Aleyrodidae) and some of its natural enemies. Journal of Economic Entomology 103: 1848–59.

Ahmed, M.Z., De Barro, P.J., Ren, S.X., Greeff, J.M. and Qiu, B.-L. (2013). Evidence for horizontal transmission of secondary endosymbionts in the *Bemisia tabaci* cryptic species complex. Plops One 8: e0053084.

Ahmed, M.Z., Li, S.J., Xue, X., Yin, X., Ren, S.X., Jiggins, F.M., Greeff, J.M. and Qiu, B.L. (2015). The intracellular bacterium *Wolbachia* uses parasitoid wasps as phoretic vectors for efficient horizontal transmission. PLoS Pathog 2015b; 10: e1004672.

Alemandri, V., De Barro, P., Bejerman, N., Arguello Caro, E.B. and Dumon, A.D. (2012). Species within the *Bemisia tabaci* (Hemiptera: Aleyrodidae) complex in soybean and bean crops in Argentina. Journal of Economic Entomology 105: 48–53.

Alemandri, V., Vaghi Medina, C.G., Dumont, A.D., Argüelles Caro, E.B., Matteo, M.F., Medina, S.G., Albertini, P.M. and Troll, G. (2015). Three members of the *Bemisia tabaci* (Hemiptera: Aleyrodidae) cryptic species complex occur sympatric ally in argentine horticultural crops. Journal Economic Entomology 108: 405–413.

Al-Musa, A. (1982). Incidence, economic importance, and control of Tomato Yellow Leaf Curl in Jordan. Plant Disease 66: 561–563.

Alomar, O., Riudavets, J. and Castane, C. (2006). *Macrolophus caliginosus* in the biological control of *Bemisia tabaci* on greenhouse melons. Biological Control 36: 154–162.

Alon, M., Benting, J., Lueke, B., Ponge, T., Alon, F. and Morin, S. (2006). Multiple origins of pyrethroids resistance in sympatric biotypes of *Bemisia tabaci* (Hemiptera: Aleyrodidae). Insect Biochemistry and Molecular Biology 36: 71–79.

Alon, M., Alon, F., Nauen, R. and Morin, S. (2008). Organophosphates' resistance in the B-biotype of *Bemisia tabaci* (Hemiptera: Aleyrodidae) is associated with a point mutation in acel-type acetylcholinesterase and overexpression of carboxylesterase. Insect Biochemistry and Molecular Biology 38: 940–949.

Alonso, R.S., Raccka-Filho, F. and Lima, A.F. (2012). Occurrences of whiteflies (Homoptera: Aleyrodidae) on cassava, *Manihot esculenta* (Cants) in the state of Ri de Janeiro, Brazil. EnomoBrasilis 5: 48–49.

Al-Shayji, Y. and Shaheen, M. (2008). Isolation of *Bacillus thuringiensis* strain from Kuwait, soil effective against whitefly nymphs. Journal of Insect Science 8(4): 53–59.

Al-Shehi, A.A. and Khan, A.J. (2013). Identification of whitefly (*Bemisia tabaci* Genn.) biotypes and associated bacterial symbionts in Oman. Journal of Plant Sciences 8: 39–44.

Al-Shihi, A.A., Al-Sadi, A.M., Al-Said, F.A., Ammara, U.E. and Deadman, M.L. (2016). Optimising the duration of floating row cover period to minimize the incidence of Tomato Yellow Leaf Curl disease and maximize the yield of tomato. Annals of Applied Biology 168: 328–336.

Alston, D. (2007). Insect pests in greenhouse and nursery crops. Utah State University Extension Utah Green Conference January 22, 2007. https://extension.usu.edu/pests/files/slideshows/insects-landscape/07sh-insects.

Al-Zyoud, F. and Sengonca, C. (2004a). Prey consumption preferences of *Serangium parcesetosum* Sicard (Coleoptera: Coccinellidae) for different prey stages, species, and parasitized prey. Journal of Pest Science 77: 197–204.

Al-Zyoud, F. and Sengonca, C. (2004). Development, longevity, and fecundity of *Bemisia tabaci* (Genn.) (Homoptera: Aleyrodidae) on different host plants at two temperatures. Mittelungen Deutschen Gesellschaft Algemeine Angewandte Entomologie 14: 373–376.

Amro, M.A., Omar, A.S., Abdel–Monism, A.S. and Yamani, K.M.M. (2007). Determination of resistance of environmental soybeans to the lima bean pod borer, *Etiella zinckenella* Treitschke and the whitefly, *Bemisia tabaci* Gennadius at Dakhla oases, New Valley, Egypt. ASS University Bulletin of Environmental Research 10(2): 57–65.

Anonymous. (1986). The Package of Practices for *Kharif* Crops. Publication of Punjab Agricultural University Ludhiana.

Anonymous. (1989). Management of whitefly, *Bemisia tabaci* Genn. on cotton. Andhra Pradesh Agriculture University, Rajendra nagar, Hyderabad.

Anonymous. (1995). Controlled atmosphere to manage whitefly. The Cut Flower Quarterly. July. p. 14–16.

Anonymous. (2004). Integrated Management for Cotton, Ministry of Agriculture, Department of Agriculture and Cooperation, Directorate of Plant Protection, Quarantine and Storage, Govt of India, IPM Package No. 25.

Anonymous. (2015). International Cotton Advisory Committee, 2015.

Anonymous. (2019). Invasive species compendium: Detailed coverage of invasive species threatening the livelihood and the environment worldwide. *Trialeurodes vaporariorum* (whitefly & Greenhouse).

Anonymous. (2019a). Pest Alert: Whiteflies/IFAS Extension, University of Florida, and Institute of the food and Agricultural Sciences. Post Box 110810, Gainsvilla, Fl1 12611, 352: 392–1761.

Anonymous. (2020). Cotton Insect Management Guide Texas A&M Agriculture life Extension.

Antignus, Y. and Cohen, S. (1987). Purification and some properties of a new strain of Cowpea mild mottle virus in Israel. Annals of Applied Biology 110: 563–569.

Antignus, Y., Mor, N., Joseph, R.B., Lapidot, M. and Cohen, S. (1996). Ultraviolet-absorbing plastic sheets protect crops from insect pests and virus diseases vectored by insects. Environmental Entomology 25: 919–924.

Antignus, Y., Lapidot, M., Hadar, D., Messika, Y. and Cohen, S. (1998). Ultraviolet-absorbing screens serve as optical barriers to protect crops from viruses and insect pests. Journal of Economic Entomology 91: 1401–1405.

Antignus, Y., Nestel, D., Cohen, S. and Lapidot, M. (2001). Ultraviolet-deficient greenhouse environment affects whitefly attraction and flight-behavior. Environmental Entomology 30: 394–399.

Antonious, G. and Snyder, J. (2008). Tomato leaf crude extracts for insects and spider mite control. pp. 269–297. *In*: Preedy, V.R. and Watson, R.R. (eds.). Tomatoes and Tomato Products: Nutritional, Medicinal and Therapeutic Properties. Science Publishers, USA.

Anwar, W., Ali, S., Nawaz, K., Iftikhar, S., Javed, M.A. and Hashem, A. (2018). Entomopathogenic fungus, *Clonostachys rosea* as a biocontrol agent against whitefly *Bemisia tabaci*. Biological Science and Technology 28: 700–750.

Ardeh, M.J. (2004). Whitefly control potential of *Eretmocerus* parasitoids with different reproductive modes. Ph.D. Thesis, Wageningen University, Wageningen, Netherlands.

Aregbesola, O.Z. (2018). Understanding the impact of climate change on whitefly, *Bemisia tabaci* (Gennadius) (Hemiptera: Aleyrodidae) Ph.D. thesis University of Catania, Italy and University of Copenhagen (Denmark).

Aregbesola, O.Z., Legg, J.P., Sigsgaard, L., Lund, O.S. and Rapisarda, C. (2019). Understanding the potential impact of climate change on whiteflies and implications for the spread of vectored viruses. Journal of Pest Science 92(2): 581–592.

Arif, M., Gog, D., Zia, K. and Hafeez, I. (2006). Impact of plant spacings and abiotic factors on the population dynamics of sucking insect pests of cotton. Pakistan Journal of Biological Sciences 9(7): 1364–1369.

Armes, N.J., Jadhav, D.R. and DeSouza, K.R. (1996). A survey of insecticide resistance in *Helicoverpa armigera* in the Indian subcontinent. Bulletin Entomological Research 86: 499–514.

Ashfaq, M., Noor Ul Ane, M., Zia, K., Nasreen, A. and Hasan, M.U.H. (2010). The correlation of abiotic factors and physiomorphic characteristics of (*Bacillus thuringiensis*) Bt transgenic cotton with whitefly, *Bemisia tabaci* (Homoptera: Aleyrodidae) and jassid, *Amrasca devastans* (Homoptera: Jassidae) populations. African Journal of Agricultural Research 5: 3102–3107.

Aslam, M. and Misbah-ul-Haq, M. (2003). Resistance of different genotypes of *Helianthus annuus* Linnaeus against *Bemisi tabaci* and *Empoasca* spp. and their correlation with yield. Asian Journal of Plant Sciences (Pakistan) 2: 220–223.

Atta, B., Mustafa, F., Adil, M., Raja, M.F. and Farooq, M.A. (2015). Impact of different transgenic and conventional cotton cultivars on population dynamics of whitefly *Bemisia tabaci*. Bulgaria Journal of Agricultural Science 3: 175–178.

Aulakh, S.S. and Butter, N.S. (2009). Conservation of natural enemies in cotton ecosystem with pre-dosing of lufenuron prior to the use of traditional insecticide. Journal of Biological Control 23(3): 233–24.

AVA. (2001). Diagnostic records of Plant Health Diagnostic Services, Singapore Plant Health Centre. Agriculture and Veterinary Authority.

Avidov, Z. (1956). Bionomics of the tobacco whitefly (*Bemisia tabaci* Gennadius) in Israel. Ktavim 7: 25–41.

Awasthi, L.P. and Kumar, P. (2008). Response of chilli genotypes/cultivars against viral diseases. Indian Phytopathol. 61: 282–284.

Baig, M.M., Dubey, A.K. and Ramamurthy, V.V. (2015). Biology and morphology of life stages of three species of whiteflies (Hemiptera: Aleyrodidae) from India. Pan-Pacific Entomologist 91: 168–183.

Baldin, E.L. and Beneduzz, R.A. (2010). Characterization of antibiosis antexenosis to Silver leaf whitefly (Homoptera: Aleyrodidae) in some squash varieties. Journal Pest Management 83: 221–227.

Baldo, L., Ayoub, N.A., Hayashi, C.Y., Russell, A.J., Stahlhut, J.K. and Werren, Z.H. (2008). Insight into the routes of *Wolbachia* invasion: high levels of horizontal transfer in the spider genus *Agelenopsis* revealed by *Wolbachia* strain and mitochondrial DNA diversity. Molecular Ecology 17: 557–569.

Bandyopadhay, U.K., Santhakumar, M.V., Das, K.K. and Saratchandra, B. (2002). Determination of economic threshold level of whitefly, *Dialeuropora decempunctata* in mulberry, *Morus alba* L. International Journal of Industrial Entomology 4(2): 133–136.

Banerjee, I., Tripathi, S.K., Roy, A.S. and Sengupta, P. (2014). Pesticide use pattern among farmers in a rural district of West Bengal, India. Journal National Science Biology Med. 5: 313–316.

Barbosa, L., Julio, B.L., Marubayashi, M., Bruno, A., Marchi, RDe, Valdir, A.A. Yuki,c Pavan, M.A., Moriones, bAn, Navas-Castillob, E.J. and Krause-Sakate, R. (2014). Indigenous American species of *Bemisia tabaci* complex still widespread in the Americas. Pest Management Science 70: 1440–1445.

Bashir, M.H., Afzal, M., Sabri, A.M. and Raza, A.B.M. (2001). Relationship between sucking insect pests and physic-morphic plant characters towards resistance/susceptibility and some new genotype of cotton. Pakistan Entomologist 23: 75–78.

Basij, M., Talebi, K., Ghadamyari, M., Hosseininaveh, V. and Salami, S.A. (2016). Status of resistance of *Bemisia tabaci* (Hemiptera: Aleyrodidae) to neonicotinoids in Iran and detoxification by cytochrome P450-dependent monooxygenases. Neotropical Entomology doi: 10.1007/s13744-016-0437-3.

Basit, M., Sayyed, A.H., Saleem, M.A. and Saeed, S. (2011). Cross-resistance, inheritance and stability of resistance to acetamiprid in cotton whitefly, *Bemisia tabaci* Genn. (Hemiptera: Aleyrodidae). Crop Protection 30: 705–712.

Basit, M., Saeed, S., Saleem, M.A., Denholm, I. and Shah, M. (2013). Detection of resistance, cross-resistance, and stability of resistance to new chemistry insecticides in *Bemisia tabaci* (Homoptera: Aleyrodidae). Journal of Economic Entomology 106: 1414–1422.

Basu, A.K. (1986). Resurgence of whitefly in cotton and strategies for its management. Proceedings of the National Symposium on Resurgence of Sucking Pests (Tamil Nadu Agricultural University. Coimbatore), 129–133.

Baumann, P. and Moran, N.A. (1985). Non-cultivable microorganisms from symbiotic associations of insects and other hosts. Anotonie Van Leeunhoek 72: 9–48.

Baumann, P. and Moran, N.A. (1997). Noncultivable microorganisms from symbiotic associations of insects and other hosts. Antonne Von Leeuwenhoek 72: 39–48.

Baumann, P. (2005). Biology of bacteriocyte-associated endosymbionts of plant sap-sucking insects. Annual Reviews of Microbiology 59: 155–189.

Baumann, P., Moran, N.A. and Baumann, L. (2006). Bacteriocyte-associated endosymbionts of insects. pp. 403–438. *In*: Dworkin, M., Falkow, S., Rosenberg, E., Schleifer, K.H. and Stackebrandt,

E. (eds.). Prokaryotes vol. 1 Symbiotic Association of Biotechnologists. Applied Microbiology Springer New York, New York, NY.

Bedford, I.D., Briddon, R.W., Brown, J.K., Rosell, R.C. and Markham, P.G. (1994). Geminivirus-transmission and biological characterization of *Bemisia tabaci* (Gennadius) biotypes from different geographic regions. Annals of Applied Biology 125: 311–325.

Begum, S., Anis, S.B., Farooqi, M.K., Rehmat, T. and Fatma, J. (2011). Aphelinid parasitoids (Hymenoptera; Aphelinidae) of whiteflies from India. Journal of Biology and Medicine 3(2) Special Issue: 222–231.

Bellotti, A., Campo, B.V.H. and Hyman, G. (2012). Cassava production and Pest Management: Present and potential threat in changing environment. Tropical Plant Biology 5: 39–72.

Bellotti, A.C. and Arias, B. (2001). Host plant resistance to whiteflies with emphasis on cassava as a case study. Crop Protection 20: 813–823.

Bellotti, A.C. (2008). Cassava pests and their management. pp. 764–794. *In*: Capinera, J.L. (ed.). Encyclopedia of Entomology. Springer Netherlands, Dordrecht, the Netherlands.

Bellows, T.S. Jr, Paine, T.D., Arakawa, K.Y., Meisenbacher, C., Leddy, P. and Kabashimo, J. (1990). Biological control sought for ash whitefly. California Agriculture 44(1): 4–6.

Bellows, T.S. Jr, Perring, T.M., Gill, R.J. and Headrick, D.H. (1994). Description of a species of *Bemisia* (Homoptera: Aleyrodidae). Annals of the Entomological Society of America 87(2): 195–206.

Bellows, T.S. and Meisenbacher, C. (2000). Biological control of giant whitefly, Aleurodicus dugesii, in California. pp. 113–16. *In*: Hoddle, M.S. (ed.). California Conference on Biological Control II. Riverside, California Agricultural University. California. Riverside.

Ben-Israel, I., Yu, G., Austin, M.B., Bhuiyan, N., Auldridge, M., Nguyen, T., Schauvinhold, I., Noel, J.P., Pichersky, E. and Fridman, E. (2009). Multiple biochemical and morphological factors underlie the production of methylketones in tomato trichomes. Plant Physiology 151: 952–1964.

Bennett, R.M., Ismael, Y., Kambhampati, U. and Morse, S. (2004). The economic impact of genetically modified cotton in India. AgBioForum 7: 96–100.

Bentz, J.A. and Larew, H.G. (1992). Ovipositional preference and nymphal performance of *Trialeurodes vaporariorum* (Homoptera: Aleyrodidae) on *Dendranthema grandiflora* under different fertilizer regimes. Journal of Economic Entomology 85: 514–517.

Bentz, J.A., Reeves, III J., Barbosa, P. and Francis, B. (1995). Nitrogen fertilizer effect on selection, acceptance, and suitability of *Euphorbia pulcherrima* (Euphorbiaceae) as a host plant to *Bemisia tabaci* (Homoptera: Aleyrodidae). Environment Entomology. 24: 40–45

Ben-Yakir, D., Antignus, Y., Offir, Y. and Shahak, Y. (2012). Colored shading nets impede insect invasion and decrease the incidences of insect-transmitted viral diseases in vegetable crops. Entomologia Experimentalis et Applicata 144: 249–257.

Berlinger, M.J., Goldberg, A.M., Dahan, R. and Cohen, S. (1983). Use of plastic covering to prevent the spread of Tomato Yellow Leaf Curl Virus in greenhouses. Hasadeh 63: 1862–1865.

Berlinger, M.J., Magal, Z. and Benzioni, A. (1983a). The importance of PH in food selection of whitefly, *Bemisia tabaci*. Phytoparasitca 11: 151 HTTP/doi;.org/10. 1007/02980686.

Berlinger, M.J. (1986). Host plant resistance to *Bemisia tabaci*. Agriculture, Ecosystems, and Environment 17: 69–92.

Berry, S.D., Fondong, V.N., Rey, C., Rogan, D., Fauque, C.M. and Brown, J.K. (2004). Molecular evidence for five distinct *Bemisia tabaci* (Homoptera: Aleyrodidae) geographic haplotypes associated with cassava plants in Sub-Saharan Africa. Annals of the Entomological Society of America 97(4): 852–859.

Bethke, J.A. and Paine, T.D. (1991). Screen hole size and barriers for the exclusion of insect pests of glasshouse crops. Journal of Entomological Science 26: 169–177.

Bethke, J.A., Paine, T.D. and Nuessly, G.S. (1991). Comparative biology, morphometrics, and development of two populations of *Bemisia tabaci* (Homoptera: Aleyrodidae) on cotton and poinsettia. Annals of the Entomological Society of America 84.4: 407–411.

Bethke, J.A., Redak, R.A. and Paine, R.A. (1994). Screens deny specific pests entry to greenhouses. California Agriculture 48: 37–40.

Bhatt, B., Giri, G.S., Karnatak, A.K. and Shivashankara. (2018). Effect of weather parameters on population fluctuations of sucking pests and their predators on okra crop. International. Journal of Current Microbiology and Applied Sciences 7(9): 757–762.

Bielza, P., Fernanadez, E., Gravalos, C. and Izquierdo, J. (2009). Testing for non-target effects of spiromesifen on *Eretmocerus mundus* and *Orius laevigatus* under greenhouse conditions. BioControl 54: 229–236.

Bielza, P., Moreno, I., Belando, A., Grávalos, C., Izquierdo, J. and Nauen, R. (2019). Spiromesifen and spirotetramat resistance in field populations of *Bemisia tabaci* Gennadius in Spain. Pest Management Science 75(1): 45–52. https://doi.org/10.1002/ps.5144.

Bing, X., Li Yang, J., Liu, S.S. and Wang, X.-W. (2012). Identification of a new secondary symbiont in the whitefly *Bemisia tabaci* (Hemiptera: Aleyrodidae). XXIV International Congress of Entomology. Abstract, S511TH11.

Bing, X.-L., Yang, J., Zchori-Fein, E., Wang, X.-W. and Liu, S.-S. (2013). Characterization of a newly discovered symbiont of the whitefly *Bemisia tabaci* (Hemiptera: Aleyrodidae). Applied Environmental Microbiology 79: 569–75.

Bing, X.L., Ruan, Y.M., Rao, Q., Wang, X.W. and Liu, S.S. (2013a). Diversity of secondary endosymbionts among different putative species of the whitefly *Bemisia tabaci*. Insect Science 20: 194–206.

Birkett, M.A., Chamberlain, K., Guerrieri, E. and Pickett, J.A. (2003). Volatiles from whitefly infested plants elicit a host locating response in parasitoid, *Encarsia formosa*. Journal of Chemical Ecology 29: 1589–1600.

Birkett, M.A. and Pickett, J.A. (2014). Prospects of genetic engineering for robust insect resistance. Current Opinion in Plant Biology 19: 59–67.

Blackburn, M.B., Domek, J.M., Gelman, D.B. and Hu, J.S. 2005. The broadly insecticidal *Photorhabdus luminescens* toxin complex a (Tca): Activity against the Colorado potato beetle, *Leptinotarsa decemlineata*, and sweet potato whitefly, *Bemisia tabaci*. 11pp. Journal of Insect Science 5: 32.

Bleeker, P.M., Diergaarde, P.J., Ament, K., Guerra, J., Weidner, M., Schutz, S., De Both, M.T.J., Harring, M.A. and Schuurink, R.C. (2009). The role of specific tomato volatiles in tomato-whitefly interaction. Plant Physiology 151: 925–935.

Bleeker, P.M., Diergaarde, P.J., Ament, K., Schutz, S., Johne, B. and Dijkink, J. (2011). Tomato-produced 7-epizingiberene and R-curcumin act as repellents to whiteflies. Phytochemistry 72: 68–73.

Bleeker, P.M., Mirabella, R., Diergaarde, P.J., Van Doorn, P.J., Tissier, A., Kant, M.R., Prins, M., de Vos, M., Haring, M.A. and Schuurink, R.C. (2012). Improved herbivore resistance in cultivated tomato with the sesquiterpene biosynthetic pathway from a wild relative. Proceedings of National Academy of Sciences, USA 109: 20124–20129.

Bloch, G. and Wool, D. (1994). Methidathion resistance in the sweet potato whitefly (Aleyrodidae: Homoptera) in Israel: Selection, heritability, and correlated changes of esterase activity. Journal of Economic Entomology 87: 1147–1156.

Bolckmans, K., Sterk, G., Eyal, J., Sels, B. and Stepman, W. (1995). PreFeRal (*Paecilomyces fumosoroseus* strain Apopka 1997), a new microbial insecticide for the biological control of whiteflies in greenhouses. Medical Faculty Landbouww University Gent 60/3a: 707–711.

Bondar, G.K. (1967). Boli du laboratorio patrol. Veg 4: 39–46.

Boopathi, T., Karuppuchamy, P., Singh, S.B., Kalyanasundaram, M., Mohankumar, S. and Ravi, M. (2015). Microbial control of the invasive spiraling whitefly on cassava with entomopathogenic fungi. Brazillian Journal Microbiology 46: 1077–1085.

Boykin, L.M., Shatters, R.G., Jr, Rosell, R.C., McKenzie, C.L., Bagnall, R.A., De Barro, P.J. and Frohlich, D.R.Y. (2007). Global relationships of *Bemisia tabaci* (Hemiptera: Aleyrodidae) revealed using Bayesian analysis of mitochondrial COI DNA sequences. Molecular Phylogenetics and Evolution 44: 1306–1319.

Boykin, L.M., Armstrong, K.F., Kubatko, L. and De Barro, P.J. (2012). Species delimitation and global biosecurity. Evolutionary Bioinformatics 8: 1–37.

Boykin, L.M. (2014). *Bemisia tabaci* nomenclature: Lessons learned. Pest Management Science 70: 1444–1459.

Boykin, L.M. and De Barro, P.J. (2014). A practical guide to identifying members of the *Bemisia tabaci* species complex: and other morphologically identical species. Frontiers of Ecology and Evolution 2: 45.

Bretschneider, T., Benet-Buchholz, J., Fischer, R. and Nauen, R. (2003). Spirodiclofen and spiromesifen: Noval acaricidal and insecticidal tetronic acid derivatives with a new mode of action. CHIMIA International Journal for Chemistry 57: 697–701.

Briddon, R.W., Pinner, M.S., Stanley, J. and Markham, P.G. (1990). Geminivirus coat protein replacement alters insect specificity. Virology 177: 85–94.

Brown, J.K. (1991). An update on the whitefly-transmitted geminiviruses in the Americas and the Caribbean Basin. FAO Plant Protection Bulletin 39: 5–23.

Brown, J.K., Coat, S.A., Bedford, I.D., Markham, P.G., Bird, J. and Frohlich, D.R.Y. (1995). Characterization and distribution of esterases electromorphs in Whitefly, *Bemisia tabaci* (Genn.) (Homoptera; Aleyrodidae). Biochemical Genetics 33: 205–214.

Brown, J.K., Frohlich, D.R. and Rosell, R.C. (1995a). The sweet potato or silverleaf whiteflies: Biotypes of *Bemisia tabaci* or a species complex? Annual Reviews of Entomology 40: 511–534.

Brown, J.K. (2000). Molecular markers for the identification and global tracking of whitefly vector Begomovirus complexes. Virus Research 71: 233–260.

Brown, J.K., Fauquet, C.M., Bridden, R.W., Zerbini, F.M., Moriones, E. and Navas-Castillo, J. (2012). Family Geminiviridae. pp. 351–373. *In*: King, A.M.Q., Adam, M.J., Carsten, E.B., Lefkowitz, E.J. (eds.). Virus Taxonomy, 9TH Report of International Committee on Taxonomy of Viruses. Elsevier, Academic Press London.

Brown, L., Brown, J.K. and Tsai, J.H. (1990). Lettuce infectious yellows virus. Plant Pathology, circular no 335, Sept 1990. Fl Dept of Agriculture Consumer Service, Division of Plant Pathology.

Bruce, T., Aradottir, G.I., Smart, L.E., Martin, J.L., Caulfield, J.C., Doherty, A., Sparks, C.A., Woodcock, C.M., Birkett, M.A., Napier, J.A., Jones, H.D. and Pickett, J.A. (2015). The first crop plant genetically engineered to release an insect pheromone for defence. Scientific Reports 5: 11183. https://doi.org/10.1038/srep11183.

Brumin, M., Kontsedalov, S. and Ghanim, M. (2011). *Rickettsia* influences thermos tolerance in the whitefly *Bemisia tabaci* B biotype. Insect Science 18: 57–66.

Burade, D.D., Khobragade, A.M. and Shinde, P.R. (2019). Influence of weather parameters on pests of okra in Parbhani Kranti variety. International Journal of Current Microbiology and Science 8: 806–812.

Burban, C., Fishpool, L.D.C., Fauquet, C., Fargette, D. and Thouvenel, J.C. (1992). Host-associated biotypes within West African populations of the whitefly *Bemisia tabaci* (Genn.) (Hom., Aleyrodidae) Journal of Applied Entomology 113: 416–423. https://doi.org/10.1111/j.1439-0418.1992.tb00682.x.

Butler, G.D., Jr. and Henneberry, T.J. (1984). *Bemisia tabaci*: effect of cotton leaf pubescence on abundance. Southwestern Entomologist 9: 91–94.

Butler, G.D. and Henneberry, T.J. (1986). *Bemisia tabaci* Gennadius. A pest of cotton in the South Western United States. Agricultural Research Service Technical Bulletin 19: 1701.

Butler, G.D., Jr., Henneberry, T.J., Stansly, P.A. and Schuster, D.J. (1993). Insecticidal effects of selected soaps, oils, and detergents on the sweetpotato whitefly: (Homoptera: Aleyrodidae). The Florida Entomologist 76(1): 161–167.

Butt, T.M. and Goettel, M.S. (2000). Bioassays of Entomogenous Fungi© CAB International 2000. Bioassays of Entomopathogenic Microbes and Nematodes (eds A. Navon and K.R.S. Ascher).

Butter, N.S. and Rataul, H.S. (1973). Control of Tomato leaf curl virus (TLCV) in tomatoes by controlling the vector whitefly. *Bemisia tabaci* Gen. by mineral oil sprays. Current Science 42(24): 864–865.

Butter, N.S. and Rataul, H.S. (1977). The virus vector relationship of Tomato leaf curl virus with its vector, *Bemisia tabaci* Gen. (Hemiptera; Aleyrodidae). Phytoparasitica 5: 173–186.

Butter, N.S. and Rataul, H.S. (1981). Nature and extent of loss in tomatoes due to LeafCurl Virus (TLCV) transmitted by Whitefly, *Bemisia tabaci* Genn (Aleyrodidae; Hemiptera). Indian Journal of Ecology 8: 299–300.

Butter, N.S. and Kular, J.S. (1986). Economic threshold of Whitefly, *Bemisia tabaci* Genn (Homoptera; Aleyrodidae) on cotton. Journal of Research Punjab Agricultural University 23: 66–70.

Butter, N.S. and Sukhija, H.S. (1987). Efficacy of flucythrinate (payoff 10EC) against bollworms (*Pectinophora gossypiella, Earias* spp., and *Helicoverpa armigera*) infesting cotton. Journal of Research Punjab Agricultural University 24: 615–622.

Butter, N.S. and Kular, J.S. (1987). Effect of cotton whitefly damage on seed germination and fiber qualities of upland cotton. Indian Journal of Ecology 14: 158–16.

Butter, N.S., Sukhija, H.S. and Kular, J.S. (1989). Resurgence of red spider mite, *Tetranychus cinnabarinus* (Boisd) on upland cotton. Journal of Research Punjab Agricultural University 26(1): 80–84.

Butter, N.S. and Vir, B.K. (1989). Morphological basis of resistance in upland cotton to whitefly, *Bemisia tabaci.* Phytoparsitica 17: 251–262.

Butter, N.S. and Vir, B.K. (1990). Sampling of Whitefly, *Bemisia tabaci* Genn in cotton. Journal of Research, Punjab Agricultural University 27: 615–619.

Butter, N.S. and Vir, B.K. (1991). Response of whitefly, *Bemisia tabaci* Genn to different cotton genotypes under glasshouse conditions. Indian Journal of Entomology 53: 115–119.

Butter, N.S., Vir, B.K., Kaur, G., Singh, T.H. and Raheja, R.K. (1992). Biochemical basis of resistance to whitefly, *Bemisia tabaci* Genn. (Aleyrodidae: Hemiptera) in cotton. Tropical Agriculture (Trinidad) 69(2): 119–122.

Butter, N.S., Kular, J.S., Singh, T.H. and Chahal, G.S. (1992a). Management of Whitefly, *Bemisia tabaci* Genn in cotton through soil application of phosphorus. Proceedings of the Vasant Rao Naik Memorial National Seminar on cotton production in India, 5th-6th December 1992, Nagpur pp. 127–135.

Butter, N.S., Brar, A.S., Kular, J.S. and Singh, T.H. (1992b). Effect of agronomic practices on the incidence of key pests of cotton under unsprayed conditions. Indian Journal of Entomology 54: 115–123.

Butter, N.S., Kular, J.S. and Singh, T.H. (1992c). Deltaphos for the control of cotton pests in Punjab. Pestology 16(12): 11–18.

Butter, N.S. and Singh, T.H. (1993). Current status of insect resistance in cotton. pp. 151–181. *In*: Dhaliwal, G.S. and Dilawari, V.K. (eds.). Advances in Host Plant Resistance to Insects.

Butter, N.S., Vir, B.K., Kular, J.S., Brar, A.S. and Nagi, P.S. (1996). Relationship of plant nutrients and Whitefly, *Bemisia tabaci* Gen. Indian Journal of Entomology 58: 1–16.

Butter, N.S., Kular, J.S. and Singh, T.H. (1997). Behavioural response of *Tetranychus cinnabarinus* (Boisd) to cotton. Indian Journal of Entomology 59: 379–384.

Butter, N.S. and Kular, J.S. (1999). Resurgence of whitefly in cotton and its management. Indian Journal of Entomology 61: 85–90.

Butter, N.S. and Dhawan, A.K. (2001). Integrated management of whitefly, *Bemisia tabaci* (Gennadius) on cotton: Indian perspective: Proceedings of National Seminar On Sustainable Cotton Production to Meet the Requirement of Industry held on October 3–4, 2001 at Mumbai.

Butter, N.S. (2018). Insect Vectors and Plant Pathogens. CRC Press, Boca Raton, Florida: Pages 496.

Byrne, D.N., Bellos, T.S. and Parella, M.P. (1990). Whiteflies in agricultural systems. pp. 227–261. *In*: Gerling, D. (ed.). Whiteflies: Their Bionomics, Pest Status, Management, and or, Uk. Intercept Limited.

Byrne, D.N. and Bellows, T.S. Jr (1991). Whitefly biology. Annual Reviews of Entomology 36: 431–457.

Byrne, F., Castle, S., Prabhaker, N. and Toscano, N. (2003). Biochemical study of resistance to imidacloprid in B biotype *Bemisia tabaci* from Guatemala. Pest Management Science 59: 347–352.

Byrne, F.J., Denholm, I., Devonshire, A.L. and Rowland, M.W. (1992). Analysis of insecticidal resistance in whitefly, *Bemisia tabaci.* pp. 165–178. *In*: Denholm, I., Devonshire, A.L. and Hollomon, D.W. (eds.). Resistance 91: Achievements and Development in Combating Pesticide Resistance. Elsevier, London.

Byrne, F.J. and Devonshire, A.L. (1993). Insensitive acetylcholinesterase and esterase polymorphism in susceptible and resistant populations of the tobacco whitefly *Bemisia tabaci* (Genn.). Biochemistry. Physiology 45: 34–42.

Byrne, F.J., Cahill, M., Denholm, I. and Devonshire, A.L. (1994). A biochemical and toxicological study of the role of insensitive acetylcholinesterase in organophosphorus resistant *Bemisia tabaci* (Homoptera: Aleyrodidae) from Israel. Bulletin of Entomology Research 84: 179–184.

Byrne, F.J., Gorman, K.J., Cahil, M., Denholm, I. and Devonshire, A.L. (2000). The role of B-type esterases in conferring insecticide resistance in the tobacco whitefly, *Bemisia tabaci* (Genn). Pest Management Science 56: 867–874.

Bharpoda, T.M. and Suther, M.D. (2015). Physiochemical bases of resistance in cotton with special reference to sucking pests. AGRES-an International e-Journal 4: 87–96.

Caballero, R., Cyman, S. and Schuster, D. (2013). Monitoring insecticide resistance in biotype B of *Bemisia tabaci* (Hemiptera: Aleyrodidae) in Florida. Florida Entomologist 96: 1243–1256.

CABI. (2017). Center for Agriculture and Biosciences International, Invasive Species Compendium. *Bemisia tabaci* (tobacco whitefly). http://www.cabi. org/iso/datasheet/8927.

CABI. (2020a). Invasive Species Compendium. Detailed coverage of Invasive species threatening livelihoods and the environment worldwide-the Greenhouse whitefly, *Trialeurodes vaporariorum.*

CABI. (2020). Invasive Species Compendium. Detailed coverage of Invasive species threatening livelihoods and the environment worldwide-the tabacco whitefly, *Bemisia tabaci.*

Cahill, M., Byrne, F.J., Denholm, I., Devonshire, A.L. and Gorman, K.J. (1994). Insecticide resistance in *Bemisia tabaci.* Pesticide Science 42: 137–139.

Cahill, M., Byrne, F.J., Gorman, K., Denholm, I. and Devonshire, A.L. (1995). Pyrethroid and organophosphate resistance in the tobacco whitefly *Bemisia tabaci* (Homoptera: Aleyrodidae). Bulleti Entomological Research 85: 181–187.

Cahill, M., Gorman, K., Day, S., Denholm, I., Elbert, A. and Nauen, R. (1996). Baseline determination and detection of resistance to imidacloprid in *Bemisia tabaci* (Homoptera: Aleyrodidae). Bulletin of Entomological Research 86(4): 343–349. doi:10.1017/S000748530003491X.

Calabrese, E.J. and Baldwin, L.A. (2003). Hormesis, the dose-response solution. Annual Reviews of Pharmacology and Toxicology 43: 175–197.

Calvert, L.A., Cuervo Mcalvertrroyave, J.A., Constantino, L.M. and Bellotti, A. (2001). Morphological and DNA marker analyses of whiteflies (Homoptera: Aleyrodidae) colonizing mitochondrial cassava and beans in Colombia. Annals of Entomological Society of America 94: 512–519.

Calvitti, M. and Buttarazzi, M. (1995). Determination of biological and demographic parameters of *Trialeurodes vaporariorum* Westwood (Homoptera Aleyrodidae) on two host plant species: Zucchini (*Cucurbita pepo*) and tomato (*Lycopersicon esculentum*). Redia 78: 29–37.

Campbell, B.C., Duffus, J.E. and Baumann, P. (1993). Determining whitefly species. Science 261: 1333–1335.

Campbell, B.C., Steffen-Campbell, J.D. and Gill, R.J. (1994). Evolutionary origin of whiteflies (Hemiptera; Sternorrhyncha; Aleyrodidae) inferred from 18S rDNA sequences. Insect Molecular Biology 3(2): 73–88.

Campbell, B.C., Steffen-Campbell, J.D. and Gill, R.J. (1996). Origin and radiation of whiteflies: an initial molecular phylogenetic assessment, in *Bemisia* 1995: Taxonomy, Biology, Damage, Control, and Management, edited by Gerling D and Mayer RT. Intercept, Andover, UK.

Cao, F.Q., Liu, W., Fan, Z., Wang, F. and Cheng, L. (2008). Behavioral responses of *Bemisia tabaci* B-biotype to three host plants and their volatiles. Acta Entomologica Sinica 51: 830–838.

Carabali, A., Bellotti, A.C., Montoya-Lerma, J. and Cuellar, M.E. (2005). Adaptation of *Bemisia tabaci* biotype B (Gennadius) to cassava, Manihot sculenta (Crantz). Crop Protection 24: 643–649.

Carabali, A., Bellotti, A.C., Montoya-Lerma, J. and Fregene, M. (2010). Resistance to the whitefly, *Aleurotrachelus socialis*, in wild populations of cassava, Manihot tristis. Journal of Insect Science 10: 170A.

Cardona, C.I. (2001). Resistencia an insecticides en *Bemisia tabaci Trialeurodes vaporariorum* (Homoptera: Aleyrodidae) en Colombia y Ecuador. Reviews of Colombian Entomology 27: 33–38.

Carriere, Y., Ellers-Kirk, C., Hartfield, K., Larocque, G., Degain, B., Dutilleul, P., Dennehy, T.J., Marsh, S.E., Crowder, D.W., Xianchun, W.C., Li, X., Ellsworth, P.C., Naranjo, S.E., Palumbo, J.C., Fournier, A., Antilla, L. and Tabashnik, B.T. (2012). Large-scale, a spatially-explicit test of the refuge strategy for delaying insecticide resistance. Proceedings of National Academy of Sciences 109: 775–780.

Carvalho, M.G., Bortolotto, O.C. and Ventura, M.U. (2017). Aromatic plants affect the selection of host tomato plants by *Bemisia tabaci* biotype B. Entomologia Experimentalis et Applicata 162: 86–92.

Casida, J.E. and Quistad, G.B. (1998). Golden age of insecticide research: past, present, or future? Annual Reviews of Entomology 43: 1–16.

Caspi-Fluger, A., Inbar, M., Mozes-Daube, N., Katzir, N., Portnoy, V., Belausov, E., Hunter, M.S. and Zchori-Fein, E. (2012). Horizontal transmission of the insect symbiont Rickettsia is plant-mediated. Proceedings of Biological Sciences 279: 1791–1806.

Castle, S., Palumbo, J., Prabhaker, N., Horowitz, A. and Denholm, I. (2010). Ecological determinants of *Bemisia tabaci* resistance to insecticides. pp. 423–465. *In*: Stnsly, P. and Naranjo, S. (eds.). Bionomics and Management of a Global Pest. Amsterdam, The Netherlands: Springer.

Castle, S.J. and Prabhaker, N. (2013). Monitoring changes in *Bemisia tabaci* (Hemiptera: Aleyrodidae) susceptibility to neonicotinoid insecticides in Arizona and California. Journal of Economic Entomology 106: 1404–1413.

Castle, S.J., Merten, P. and Prabhaker, N. (2013). Comparative susceptibility of *Bemisia tabaci* to imidacloprid in field-and laboratory-based bioassays. Pest Management Science 70: 1538–1546.

Caterino, M.S., Cho, S. and Sperling, F.A.H. (2000). The current state of insect molecular systematics: a thriving tower of Babel. Annual Review of Entomology 45: 1–54.

Cave, R.D. (2008). Biocontrol of whitefly on coconut palms in Comoros. Biocontrol News Information 29: 1–18.

Chaba, A.A. (2018). Pest management: Punjab cotton farmers conquer whitefly. The Indian Express, August 30, 2018.

Chandi, R.S. and Kular, J.S. (2020). A sampling plan for whitefly, *Bemisia tabaci* (Gennadius) population on Bt cotton. Agricultural Research Journal 57(1): 53.

Chandrashekar, K. and Shashank, P.R. (2017). Indian contribution to whitefly (*Bemisia tabaci*) research. pp. 563–580. A Century of Plant Virology in India. Springer, Singapore.

Chaubey, R., Andrew, R.J., Naveen, N.C., Rajagopal, R., Ahmed, B. and Rammamurthy, V.V. (2015). Morphometric analysis of three putative species of *Bemisia tabaci* (Hemiptera: Aleyrodidae) species complex from India. Annals of Entomological Society of America 108: 600–612.

Chen, W., Hasegawa, D.K., Kaur, N., Kliot, A., Pinheiro, P.V., Luan, J., Stensmyr, M.C., Zheng, Y., Liu, W. and Sun, H. (2016). The draft genome of whitefly *Bemisia tabaci* MEAM1, a global crop pest, provides novel insights into virus transmission, host adaptation, and insecticide resistance. BMC Biology 14: 110. (DOI: 10.1186/s12915-016-0321-y).

Chiel, E., Gottlieb, Y., Zchori-Fein, E., Mozes-Daube, N., Katzir, N., Inbar, M. and Ghanim, M. (2007). Biotype-dependent secondary symbiont communities in sympatric populations of *Bemisia tabaci*. Bulletin of Entomology Research 97: 407–413.

Chiel, E., Inbar, M., Mozes-Daube, N., Jennifer, A., White, Martha, S., Hunter and Einat Zchori-Fein. (2009). Assessments of fitness effects by the facultative symbiont Rickettsia in the sweet potato whitefly (Hemiptera: Aleyrodidae). Annals of Entomological Society of America 102: 413–418.

Choi, Y., Seo, J., Whang, I., Kim, G., Choi, B. and Jeong, T. (2015). Effects of eggplant as a trap plant attracting *Bemisia tabaci* Genn. (Hemiptera: Aleyrodidae) adults available on tomato greenhouses. Korean Journal of Applied Entomology 54: 311–316.

Choi, Y., Hwang, I., Lee, G. and Kim, G. (2016). Control of *Bemisia tabaci* Genn. (Hemiptera: Aleyrodidae) adults on tomato plants using trap plants with a systemic insecticide. Korean Journal of Applied Entomology 55: 109–117.

Chowda-Reddy, R.V., Kirankumar, M., Seal, S.E., Muniyappa, V. and Valand, G.B. (2012). *Bemisia tabaci* phylogenetic groups in India and the relative transmission efficacy of Tomato leaf curl Bangalore virus by an indigenous and an exotic population. Journal Integrated Agriculture 11: 235–248.

Chu, C.C., Natwick, E.T., Perkins, H.H., Brushwood, D.E., Henneberry, T.J., Castle, S.J., Cohen, A.C. and Boykin, M.A. (1998). Upland cotton susceptibility to *Bemisia argentifolii* (Homoptera: Aleyrodidae) infestations. Journal of Cotton Science 2: 1–9.

Chu, C.C., Cohen, A.C., Natwick, E.T., Simmons, G.S. and Henneberry, T.J. (1999). *Bemisia tabaci* (Hemiptera: Aleyrodidae) biotype B colonisation and leaf morphology relationships in upland cotton cultivars. Australian Journal of Entomology 38: 127–131.

Chu, C.C., Freeman, T.P., Buckner, J.S., Henneberry, T.J., Nelson, D.R., Walker, G.P. and Natwick, E.T. (2000). *Bemisia argentifolii* (Hemiptera: Aleyrodidae) colonization on upland kinds of cotton and relationships to leaf morphology and leafage. Annals of Entomological Society of America 93: 912–919.

Chu, C.C., Freeman, T.P., Buckner, J.S. and Henneberry, T.J. (2001). Susceptibility of upland cotton cultivars to *Bemisia tabaci* biotype B (Homoptera: Aleyrodidae) about leaf age and trichome density. Annals of Entomological Society of America 94: 743–749.

Chu, C.C., Natwick, E.T. and Henneberry, T.J. (2002). *Bemisia tabaci* (Homoptera: Aleyrodidae) biotype B colonization on Okra- and Normal-Leaf upland cotton strains and cultivars. Journal of Economic Entomology 95(4): 733–738.

Chu, C.C., Natwick, E.T., Chen, T.Y. and Hennebury, T.J. (2003). Analysis of cotton leaf characteristics affects on *Bemisia tabaci* (Homoptera: Aleyrodidae) biotype B colonization on cotton in Arizona and California. Southwestern Entomologist 28: 235–240.

Chu, D., Wan, F.-H., Tao, Y.-L., Liu, G.-X., Fan, Z.-X. and Bi, Y.-P. (2008). Genetic differentiation of *Bemisia tabaci* (Gennadius) (Hemiptera: Aleyrodidae) biotype Q based on mitochondrial DNA markers. Insect Science 15: 115–123.

Chu, D., Gao, C.S., De Barro, P., Zhang, Y.J., Wan, F.H. and Khan, I.A. (2011). Further insights into the strange role of bacterial endosymbionts in whitefly, *Bemisia tabaci*: comparison of secondary symbionts from biotypes B and Q in China. Bulletin Entomological Research 101: 477–486.

Clark, E., Karley, A.J. and Hubbard, S.F. (2010). Insect endosymbionts: manipulators of insect herbivore trophic interactions? Protoplasma 244: 25–51.

Clark, M.A., Baumann, L., Munson, M.A., Baumann, P., Campbell, B.C., Duffus, J.P., Osborne, L.S. and Moren, N.A. (1992). Eubacterial endosymbionts of whiteflies (Homoptera: Aleyrodidae constitute a lineage distinct from the endosymbionts of aphids and mealybugs. Current Microbiology 25: 119–123.

CMRAB (California Melon Research Advisory Board). (2003). Management strategic plan—Cantaloupe, Honeydew, and Mixed Melon Production in California. http://pest-data.ncsu.edu/pmsp/pdf/CAMelon.pdf.

Cock, M.J.W. (1986). *Bemisia tabaci*—A Literature Survey on the Cotton Whitefly with an Annotated Bibliography. CAB International Institute of Biological Control, UK.

Cock, M.J.W. (1993). *Bemisia tabaci* An update 1986–1992 on Cotton Whitefly with an Annotated Bibliography. CABI, Silwood Park, UK: 78.

Cockerell, T.D.A. (1902). The classification of the Aleyrodidae. Proceedings of the Academy of Natural Sciences of Philadelphia 54: 279–283.

Coffey, J.L., Simmons, A.M. and Merle, B. (2015). Potential sources of whitefly (Hemiptera: Aleyrodidae) resistance in Desert watermelon *Citrullus colocynthis* germplasm. Hortscience 50: 13–17.

Cohen, S., Kern, J., Herpaz, I. and Ben-Joseph, R. (1988). Epidemiological studies of Tomato yellow leaf curl virus (TYLCV) in the Jordan valley, Israel. Phytoparasitica 16: 259–270.

Colinet, D., Kummert, J. and Lepoivre, P. (1996). Molecular evidence that the whitefly-transmitted Sweet potato Mild Mottle Virus belongs to a distinct genus of the Potyviridae. Archives of Virology 141(1): 125–35.

Cook, S.M., Khan, Z.R. and Pickett, J.A. (2007). The use of push-pull strategies in integrated pest management. Annual Reviews of Entomology 52: 375–400.

Corbett, G.H. (1936). New Aleurodidae (Hemiptera). Proceedings of Royal Entomological Society of London (B) 5: 18–22.

Cossa, N.S. (2011). Epidemiology of Cassava mosaic disease in Mozambique. Unpublished Ph.D. thesis, Johannesburg, South Africa: University of Witwatersrand.

Costa, H.S. and Brown, J.K. (1991). Variation in biological characteristics and esterase patterns among populations of *Bemisia tabaci* (Genn) and the association of one population with Silverleaf symptom development. Entomologic Experimentalism et Applicata 61: 211–219.

Costa, H.S., Brown, J.K., Sivasupramaniam, S. and Bird, I. (1993a). Regional distribution, insecticide resistance, and reciprocal crosses between the A and B biotypes of *Bemisia tabaci*. Insect Science Application 14: 127–138.

Costa, H.S., Ullman, D.E., Johnson, M.W. and Tabashnik, B.E. (1993). Squash Silverleaf symptoms induced by immature, but not adult, *Bemisia tabaci*. Phytopathology 83(7): 763–766.

Costa, H.S., Westcot, D.M. and Ullman, D.E. (1995). Morphological variation in *Bemisia* endosymbionts. Protoplasma 189: 194–202.

Costa, H.S., Toscano, N.C. and Henneberry, T.J. (1996). Mycetocyte inclusion in the oocyte of *Bemisia argentifolii* (Homoptera: Aleyrodidae). Annals of Entomological Society of America 89: 694–699.

Costa, H.S. and Robb, K.L. (1999). Effects of ultraviolet-absorbing greenhouse plastic films on the flight behavior of *Bemisia argentifolii* (Homoptera: Aleyrodidae) and *Frankliniella occidentalis* (Thysanoptera: Thripidae). Journal of Economic Entomology 92: 557–562.

Coudriet, D.L., Parbhaker, N., Krishna, A.N. and Meyerdirk, D.E. (1985). Variation in development rate on different hosts and overwintering of the sweet potato whitefly, *Bemisia tabaci* (Homoptera: Aleyrodidae). Environmental Entomology 14: 516–519.

Crowder, D.W. and Carriere, Y. (2009). Comparing the refuge strategy for managing the evolution of insect resistance under different reproductive strategies. Journal of Theoretical Biology 261: 423–430.

Csizinszky, A.A., Schuster, D.J. and Kring, J.B. (1995). Color mulches influence yield and insect pest populations in tomatoes. Hortscience 120: 778–784.

Csizinszky, A.A., Schuster, D.J. and Polston, J.E. (1999). Effect of Ultraviolet-reflective mulches on tomato yields and the Silverleaf whitefly. Hortscience 34(5): 911–914.

Cui, X., Wan, F., Xie, M. and Liu, T. (2008). Effect of heat shock on the survival and reproduction of two species *Trialeurodes vaporariorum* and *Bemisia tabaci*, biotype B. Journal of Insect Science 8(1): 24.

Cui, X., Chen, Y., Xie, M. and Wan, F. (2010). Survival characteristics of *Bemisia tabaci* B-biotype and *Trialeurodes vaporariorum* (Homoptera:Aleyrodidae) after exposure to adverse temperature conditions. Europe PMC 1232–1238.

Curnutte, L.B., Simmons, A.M. and Abd-Rabou, S. (2014). Climate change and *Bemisia tabaci* (hemiptera: Aleyrodidae) impacts of temperature and carbon dioxide on life history. Annals of Entomological Society of America 107: 933–945.

Curry, J.P. and Pimentel, D. (1971). Evaluation of tomato varieties for resistance to greenhouse whitefly. Journal of Economic Entomology 64(4): 1333–1334.

Cuthbertson, A.G.S. and Vanninen, I. (2015). The importance of maintaining protected zone status against *Bemisia tabaci*. Insects 6: 432–441.

Czosnek, H. and Ghanim, M. (2012). Back to basics: Are begomoviruses whitefly pathogens? Journal of Integrated Agriculture 11: 225–234.

Czosnek, H., Hariton-Shalev, A., Sobol, I., Gorovits, R. and Ghanim, M. (2017). The incredible journey of begomoviruses in their whitefly vector. Viruses 273.

Dadd, R.H. (1985). Nutrition: Organisms GAKerkut and LI Gerbert Comprehensive Insect Physiology Biochemistry and Pharmacology person Press, Inc, Elmsford NY 4: 315–319.

Dai, P., Ruan, C.C., Zang, L.S., Wan, F.H. and Liu, L.Z. (2014). Effects of rearing host species on the host-feeding capacity and parasitism of the whitefly parasitoid *Encarsia formosa*. Journal of Insect Science 14: 489–492.

Dale, C. and Moran, N.A. (2006). Molecular interactions between bacterial symbionts and their hosts. Cell 126: 453–465.

Dalton, R. (2006). The Whitefly infestations; The Christmas invasion. Nature 443: 898–900.

Dangelo, R.A.C., Michereff-Filho, M., Campos, M.R., da Silva, P. and Guedes, R.N.C. (2018). Insecticide resistance and control failure likelihood of the whitefly *Bemisia tabaci* (MEAM1; B biotype): a Neotropical scenario. Annals of Applied Biology 172: 88–99.

Dara, S.K. (2017). Entomopathogenic microorganisms: Modes of action and role in IPM. E-Journal of Entomology and Biologicals, UCANR.

Darshnee, H., Chunila and Tong-Kian, L. (2016). Potential for using semiochemicals to manage whiteflies (Hemiptera: Aleyrodidae) in agricultural fields. Journal of Plant Protection 43: 17.

Darshanee, H.L.C., Ren, H., Ahmed, N., Zhang, Z.F., Liu, Y.H. and Liu, T.X. (2017). Volatile-mediated attraction of greenhouse whitefly *Trialeurodes vaporariorum* to tomato and eggplant. Frontiers of Plant Sciences 8: 1285.

Das, S., Rahman, M.M., Kamal, M.M. and Shishir, A. (2017). Species richness of thrips and whiteflies and their predators in mustard fields. Journal of Bangladesh Agricultural University 15(1): 7–14.

David, B.V. and Subramaniam, T.R. (1976). Status of some Indian Aleyrodidae. Records of Zoological Survey of India 70: 133–233.

David, B.V. (2020). Invasive Whiteflies (Aleyrodidae:Hemiptera) International Seminar on Transboundary Pest Management held at Coimbatore on March 4–5, 2020.

De Bach, P. and Rose, M. (1976). Biological control of Wooly Whitefly. California Agriculture 30: 4–7.

De Barro, P. and Ahmed, M.Z. (2011). Genetic networking of the *Bemisia tabaci* cryptic species complex reveals a pattern of biological invasions. PLoS ONE 6(10): e25579. DOI:10.1371/journal.pone.0025579.

De Barro, P.J. and Driver, F. (1997). Use of RAPID PCR to distinguish the B biotypes from other biotypes of *Bemisia tabaci* Gennadius (Homoptera; Aleyrodidae). Australian Journal of Entomology 36: 1149–1152.

De Barro, P.J., Liebregts, W. and Carver, M. (1998). Distribution and identity of biotypes of *Bemisia tabaci* Genn (Hemiptra: Aleyrodidae) in member countries of Secretariat Pacific Community. Australian Journal of Entomology 37: 214–218.

De Barro, P.J., Driver, F., Trueman, J.H. and Curran, J. (2000). Phylogenetic relationships of world populations of *Bemisia tabaci* (Gennadius) using ribosomal ITS1. Molecular Physiology Evolution 16(1): 29–36.

De Barro, P.J., Scott, K.D., Graham, G.C., Lange, C.L. and Schutze, M.K. (2003). Isolation and characterization of microsatellite loci in *Bemisia tabaci*. Molecular Ecology Notes 3: 40–43.

De Barro, P.J. (2005). Genetic structure of the whitefly *Bemisia tabaci* in the Asia Pacific region revealed using microsatellite markers. Molecular Ecology 14: 3695–718.

De Barro, P.J. (2007). *Bemisia tabaci*—From molecular to landscape; Commonwealth Scientific and Industrial Research Organisation (CSIRO) – Entomology 120 Meiers Road, Indooroopilly, Queensland 4068 Australia.

De Barro, P.J., Liu, S.-S., Boykin, L.M. and Dinsdale, A.B. (2011). *Bemisia tabaci*: a statement of species status. Annual Reviews of Entomology 56: 1–19.

De Cock, A. and Degheele, D. (1998). Buprofezin: A novel chitin synthesis inhibitor affecting specifically planthoppers, whiteflies, and scale insects. *In*: Ishaaya. I. and Degheele, D. (eds.). Insecticides with Novel Modes of Action. Applied Agriculture, Springer, Berlin, Heidelberg. DOi https://DOi.Org/10.1007/978-3-662-03565-8_5.

De Moraes, L.A., Marubayashi, J.M., Yuki, V.A., Ghanim, M., Bello, V.H., De Marchi, B.R., da Fonseca Barbosa, L., Boykin, L.M., Krause-Sakate, R. and Pavan, M.A. (2017). New invasion of *Bemisia tabaci* Mediterranean species in Brazil associated to ornamental plants. Phytoparasitica 45: 517–525.

De Moraes, L.A., Muller, C., Bueno, R.C.O.F., Santos, A., Bello, V.H., De Marchi, B.R., Watanabe, L.M., Marubayashi, J.M., Santos, B.R., Yuki, V.A., Hélio Minoru Takada, Danielle Ribeiro de Barros, Carolina Garcia Neves, Fábio Nascimento da Silva, Mayra Juline Gonçalves, Murad Ghanim, Laura Boykin, Marcelo Agenor Pavan and Renate Krause-Sakate. (2018). Distribution and phylogenetics of whiteflies and their endosymbiont relationships after the Mediterranean species invasion in Brazil. Science Reporter 8: 14589.

Deepak, S., Reddy, C.N. and Shashibhushan, V. (2017). Bioefficacy and dissipation of β-cyfluthrin against whitefly *Bemisia tabaci* Genn in okra (*Abelmoschus esculentus* L.). Journal of Applied and Natural Science 9(2): 950–953.

Degenhardt, J., Gershenzon, J., Baldwin, I.T. and Kessler, A. (2003). Attracting friends to feast on foes: engineering terpene emission to make crop plants more attractive to herbivore enemies. Current Opinion in Biotechnology 14: 169–176.

Deletre, E., Schatz, B., Bourguet, D., Chandre, F., Williams, L. and Ratnadass, A. (2016). Prospects for repellent in pest control: Current developments and future challenges. Chemoecology 26: 127–142.

Denholm, I., Cahill, M., Dennehy, T.J. and Horowitz, A.R. (1998). Challenge with managing insecticide resistance in agricultural pests exemplified by the Whitefly, *Bemisia tabaci*. Philosophical Transactions of the Royal Society of London 853: 1757–1767.

Denholm, I., Germam, K. and Williamson, M. (2008). Insecticide resistance in *Bemisia tabaci*: A global perspective. Journal of Insect Science 8(4): 16–16.

Dennehy, T.J., Degain, B.A., Harpold, V.S., Brown, J., Morin, S., Fabrick, J.A., Byrne, F. and Nichols, R.L. (2005). New challenges to management of whitefly resistance to insecticides in Arizona. Vegetable Report, Series p-144.

Dennehy, T.J., Degain, B.A., Harpold, V.S., Zaborac, M., Morin, S., Fabrick, J.A., Nichols, R.L., Brown, J.K., Byrne, F.J. and Li, X. (2010). Extraordinary resistance to insecticides reveals exotic Q biotype of *Bemisia tabaci* in the New World. Journal of Economic Entomology 103: 2174–2186.

Devine, G.J.I., Ishaaya, I., Horowitz, A.R. and Denholm, I. (1999). The response of pyriproxyfen-resistant and susceptible *Bemisia tabaci* Genn (Homoptera: Aleyrodidae) to pyriproxyfen and fenoxycarb alone and in combination with piperonyl butoxide. Pesticide Science 55: 405–411.

Dhaliwal, M.S., Jindal, S.K. and Cheema, D.S. (2013). Punjab Sindhuri and Punjab Tej: new varieties of chilli. Journal of Research Punjab Agricultural University 50: 79–81.

Dhaliwal, M.S., Jindal, S.K. and Cheema, D.S. (2015). CH-27: a multiple disease resistant chilli hybrid. Agriculture Research Journal 52: 127–129.

Dhawan, A.K. (1990). Management of cotton bollworms. pp. 153–171. In Proceedings Summer Institute on Key insect pests of India: Their Bioecology with Special Reference to Integrated Pest Management. June 6–15, 1990, PAU Ludhiana.

Dhawan, A.K. and Simwat, G.S. (1997). Insecticide induced resurgence of sucking pests in upland cotton in Punjab, India. *In*: Abstr. Third International Congress of Entomological Sciences, March, 1820, 1997.

Dhawan, A.K. and Simwat, G.S. (1998). Evaluation of different biopesticides against cotton bollworms, *Helicoverpa armigera* (Hubner). *In*: Dhaliwal, G.S., Randhawa, N.S., Arora, R. and Dhawan, A.K. (eds.). Ecological Agriculture and Sustainable Development, IES CRID Chandigarh 274–280.

Dhawan, A.K. and Saini, H.K. (1998). Impact of synthetic pyrethroids on the biology of whitefly, *Bemisia tabaci* (Genn.). Proc. National Seminar on Entomology in 21st Century, College of Agriculture, Udaipur, India.

Dhawan, A.K. (1999). Major insect pests of cotton and their integrated management. pp. 165–225. *In*: Upadadhyay, R.K., Mukerji, K.G. and Rajak, R.L. (eds.). IPM System in Agriculture Vol 6. Cash Crops, Aditya Books Pvt Limited, New Delhi.

Dhawan, A.K. and Simwat, G.S. (1999). Population dynamics of whitefly, *Bemisia tabaci* (Genn) on cotton An ecological approach. pp. 335–348. *In*: Dhaliwal, G.S., Randhawa, N.S., Ramesh Arora and Dhawan, A.K. (eds.). Ecological Agriculture and Sustainable Development. Vol 1; IES and CRID, Chandigarh.

Dhawan, A.K., Dhaliwal, G.S. and Chelliah, S. (2000). Insecticide induced resurgence of insect pests in crop plants. pp. 86127. *In*: Dhaliwal, G.S. and Singh, B. (eds.). Pesticides and the Environment. Commonwealth Publishers, New Delhi.

Dhawan, A.K. (2016). Integrated pest management in cotton. pp. 499–575. *In*: Dharam P. Abrol (ed.). Integrated Pest Management in the Tropics. New India Publishing Agency, New Delhi (India).

Dhooria, M.S. and Butter, N.S. (1990). Effect of pesticides on the incidence and development of red spider mite, *Tetranychus cinnabarinus* on cotton. Acarology Newsletter 17/18: 23–24.

Dickey, A.M., Osborne, L.S., Shatters, R.G. and Mckenzie, C.l. (2013). Identification of the MEAM1 crypyic species of *Bemisia tabaci* (Hemiptera; Aleyrodidae) by loop mediated isothermal amplification. Florida Entomologist 96: 756–764.

Dietz, H.F. and Zetek, J. (1920). The Blackfly of citrus and other sub-tropical plants. USDA Bulletin 885: 1–55.

Dinsdale, A., Cook, Riginos, C., Buckley, Y.M. and De Barro, P.J. (2010). Refined global analysis of *Bemisia tabaci* (Hemiptera: Sternorrhyncha: Aleyrodoidea: Aleyrodidae) mitochondrial cytochrome oxidase 1 to identify species level genetic boundaries. Annals of the Entomological Society of America 103: 196–208.

Dittrich, V., Hassan, S.D. and Ernst, G.H. (1985). Sudanese cotton in whitefly a case of study of the emergence of new primary pest. Crop Protection 4: 161–176.

Dittrich, V., UK, S. and Ernst, G.H. (1990). Chemical control and insecticide resistance in whiteflies. pp. 263–285. *In*: Gerling, D. (ed.). Whiteflies: Their Bionomics, Pest Status, and Management. Intercept, Hants, UK.

Dittrich, V., Ernest, G.H., Ruesch, O. and UK, S. (1990a). Resistance mechanisms in Sweet Potato whitefly (Homoptera: Aleyrodidae). Populations from Sudan, Turkcy, Guatemala, and Nicaragua. Journal of Economic Entomology 83: 1665–1670.

Do Nascimento Silva, J., Mascarin, G.M., de Paula Vieira de Castro, R., Castilho, L.R. and Freire, D.M. (2019). A novel combination of a biosurfactant with entomopathogenic fungi enhances efficacy against *Bemisia* whitefly. Pest Management Science 2019 Nov; 75(11): 2882–2891. DOI: 10.1002/ps.5458.

Domenico, B., Loria, A., Chiara, S. and Cenis, J.L. (2006). PCR-RFLP identification of *Bemisia tabaci* biotypes in the Mediterranean Basin, Phytoparasitica 34(3): 243–251.

Domingos, G.M., Baldin, E.L.L., Canassa, V.F. et al. (2018). Resistance of collard green genotypes to *Bemisia tabaci* biotype B: Characterization of antixenosis. Neotrop. Entomol. 47: 560–568. https://doi.org/10.1007/s13744-018-0588-5.

Dong, T., Zhang, B., Jiang, Y. and Hu, Q. (2016). Isolation and classification of fungal whitefly Entomopathogenic from soils of Qinghai-Tibet Plateau in Gansu Corridor in China. PLoS One 11(5): e 0156087. Doi:10.137/journal. one.1.

Dooley, J.W. (2006). Whitefly pupa of the world: Compendium and Key to the genera of the Aleurodicinae & the Aleyrodinae USDA, APHIS, PPQ, South San Francisco, CA. http://keys.lucidcentral.org/keys/v3/whitefly/Default.htm.

Douglas, A.E. (1998). Nutritional interaction in insect-microbial symbiosis aphid and their symbiotic bacteria *Buchnera*. Annual Reviews of Entomology 43: 17–37.

Doukas, D. and Payne, C.C. (2007). Greenhouse whitefly (Homoptera: Aleyrodidae) dispersal under different UV light environments. Journal of Economic Entomology 100: 389–397.

Dowell, R.V. 1982. Biology of *Tetraleurodes acaciae* (Quaintance) (Homoptera: Aleyrodidae). Pan-Pacific Entomologist 58: 321–318.

Drost, Y.C., Qiu, Y.T., Posthuma-Doodeman, C.J.A.M. and van Lenteren, J.C. (1999). Life-history and oviposition behavior of *Amitus bennetti*, a parasitoid of *Bemisia argentifolii*. Entomologica Experimentalis et Applicata 90: 183–89.

Du, W.X., Han, X.Q., Wang, Y.B. and Qin, Y.C. (2016). A primary screening and applying of plant volatiles as repellents to control whitefly *Bemisia tabaci* (Gennadius) on tomato. Scientific Reports 6: 22140.

Dubey, A.K. and Ko, C.C. (2008). Whitefly (Aleyrodidae) host plants from India. Oriental Insects 42: 49–102.

Dubey, A.K., Ko, C.-C. and David, B.V. (2009). The Genus Lipaleyrodes Takahashi, a junior synonym of *Bemisia* Quaintance and Baker (Hemiptera: Aleyrodidae): A revision based on morphology. Zoology Studies 48: 539–557.

Dutcher, J.D. (2007). A review of resurgence and replacement causing pest outbreaks in IPM. pp. 27–43. *In*: Ciancio, A. and Mukerji, K.G. (eds.). General Concepts in Integrated Pest and Disease Management. Springer, Netherlands.

Elbaz, M., Lahav, N. and Morin, S. (2010). Evidence for a pre-zygotic reproductive barrier between the B and Q biotypes of *Bemisia tabaci* (Hemiptera: Aleyrodidae). Bulletin of Entomological Research 100: 581–590.

Elbert, A. and Nauen, R. (2000). The resistance of *Bemisia tabaci* (Homoptera: Aleyrodidae) to insecticides in southern Spain with special reference to neonicotinoids. Pest Management Science 56: 60–64.

Elbert, A., Haas, M., Springer, B., Thielert, W. and Nauen, R. (2008). Applied aspects of neonicotinoid uses in crop protection. Pest Management Science 64: 1099–1105.

El-Gendi, S.S., Adam, K.M. and Bachatly, M.A. (1997). Effect of the planting date of tomato on the population density of *Bemisia tabaci* (Genn.) and *Heliothis armigera* (HB), viral infection and yield. Arab Universities Journal of Agricultural Sciences 5: 135–144.

El-Khidir, E. (1965). Bionomics of cotton whitefly (*Bemisia tabaci* Genn.) in Sudan and effects of irrigation on population density of whitefly. Sudan Agriculture Journal 2: 8–22.

Ellango, R., Shalini Thakur Singh, Vipin Singh Rana, Gayatri Priya, N., Harpreet Raina, Rahul Chaubey, Naveen, N.C., Riaz Mahmood, Ramamurthy, V.V., Asokan, R. and Rajagopal, R. (2015). Distribution of *Bemisia tabaci* genetic groups in India. Environmental Entomology 44: 1258–1264.

El-Latif, A.O.A. and Subrahmanyam, B. (2010). Pyrethroids resistance and esterase activity in three strains of the cotton bollworm, Helicoverpa armigera (Hübner). Pesticide Biochemistry and Physiology 96: 155–159.

Ellsworth, P.C., Diehl, J.W., Dennehy, T.J. and Naranjo, S.E. (1995). Sampling sweet potato whiteflies in cotton, IPM series 2, The University of Arizona, Cooperative Extension Publication 19423, Tucson, 2pp.

Ellsworth, P.C. and Diehl, J.W. (1996). Whiteflies in Arizona No. 6, Commercial-scale Trial 1995 (Rev 5/97), The University of Arizona, Cooperative Extension Publication, 4pp.

Ellsworth, P.C., Dennehy, T.J. and Nichols, R.C. (1996). Whitefly Management in Arizona cotton, 1996 IPM series 3. The University of Arizona, Cooperative Extension Publication 196004, Tucson, AZ, 2pp.

Ellsworth, P.C., Diehl, J.W., Dennehy, T.J. and Naranjo, S.E. (1996). Sampling sweet potato whiteflies in cotton, IPM series 6, The University of Arizona, Cooperative Extension Publication 196006, Tucson, AZ, 2pp.

Ellsworth, P.C. and Martinez-Carrillo, J.L. (2001). IPM for *Bemisia tabaci*: a case study from North America. Crop Protection 20: 853–869.

El-Serwiy, S.A., Ali, A.A. and Razoki, I.A. (1987). Effect of intercropping of some host plants with tomato on the population density of tobacco whitefly, *Bemisia tabaci* (Genn.), and the incidence of Tomato Yellow Leaf Curl Virus (TYLCV) in plastic houses. Journal of Agriculture and Water Resources Research, Plant Production 61: 81–89.

EPPO. (1998). Specific quarantine requirements. EPPO Technical Documents, No. 1008. Paris, France: European and Mediterranean Plant Protection.

EPPO. (2006). Datasheets on quarantine organisms No. 178, *Bemisia tabaci*. Bulletin OEPP/EPPO Bulletin 19: 733–737.

EPPO. (2014). QPR database, Paris France, European and Mediterranean plant Protection Organization. HTTP/www. eppo. int/DATABASES/QPR/pqr him.

EPPO/CABI. (1996). Bean golden mosaic bigeminivirus; Lettuce infectious yellows Closterovirus; Squash Leaf Curl geminivirus; Tomato Mottle geminivirus; Tomato Yellow Leaf Curl Bigeminivirus.

In: Smith, I.M., McNamara, D.G., Scott, P.R. and Holderness, M. (eds.). Quarantine Pests for Europe 2nd edition.

Erdogan, C., Moores, G.D., Oktay Gurkan, M., Gorman, K.J. and Denholm, I. (2008). Insecticide resistance and biotype status of populations of the tobacco whitefly *Bemisia tabaci* (Hemiptera: Aleyrodidae) from Turkey. Crop Protection 27: 600–605.

Erdogan, C., Velioglu, A.S., Oktay Gurkan, M., Denholm, I. and Moores, G.D. (2012). Chlorpyrifos ethyl-oxon sensitive and insensitive acetylcholinesterase variants of greenhouse whitefly *Trialeurodes vaporariorum* (Westw.) (Hemiptera: Aleyrodidae) from Turkey. Physiology 104: 273–276.

Eveleens, K.G. (1983). Cotton-insect control in the Sudan Gezira: Analysis of a crisis. Crop Protection 2: 273–287.

Everett, K.D., Thao, M., Horn, M., Dyszynski, G.E. and Baumann, P. (2005). Novel chlamydiae in whiteflies and scale insects: endosymbionts 'Candidatus ea. *bemisiae*' strain Falk and *'Candidatus Fritschea eriococci'* strain Elm. International Journal of Systematic and Evolutionary Microbiology 55: 1581–1587.

Fahmy, I.F. and Abou-Ali, R.M. (2015). Studying genetic diversity of whitefly, *Bemisia tabaci* Egyptian isolates about some worldwide isolates. Journal of Genetic Engineering and Biotechnology 13: 87–92.

Fang, Y., Wang, J., Luo, C. and Wang, R. (2018). Lethal sub-lethal effects of clothianidin on its development and reproduction of *Bemisia tabaci* (Hemiptera: Aleyrodidae) MED and MEAM1. Journal of Insect Science 18: 37. DOI: 10.1093/jisesa/iey025.

Farag-El-Shafie, H.A. (2019). Insect pest management in organic farming system. >DOI10.5772/I techopen 84483.

Fargues, J., Smits, N., Rougcer, M., Boulard, T., Rdray, G., Lasier, J., Jeonnequin, B., Fatnassi, H. and Mermier, M. (2005). Effect of microclimate heterogeneity and ventilation system on entomopathogenic hyphomycete infection of *Trialeurodes vaporariorum* (Homoptera: Aleyrodidae) in Mediterranean greenhouses. Biological Control 32: 461–472.

Faria, M. and Wraight, S.P. (2001). Biological control of *Bemisia tabaci* with fungi. Crop Protection 767–778.

Feldhaar, H. (2011). Bacterial symbionts as mediators of ecologically important traits of insect hosts. Ecology Entomology 36: 533–543.

Fengming, Y. (2008). Research on *Bemisia* and other whiteflies in China. Journal of Insect Science January 8(4): 53. Doi.10.1673/031 008 04a7.

Fernandez, E., Grávalos, C., Haro, P.J., Cifuentes, D. and Bielza, P. (2009). Insecticide resistance status of *Bemisia tabaci* Q-biotype in south-eastern Spain. Pest Management Science 65: 885–891.

Ferro, D.N., Mackenzie, J.D. and Margolies, D.C. (1980). Effect of mineral oil and a systemic insecticide on-field spread of aphid-borne Maize Dwarf Mosaic Virus in sweet corn. Journal of Economic Entomology 73: 730–734.

Firdaus, S., Van Heusden, A.W., Hidayati, N., Supena, E.D.J., Visser, R.G.F. and Vosman, B. (2012). Resistance in *Bemisia tabaci* in tomato wild relatives. Euphytica 187: 31–45.

Firdaus, S., Vosman, B., Hidayati, N., Supena, E.D.J., Visser, R.G. and van Heusden, A.W. (2013). The *Bemisia tabaci* species complex: additions from different parts of the world. Insect Science 20: 723–733.

Firdaus, S., Van Heusden, A.W., Hidayati, N., Darmas, E., Supena, E.D.J., Mumm, R., de Vos, R.C.H., Visser, R.G.F. and Vosman, B. (2013a). Identification and QTL mapping of whitefly resistance Components in *Solanum galapagense*. Theoretical Applications in Genetics 126: 1487–1501.

Fishpool, L.D.C. and Burban, C. (1994). *Bemisia tabaci*: the whitefly vector of African cassava mosaic *Geminivirus*. Tropical Science 34(1): 55–72.

Flint, H.M. and Parks, N.J. (1989). Effect of azadirachtin from the neem tree on immature sweet potato whitefly, *Bemisia tabaci*, (Homoptera: Aleyrodidae) and other selected pest species on cotton. Journal of Agricultural Entomology 6(4): 211–215.

Forero, D. (2008). The systematics of the Hemiptera. Reviews of Colombian Entomology 34: 1–21.

Frank, D.L., Oscar, E. and Liburd. (2005). Effect of living and synthetic mulches on the population dynamics of whiteflies and aphids; their associated natural enemies and insect-transmitted plant diseases in Zucchini. Environmental Entomology 34(4): 857–865.

Fridman, E., Wang, J., Iijima, Y., Froehlich, J.E., Gang, D.R., Ohlrogge, J. and Pichersky, E. (2005). Metabolic, genomic, and biochemical analyses of glandular trichomes from the wild methylketones. Plant Cell 17: 1252–1267.

Frohlich, D.R.Y. and Brown, J.K. (1994). Mitochondrial 16S ribosomal subunit as a molecular marker in Bemisia tabaci and implications for population variability. Bemisia Newsletter 8: 3. https:// www. cabi. org/is/abstract/19941105410.

Frohlich, D.R.Y., Torres-Jerez, II, Bedford, I.D., Markham, P.G. and Brown, J.K. (1999). A phylogeographical analysis of the Bemisia tabaci species complex based on mitochondrial DNA markers. Molecular Ecology 8: 1683–1691.

Funderburk, J., Casuo, N., Leppla, N. and Donahoe, M. (2019). Insect and mite integrated Pest Management in Florida cottons. EDIS (Electronic Data based Information System).

Fuog, D., Fergusson, S.J. and Flückiger, C. (1998). Pymetrozine: a novel insecticide affecting aphids and whiteflies. pp. 40–49. In Insecticides with Novel Modes of Action. Springer, Berlin, Heidelberg.

Gameel, O.I. (1978). The cotton whitefly, Bemisia tabaci (Gennadius) in the Sudan Gezira. Proceedings of the Third Ciba-Geigy Seminar on the Strategy for Cotton Pest Control in the Sudan Basel, 8–10 May 1978, London: 111–131.

Gangwar, R.K. and Gangwar, C. (2018). Lifecycle, distribution, nature of the damage and economic importance of whitefly, Bemisia tabaci (Gennadius). Acta Scientific Agriculture 2(4): 36–39.

Gauthier, N., Clouet, C., Perrakis, A., Kapantaidaki, D., Peterschmitt, M. and Tsagkarakou, A. (2014). Genetic structure of Bemisia tabaci med populations from home-range countries, inferred by nuclear and cytoplasmic markers: Impact on the distribution of the insecticide resistance genes. Pest Management Science 14(70): 1477–1491.

Gawel, N.J. and Barlett, A.C. (1993). Characterization of differences between whiteflies using RAPD–PCR. Insect Molecular Biology 2: 33–38.

Gelman, D.B. and Hu, J.S. (2007). Critical feeding periods for last instar nymphal and pharate adults of the whiteflies, Trialeurodes vaporariorum and Bemisia tabaci. Journal of Insect Science 7: 33. doi: 10.1673/031.007.3301.

Gelman, D.B., Hu, J.S., Martin, P.A.W., Blackburn, M.B. and Salvucci, M.K. (2008). Novel candidates for the development of biopesticides to control whitefly pests. Journal of Insect Science 8(4): 19–20.

Gennadius, P. (1989). Disease of tobacco plantation in Trikunza, the aleyrodid of tobacco. Ellenike Georgia 5: 1–3.

Gerling, D. and Horowitz, A.R. (1984). Yellow sticky traps for evaluating the population levels and dispersal pattern of Bemisia tabaci. Annals of Entomological Society of America 77: 753–759.

Gerling, D. (1992). Approaches to biological control of whiteflies: biological control workshop-91. Florida Entomologist 75: 446–456.

Gerling, D. and Mayer, R.T. (1996). Bemisia; Taxonomy Biology, Damage, Control, and Management. Intercept, Andover, Hants, UK.

Gerling, D., Alomar, O. and Arno, J. (2001). Biological control of Bemisia tabaci using predators and parasitoids. Crop Protection 20: 779–799.

Gerling, D., Rottenberg, O. and Bellows, T.S. (2004). Role of natural enemies and other factors in the dynamics of field populations of the whitefly Siphoninus phillyreae (Haliday) is introduced and native environments. Biological Control 31: 199–209.

Ghahari, H., Abd-Rabou, S., Zahradnik, J. and Ostovan, H. (2013). Annotated catalog of whiteflies (Hemiptera: Sternorrhyncha: Aleyrodidae) from Arasbaran, Northwestern Iran. International Journal of Nematology and Entomology 1: 42–52.

Ghanim, M., Rosell, R.C., Campbell, L.R., Czosnek, H., Brown, J.K. and Ulman, D.E. (2001). Digestion, salivary and reproductive organs of Bemisia tabaci, Gennadius (Homoptera; Aleyrodidae) B–type. Journal Morphology 248: 22–40.

Ghanim, M., Morin, S. and Czosnik, H. (2001a). Rate of Tomato Yellow Leaf Curl Virus in the circulative transmission pathways of its vector, the whitefly, Bemisia tabaci. Phytoparasitica 91: 188–196.

Ghanim, M. and Kontsedalov. (2009). Susceptibility to insecticides in the Q biotype of Bemisia tabaci is correlated with bacterial symbiont densities. Pest Management Science 65: 939–942.

Ghanim, M., Zchori-Fein, E. and Fleury, F. (2010). Endosymbiont metacommunities, mtDNA diversity and the evolution of the Bemisia tabaci (Hemiptera: Aleyrodidae) species complex. Molecular Ecology 19: 4365–4376.

Ghanim, M. (2014). A review of the mechanisms and components that determine the transmission efficiency of Tomato Yellow Leaf Curl Virus (Geminiviridae; *Begomovirus*) by its whitefly vector. Virus Research 186: 47–54.

Ghosal, A. and Chaterjee, M. (2018). Insecticide induced resistance study of whitefly in cotton and tomato. University Journal of Animal Sciences 2: 1–6.

Ghosh, S., Bouvaine, S. and Maruthi, M.N. (2015). Prevalence and genetic diversity of endosymbiotic bacteria infecting cassava whiteflies in Africa. BMC Microbiology 15: 93.

Ghosh, S., Bouvaine, S., Richardson, S0.C.W., Ghanim, M. and Maruthi, M.N. (2018). Fitness costs associated with infections of secondary endosymbionts in the cassava whitefly species *Bemisia tabaci*. Journal Pest Science 91: 17–22.

Gibson, R.M., Rice, A.D., Minks, A.K. and Harrewijn, P. (1989). Modifying aphid behavior. pp. 209–224. Aphids: Their Biology, Natural Enemies, and Control. Amsterdam: Elsevier.

Gilbertson, R.L., Rojas, M.R., Kon, T. and Jaquez, J. (2007). Introduction of Tomato Yellow Leaf Curl Virus into the Dominican Republic: the Development of a Successful Integrated Pest Management Strategy in Tomato Yellow Leaf Curl Virus Disease by Henry k Czosnek Pages 279–303, Springer Netherlands.

Gill, R.J. (1990). The morphology of whiteflies. pp. 13–46. *In*: Gerling, D. (ed.). Whiteflies: Their Bionomics, Pest Status, and Management. Intercept, Andover, UK.

Gill, S. (2000). Pest control: whitefly control for cut flower growers. The Cut Flower Quarterly 12(1): 26–30.

Glas, J.J., Schimmel, B.C.J., Alba, J.M., Escobar-Bravo, R., Schuurink, R.C. and Kant, M.R. (2012). Plant glandular trichomes as targets for breeding or engineering of resistance to herbivores. International Journal of Molecular Sciences 13: 17077–17103.

Glowska, E., Dragun-Damian, A., Dabert, M. and Gerth. M. (2015). New *Wolbachia* supergroup detected in quill mites (Acari: Syringophilidae). Infect Genetic Evolution 30: 140–146.

Gloyd, R.A. (1999). Know your friends: *Delphastus pusillus*: whitefly predator. Midwest Biological Control News. October. p. 3.

Gogi, M.D., Sarfraz, R.M., Dosdall, L.M., Arif, M.J., Keddle, A.B. and Ashfaq, M. (2006). Effectiveness of two insect growth regulators against *Bemisia tabaci* (Gennadius) (Homoptera: Aleyrodidae) and *Helicoverpa armigera* (Hübner) (Lepidoptera: Noctuidae) and their impact on population densities of arthropod predators in cotton in Pakistan. Pest Management Science 62: 982–990.

Gogo, E.O., Saidi, M., Itulya, F.M., Martin, T. and Ngouajio, M. (2014). Ecofriendly nets and floating row covers reduce pest infestation and improve tomato (*Solanum lycopersicum* L.) yields for smallholder farmers in Kenya. Agronomy 4: 1–12.

Gold, C.S., Altieri, M.A. and Bellotti, A.C. (1990). Direct and residual effects of short duration intercrops on the cassava whiteflies *Aleurotrachelus socialis* and *Trialeurodes variabilis* (Homoptera: Aleyrodidae) in Colombia. Agriculture Ecosystem and Environment 32: 57–67.

Gomez-Diaz, J.S., Montoya-Lema, J. and Muñoz-Valencia, V. (2019). Prevalence and diversity of endosymbionts in Cassava whiteflies (Hemiptera: Aleyrodidae) from Colombia. Journal of Insect Science 19(3): 12. 22. DOI: 10.1093/ Jessa/ie 047.

Gonella, E., Pajoro, M., Marzorati, M., Crotti, E., Mandrioli, M., Pontini, M., Bulgari, D., Negri, I., Sacchi, L., Chouaia, B., Daffonchio, D. and Alma, A. (2015). Plant-mediated interspecific horizontal transmission of an intracellular symbiontininsects. Scientific Reports 5: 15811. https://doi.org/10.1038/srep15811.

Goodell, P.B., Davis, R.M., Godfrey, L.D., Hutmacher, R.B., Roberts, P.A., Wright, S.D., Barlow, V.M., Haviland, D.R., Munier, D.J. and Natwick, E.T. (2015). UC IPM Pest Management Guidelines Cotton. UC ANR Publication 3444. Oakland, CA.

Gorman, K., Russell Slater, James, D. Blande, Alison Clarke, Jodie Wren, Alan McCaffery and Ian Denholm. (2010). Cross-resistance relationships between neonicotinoids and pymetrozine in *Bemisia tabaci* (Hemiptera: Aleyrodidae). Pest Management Science 66: 1186–1190.

Gorsane, F., Ben Halima, A., Ben Khalifa, M., Bel-Kadhi, M.S. and Fakhfakh, H. (2011). Molecular characterization of *Bemisia tabaci* populations in Tunisia: genetic structure and evidence for multiple acquisitions of secondary symbionts. Environmental Entomology 40: 809–817.

Gottlieb, Y., Ghanim, M., Chiel, E., Gerling, D., Portnoy, V., Steinberg, S., Tzuri, G., Horowitz, A.R., Belausov, E. and Mozes-Daube, N. (2006). Identification and localization of a Rickettsia sp. in *Bemisia tabaci* (Homoptera: Aleyrodidae). Applied Environmental Microbiology 72: 3646–3652.

Gottlieb, Y., Ghanim, M., Gueguen, G., Kontsedalov, S., Vavre, F., Fleury, F. and Zchori-Fein, E. (2008). Inherited intracellular ecosystem; Symbiotic bacteria share bacteriocytes in whiteflies. FASEB Journal 22: 2591–2599.

Gottlieb, Y., Zchori-Fein, E., Mozes-Daube, N., Kontsedalov, S., Skaljac, M., Brumin, M., Sobol, I., Czosnek, H., Vavre, F. and Fleury, F. (2010). The transmission efficiency of tomato yellow leaf curl virus by the whitefly *Bemisia tabaci* is correlated with the presence of a specific symbiotic bacterium species. Journal of Virology 84: 9310–9317.

Gotz, M., Popovaski, K., Knollenberg, M., Gorovitz, R., Brown, J.K., cicero, J.M., Czosnek, H., Winter, S. and Ghanim, M. (2012). Implications of *Bemisia tabaci* heat shock protein 70 in Begomovirus-whitefly interactions. Journal of Virology 86132–1341.

Gould, H.J., Parr, W.J., Woodville, H.C. and Symmonds, S.P. (1975). Biological control of glasshouse whitefly (*Trialeurodes vaporariorum*) on cucumbers. Entomophaga 20: 285–292.

Greer, L. (2000). ATTRA; Greenhose IPM: Sustainable Whitefly Control. Pages 1–12. http/www.attra org/attar=pub/gh-whitefly.html.

Greer, L. and Dole, J.M. (2003). Aluminum foil, Aluminum-painted, plastic, and degradable mulches increase yields and decrease insect-vectored viral diseases of vegetables Aluminum foil and Aluminum-painted mulches. Horticultural Technology 13. Do-10.21273/HORTTECH.13.2.0276.

Gueguen, G., Vavre, F., Gnankine, O., Peterschmitt, M., Charif, D., Chiel, E., Gottlieb, Y., Ghanim, M., Zchori-Fein, E. and Fleury, F. (2010). Endosymbiont metacommunities, mtDNA diversity and the evolution of the *Bemisia tabaci* (Hemiptera: Aleyrodidae) species complex. Molecular Ecology 19: 4365–4378.

Guerrieri, E. (1997). Flight behavior of *Encarsia formosa* in response to plant and host stimuli. Entomological Experimentalis et Applicata 82: 129–33.

Gulluoglu, L., Kurt, C., Arioglu, H., Zaimoglu, B. and Aslan, M. (2010). The researches on soybean (*Glycine max* Merr.) variety breeding for resistance to whitefly in Turkey. Turkish Journal of Field Crops 15(2): 123–127.

Gunning, R.V., Byrne, F.J., Conde, B.D., Connelly, M.I., Hergstron, K. and Devonshire, B.L. (1995). First report of B-biotype, *Bemisia tabaci* (Gennadius) (Hemiptera: Aleyrodidae) in Australia. Australian Journal of Entomology Banner 14: 116.

Guo, J., Cong, L. and Wan, F. (2013). Multiple generation effects of high temperature on the development and fecundity of *Bemisia tabaci* (Gennadius) (Hemiptera: Aleyrodidae) biotype B. Insect Science 20: 541–549.

Gutierrez, A.P., Ponti, L., Herren, H.R., Baumgärtner, J. and Kenmore, P.E. (2015). Deconstructing Indian cotton: weather, yields, and suicides. Environmental Sciences. Europe 27: 1–17.

Habibi, J. (1975). The cotton whitefly, *Bemisia tabaci* Genn; Bioecology and methods of control. Entomologie et phytopathology appliques 38: 3–4.

Hadi, M.Z. and Bidgen, M.P. (1996). Somaclonal variation as a tool to develop new pest-resistant plants of Torana Fournier, Compacta Blue. Plant Cell Tissue 46: 43–50.

Hadjistylli, M., Roderik, G.K. and Brown, J.K. (2016). Global population structure of a worldwide pest and virus vector genetic diversity to the population history of the *Bemisia tabaci* sibling species group. PLoS One 2076: 11: e 0165105.

Hagler, R., Jackson, C.G., Issacs, R. and Machtley, S.A. (2004). Foraging behavior and prey interactions by a guild of predators on various life stages of *Bemisia tabaci*. Journal of Insect Science 4(1): 13.

Hai, Lin, Y., Xiang Yong, L., Li Meng, Z., Yan Qiong, Xuc and Qing, Z. (2014). Optimization for *Bemisia tabaci* egg development conditions using orthogonal design. South China Journal of Agricultural Sciences 45: 1970–1975.

Hanan, A., He, X.Z., Shakeel, M., Khetran, M.A. and Wang, Q. (2015). Eretmocerus *warrae* prefer to attack mid-aged hosts to gain fitness for both adults and their offspring. Biological Control 91: 104–9644. https://doi.org/10.1016/j.biocontrol.2015.07.005.

Harris, K.F., Esbroeck, Z.P.V. and Duffus, J.E. (1996). Morphology of Sweet potato whitefly (Homoptera, Aleyrodidae) relation to virus transmission. Zoomorphology 116: 143–156.

Hart, W.G., Selhime, A., Harlan, D.P., Ingle, S.J., Sanchez, R.N., Rhode, R.H., Garcia, C.A., Caballero, J. and Garcia, R.L. (1978). The introduction and establishment of parasites of citrus black fly, *Aleurocanthus woglumi* in Florida. (Hemiptera:Aleyrodidae). Entomophaga 23: 361–366.

Hasanuzzaman, A.T.M., Islam, N.M., Zhang, Y., Zhang, C.U., Liu, C.Y. and Liu, T.X. (2016). Leaf morphology character can be a factor for Intra varietal performance of whitefly, *Bemisia tabaci*. PLoS One 11(4): e 153 880; DOI; 10.1371/2 pone 0153 880.

Hasanuzzaman, A.T.M., Islam, Md.N., Liu, F.H., Cao, H.H. and Liu, T.X. (2018). Leaf chemical compositions of different eggplant varieties affect performance of *Bemisia tabaci* (Hemiptera: Aleyrodidae) nymphs and adults. Journal of Economic Entomology 111: 445–453.

Hassel, M.P. and Southwood, T.R.E. (1978). Foraging strategies of insects. Annual Reviews of Ecology and Systematics 9:75–98.

He, Y., Weng, Q., Huang, J., Liang, Z., Lin, G. and We, D. (2007). Insecticide resistance of *Bemisia tabaci* field populations. Chinese Journal of Applied Ecology 18: 1578–1582.

He, Y., Liu, Y., Wang, K., Zhang, Y., Wu, Q. and Wang, Y. (2019). Development and fitness of the parasitoid, *Encarsia formosa* (Hymenoptera: Aphelinidae), on the B and Q of the Sweetpotato Whitefly (Hemiptera: Aleyrodidae). Journal of Economic Entomology 112: 2597–2603, https://doi.org/10.1093/jee/toz200.

Hebert, P.D., Cywinska, A., Ball, S.L. and de Waard, J.R. (2003). Biological identifications through DNA barcodes. Proceedings of Biological Sciences 270: 313–321.

Henderson, C.F. and Tilton, E.W. (1955). Tests with acaricides against the brown wheat mite. Journal of Economic Entomology 48: 157–161.

Hendry, T.A., Hunter, M.S. and Baltrus, D.A. (2014). The facultative symbiont Rickettsia protects invasive whitefly against entomopathogenic *Pseudomonas syringae* strains. Applied Environmental Microbiology (AEM): 02447.

Henry, M., Béguin, M., Requier, F., Rollin, O., Odoux, J.F. and Aupinel, P. (2012). A common pesticide decreases foraging success and survival in honey bees. Science 336: 348–350.

Henry, S., Gururajan, K.N., Natrajan, K. and Krishnamurthy, R. (1990). Kanchan—a whitefly tolerant stable cotton. Indian Farming 40: 25–26.

Hequet, E., Henneberry, T.J. and Nicholis, R.I. (2007). Sticky cotton: Causes, effects, and prevention, USDA. Agricultural Research Service, Technical Bulletin 1915.

Hilje, L., Costa, H.S. and Stansly, P.A. (2001). Cultural practices for managing *Bemisia tabaci* and associated viral diseases. *In*: Naranjo, S. and Ellsworth, P. (eds.). Special Issue: Challenges and Opportunities for Pest Management of *Bemisia tabaci* in the New Century. Crop Protection 20: 801–812.

Hilje, L. and Stansly, P.A. (2008). Living ground covers for management of *Bemisia tabaci* (Gennadius) (Homoptera: Aleyrodidae) and Tomato Yellow Mottle Virus (ToYMoV) in Costa Rica. Crop Protection 27: 10–16.

Hilje, L. and Morales, F.J. (2008). Whitefly bioecology and management in Latin America. Encyclopaedia of Entomology 257–262.

Himler, A.G., Adachi-Hagimori, T., Bergen, J.E., Kozuch, A., Kelly, S.E. and Tabashnik, B.E. (2011). Rapid spread of a bacterial symbiont in an invasive whitefly is driven by fitness benefits and female bias. Science 332: 254–256.

Hirunkanokpun, S., Thepparit, C. and Foil, L.D. (2011). Horizontal transmission of Rickettsia felis between cat fleas, *Ctenocephalides felis*. Molecular Ecology 20: 4577–86.

Hoddle, M.S., Van Driesche, R.G. and Sanderson, J.P. (1998). Biology and use of parasitoid *Encarsia formosa*. Annual Reviews of Entomology 43: 645–669.

Hoddle, M.S. (2006). Phenology, life tables, and reproductive biology of *Tetraleurodes perseae* (Hemiptera: Aleyrodidae) on California avocados. Annals of Entomological Society of America 99: 553–559.

Hoddle, M.S. (2013). Biology of the red-banded whitefly, *Tetraleurodes perseae* Nakahara (Homoptera: Aleyrodidae). Consultado: December 2015.

Hoddie, M.S., Van Driesche, R.G., Lyon, S.M. and Sanderson, J.P. (2001). Compatibility of insect growth regulators with *Eretmocerus eremicus* (Hymenoptera; Aphelinidae) for whitefly (Homoptera; Aleyrodidae) control in Poinsettias 1. Laboratory assays. Biological Control 20: 122–131.

Hodges, G. and Evans, G. (2005). An identification guide to the whiteflies (Hemiptera: Aleyrodidae) of the Southeastern United States. Florida Entomologist 88: 518–534.

Hoelmer, K. and Goolsby, J. (2003). Release, establishment and monitoring of *Bemisia tabaci* natural enemies in the United States. 1st International Symposium on Biological Control of Arthropods USDA Forest Service PHTET, June 3–5, 2003.

Hoelmer, K.A., Osborne, L.S. and Yokomi, R.K (1991). Foliage disorders in Florida associated with feeding by sweet potato whitefly, *Bemisia tabaci*. Florida Entomologist 74: 162–166.

Horowitz, A.R., Forer, G. and Ishaaya, I. (1994). Managing resistance in *Bemisia tabaci* in Israel with emphasis on cotton. Pesticide Science 42: 113–122.

Horowitz, A.R. and Ishaaya, I. (1996). Chemical control of *Bemisia* management and application. pp. 537–556. *In*: Gerling, D. and Mayer, R.T. (eds.). *Bemisia*, Taxonomy, Biology, Damage, Control and Management. Intercept Ltd., Andover.

Horowitz, A.R., Mendelson, Z., Weintraub, P.G. and Ishaaya, I. (1998). Comparative toxicity of foliar and systemic applications of acetamiprid and imidacloprid against cotton whitefly *Bemisia tabaci* (Hemiptera; Aleyrodidae). Bulletin of Entomological Research 88: 437–442.

Horowitz, A.R., Kontsedalov, S., Denholm, I. and Ishaaya, I. (2002). Dynamics of insecticide resistance in *Bemisia tabaci*—a case study with an insect growth regulator. Pest Management Science 58: 1096–1100.

Horowitz, A.R. (2003). Novaluron (Rimon), a novel IGR: potency and cross-resistance. Archives of Insect Biochemistry and Physiology 54: 157–164.

Horowitz, A.R., Kontsedalov, S., Khasdan, V. and Ishaaya, I. (2005). Biotypes B and Q of *Bemisia tabaci* and their relevance to neonicotinoid and pyriproxyfen resistance. Archives of Insect Biochemistry Physiology 58: 216–225.

Horowitz, A.R., Kontsedalov, S., Khasdan, V., Breslauer, H. and Ishaaya, I. (2008). The biotypes B and Q of *Bemisia tabaci* in Israel-Distribution, resistance to insecticides, and implications for pest management. Journal of Insect Science 8: 23–24.

Horowitz, A.R., Antignus, Y. and Gerling, D. (2011). Management of *Bemisia tabaci*. pp. 295–322. *In*: Thomson, W.M.O. (ed.). The Whitefly, *Bemisia tabaci* (Homoptera: Aleyrodidae) Interaction with Geminiviruses Infected Host Plants. Springer, Dordrecht.

Horowitz, A.R. and Ishaaya, I. (2014). Dynamics of biotypes B and Q of the whiteflies and its impact on insecticide resistance. Pest Management Science 70: 1568–1572.

Horowitz, A.R., Ellsworth, P.C., Mesah, R. and Ishaaya, I. (2018). Integrated management of whiteflies in cotton. International Congress on Cotton and Other Fiber Crop, Feb. 20–23, 2018 at ICAR Research Complex for NEU Region, Umiam (Barapani), Meghalaya, India.

Houndete, T.A., Ketoh, G.K., Hema, O.S.A. and Bravault, T. (2010). Insecticide resistance in field populations of *Bemisia tabaci* (Hemiptera: Aleyrodidae) in West Africa. Pest Management Science 66(11): 1181–1185.

Hsieh, C.H., Wang, C.H. and Ko, C.C. (2006). Analysis of *Bemisia tabaci* (Hemiptera: Aleyrodidae) species complex and distribution in Eastern Asia based on mitochondrial DNA markers. Annals of Entomological Society of America 99: 768–775.

Hsieh, C.H., Wang, C.H. and Ko, C.C. (2007). Evidence from molecular markers and population genetic analyses suggests recent invasions of the Western North Pacific regions by biotypes B and Q of *Bemisia tabaci* (Gennadius). Environmental Entomology 36: 952–961.

Hu, J., De Barro, P., Zhao, H., Wang, J. and Nardi, F. (2011). An extensive field survey combined with a phylogenetic analysis reveals rapid and widespread Invasion of two alien whiteflies in China. PLoS ONE 6(1): e16061. doi:10.1371/journal.pone.0016061.

Hu, J., Zhang, X., Jiang, Z., Zhang, F., Liu, Y. and Li, Z. (2018). New putative cryptic species detection and genetic network analysis of *Bemisia tabaci* (Hemiptera: Aleyrodidae) in China based on mitochondrial COI sequences. Mitochondrial DNA Part A, DNA Mapping, Sequencing, and Analysis 29: 474–484.

Hunter, M.S., Perlman, S.J. and Kelly, S.E. (2003). A bacterial symbiont in the Bacteroidetes induce cytoplasmic incompatibility in the parasitoid wasp *Encarsia pergandiella*. Proceedings of Biological Sciences 270: 2185–2190.

Husain, M.A. and Trehan, K.N. (1933). Observations on life history, bionomics and control of cotton whitefly, *Bemisia gossypiperda* (M and L). Indian Journal of Agricultural Sciences 6: 701–75.

Husain, M.A. and Trehan, K.N. (1940). Final report on the scheme of investigations on the whitefly cotton in Punjab. Indian Journal of Agricultural Sciences 10: 101–119.

Husain, M.A. and Trehan, K.N. (1942). The nature and extent of damage caused by *Bemisia gossypiperda* M. and L. the whitefly of cotton in Punjab. Indian Journal. Agriculltural Sciences 12(8): 793–821.

Idris, A.M. and Brown, J.K. (2004). Cotton leaf crumple virus is a distinct Western Hemisphere begomovirus species with complex evolutionary relationship indictive of recombination and reassortment. Phytopathology 94: 1068–1074.

Ioannou, N. (1987). Cultural management of Tomato Yellow Leaf Curl disease in Cyprus. Plant Pathology 36: 367–373.

Ishaaya, I., Cock, A.D. and Degheele, D. (1994). Pyriproxyfen a potent suppression of egg hatch and adult formation of greenhouse whitefly (Homoptera: Aleyrodidae). Journal of Economic Entomology 87: 1185–1189.

Ishaaya, I. and Horowitz, A.R. (1995). Pyriproxyfen, a novel insect growth regulator for controlling whiteflies: Mechanical and resistance management. Pesticide Science 43: 227–272.

Ishaaya, I., Kontsedalov, S. and Horowitz, A.R. (2003). Novaluron (Rimon), a Novel IGR: Potency and cross-resistance. Archives of Insect Biochemistry and Physiology 54: 157–164.

Islam, M.N., Hasanuzzaman, A.T.M., Zhang, Z.-F., Zhang, Y. and Liu, T.-X. (2017). High level of nitrogen makes tomato plants releasing less volatiles and attracting more *Bemisia tabaci* (Hemiptera: Aleyrodidae). Frontier of Plant Science 8: 466.

Issacs, R., Wills, M.A. and Byrne, D.N. (1999). Modulation of whitefly take off and flight orientation by wind speed and visual cues. Physiology Entomology 24: 31–38.

Jackson, D.M., Farnham, M.W., Simmons, A.M., Van Giessen, W.A. and Elsey, K.D. (2000). Effects of planting pattern of collards on resistance to whiteflies (Homoptera: Aleyrodidae) and on parasitoid abundance. Journal of Economic Entomology 93: 1227–1236.

Jaenike, J., Polak, M. and Fiskin, A. (2007). Interspecific transmission of endosymbiotic Spiroplasma by mites. Biological Letters 3: 23–25.

Jahan, S.M.H., Lee, G.S., Lee, S. and Lee, K.Y. (2014). Acquisition of Tomato Yellow Leaf Curl Virus enhances the attraction of *Bemisia tabaci*, diodes green light. Journal of Asian Pacific Entomology 17: 79–82.

Jakkawa, Y., Nonomura, T., Kusakari, S.I., Okada, K., Kinbura, J., Osanura, k. and Toyoda, U. (2015). Electrostatic insect sweeper for eliminating whiteflies colonizing on host plants: Acomlementay pest control device in an electric field screen–guarded greenhouse. Insects 6: 442–454.

Janu, A. and Dahiya, K.K. (2017). Influence of weather parameters on the population of whitefly, *Bemisia tabaci* in American cotton (*Gossypium hirsutum*). Journal of Entomology and Zoology Studies 5(4): 649–654.

Jauset, A.M., Sarasua, M.J., Avilla, J. and Albajes, R. (2000). Effect of nitrogen fertilization level applied to tomato on the greenhouse whitefly. Crop Protection 19: 255–261.

Javaid, M., Arif, M.J., Gogi, M.D., Shahid, M.R., Iqbal, M.S. and Shehzad, M.A. (2012). Relative resistance in different cultivars of Pakistani cotton against cotton whitefly. Academic Journal of Entomology 5(3): 143–146.

Jayaraj, S. and Regupathy, A. (1986). Studies on the resurgence of sucking pests of crops in Tamil Nadu. pp. 225–240. Jayaraj, S. (ed.). Proceedings of National Symposium on Resurgence of Sucking Pests. TNAU, Coimbatore.

Jayaraj, S., Rangarajan, A.V., Murugesan, S., Santhramj, G., Jayaraghovan, S.V. and Thangaraj, A. (1986). Studies on the outbreak of Whitefly, *Bemisia tabaci* (Gennadius) on cotton in Tamil Nadu. pp. 103–115. *In*: Jayaraj, S. (ed.). Resurgence of Sucking Pests. Proceeding of the National Symposium, Tamil Nadu Agricultural University, Coimbatore, India.

Jayaraj, S. (1987) (ed.). The resurgence of Sucking Pests. Tamil Nadu Agricultural University, Coimbatore, 262 pp.

Jayaswal, A.P. (1989). Whitefly on cotton and its management. A review. Journal of Cotton Research and Development 3: 10–22.

Jeevananandham, N., Marimuthu, M., Natesan, S., Mukhaiyah, S. and Appachi, S. (2018). Levels of plant resistance in chilies *Capsicum* spp. against whitefly, *Bemisia tabaci*. International Journal of Current Microbiology and Applied Science 7: 1429–144.

Jeon, H.Y., Kim, H.H., Yang, C.Y., Kang, T.J. and Kim, D.S. (2009). A tentative economic threshold level for greenhouse whitefly on cucumber plants in the protective cultivation. Korean Journal of Science and Technology 27: 81–85.

Jervis, M.A. and Heimpel, G.E. (2005). Phytophagy. pp. 525–550. *In*: Jervis, M.A. (ed.). Insects as Natural Enemies. Springer, Dordrecht, The Netherlands.

Jeschke, P., Nauen, R., Schindler, M. and Elbert, A. (2010). Overview of the status and global strategy for neonicotinoids. Journal of Agriculture and Food Chemistry 59: 2897–2908.

Jetter, K.M., Alston, J.M. and Farquharson, R.J. (2001). The case of silver leaf whitefly in California. UC Ag Issues Center. 2001. http://aic.ucdavis.edu/oa/whitefly.pdf.

Jha, S.K. and Kumar, M. (2017). Effect of weather parameters on incidence of whitefly *Bemisia tabaci* (Gennadius) on tomato. Journal of Entomology and Zoology, Studies 5: 304–306.

Jhang, P.J., Ku, C.T., Jhang, J.M., lu, V.B., Wei, J.N., Liu, Y.Q., David, A., Boland, W. and Turlings, T.C.J. (2013). Phloem feeding whiteflies can fool their host plants, but not their parasitoids. Functional Ecology 27: 1304–1312.

Jindal, V., Dhaliwal, G.S. and Dhawan, A.K. (2007). Mechanisms of resistance in cotton to whitefly *Bemisia tabaci* (Homoptera: Aleyrodidae): Antibiosis. International Journal of Tropical Insect Science 27(3/4): 216–222.

Jindal, V. and Dhaliwal, G.S. (2009). Elucidating resistance in cotton genotypes to whitefly, *Bemisia tabaci* by population build-up studies. Phytoparasitica 37: 137–145.

Jindal, V., Dhaliwal, G.S., Dhawan, A.K. and Dilawari, V.K. (2009). Mechanisms of resistance in cotton to whitefly *Bemisia tabaci*. Tolerance Phytoparasitica 37(3): 249–254.

Jindal, V. and Dhaliwal, G.S. (2011). Mechanisms of resistance in cotton to whitefly (*Bemisia tabaci*): Antixenosis. Phytoparasitica 39: 129–136. DOI:10.1007/s12600-011-0144-x.

John, J.L. (2008). Encyclopedia of Entomology Springer Science and Business Media, pp 2944-AFSN 978-1-4020-6242-1.

Johnson, M.W., Toscano, N.C., Reynolds, H.T., Sylvester, E.S., Kido, K. and Natwick, E.T. (1982). Whiteflies cause problems for southern California growers. California Agriculture 36: 24–26.

Jones, D.R. (2003). Plant viruses transmitted by whiteflies. European Journal of Plant Pathology 109: 195–219.

Joyce, A.L., Bellows, T.S. Jr and Headrick, D.H. (1999). Reproductive biology and search behavior of Amicus Bennett (Hymenoptera: Platygasteridae), a parasitoid of *Bemisia argentifolii* (Homoptera: Aleyrodidae). Environmental Entomology 28: 282–89.

Kady, H.E. and Devine, G.A. (2003). Insecticide resistance in Egyptian populations of the cotton whitefly, *Bemisia tabaci* (Hemiptera: Aleyrodidae). Pest Management Science 59: 865–871.

Kajita, H. (1986). Parasitism of the greenhouse whitefly, *Trialeurodes vaporariorum* (Westwood) (Homoptera: Aleyrodidae) by *Encarsia formosa* Gahan (Hymenoptera: Aphelinidae) in a greenhouse covered with near-ultraviolet absorbing vinyl film. Proceedings of the Association for Plant Protection of Kyushu 155–157.

Kakutani, K., Matsuda, Y., Nonomura, T., Toyoda, H., Kimbara, J., Osamura, K. and Kusakari, S. (2012). Practical application of an electric field screen to an exclusion of flying insect pests and airborne fungal conidia from greenhouses with a good air penetration. Journal of Agricultural Sciences 4: 51–60.

Kanakala, S. and Ghanim, M. (2015). Advances in the genomics of the whitefly *Bemisia tabaci*: an insect pest and a virus vector. *In*: Raman, C., Goldsmith, M. and Agunbiade, T. (eds.). Short Views on Insect Genomics and Proteomics. Entomology in Focus, vol 3. Springer, Cham. https://doi.org/10.1007/978-3-319-24235-4_2.

Kanakala, S. and Ghanim, M. (2018). Whitefly-transmitted begomoviruses and advances in the control of their vectors. pp. 201–220. *In*: Pail, B. (ed.). Genes, Genetics, and Transgenics for Virus Resistance in Plants. United Kingdom: Caister Academic Press.

Kanakala, S. and Ghanim, M. (2019). Global genetic diversity and geographical distribution of *Bemisia tabaci* and its bacterial endosymbionts. Plos One 14: e o213948. https/doi.org/101351 Journal pone 0213948.

Kapadia, M.N. and Puri, S.N. (1990). Development and relative proportions and emergence of *Encarsia transvena*, and *Eretmocerus mundus*, important parasitoids of *Bemisia tabaci* Genn. Entomon. 15: 555–559.

Kapantaidaki, D.E., Ovčarenko, I., Fytrou, N., Knott, K.E., Bourtzis, K. and Tsagkarakou, A. (2015). Low levels of mitochondrial DNA and symbiont diversity in the worldwide agricultural pest, the greenhouse whitefly *Trialeurodes vaporariorum* (Hemiptera: Aleyrodidae). Journal Heredity 106: 80–92.

Karatolos, N., Williamson, M.S., Denholm, I., Gorman, K. and Nauen, R. (2012). Resistance to spiromesifen in *Trialeurodes vaporariorum* is associated with a single amino acid replacement in its target enzyme acetyl-coenzyme A carboxylase. Insect Molecular Biology 21: 327–334.

Karthikeyan, C., Patil, B.L., Borah, B.K., Resmi, T.R., Turco, S., Pooggin, M.M., Hohn, T. and Veluthambi, K. (2016). The emergence of a Latent Indian cassava mosaic virus from cassava which recovered from infection by a non-persistent Sri Lankan Cassava Mosaic Virus. Viruses 8: 264.

Karunker, I., Benting, J., Lueke, B., Ponge, T., Nauen, R., Roditakis, E., Vontas, J., Gorman, K., Denholm, I. and Morin, S. (2008). Over-expression of cytochrome P450 CYP6CM1 is associated with high resistance to imidacloprid in the B and Q biotypes of *Bemisia tabaci* (Hemiptera: Aleyrodidae). Insect Biochemistry and Molecular Biology 38: 634–644.

Kaur, P., Singh, H. and Butter, N.S. (2009). Formulation of weather-based criteria rules for the prediction of sucking pests in cotton (*Gossypium hirsutum*) in Punjab. Indian Journal of Agricultural Sciences 79: 375–380.

Kayser, H., Kaufmann, L. and Schurmann, F. (1994). Pymetrozine (CGA215944): a novel compound for aphid and whitefly control. An overview of its mode of action. Proceedings, 1994 Brighton Crop Protection Conference—Pests and Diseases 2: 737–742.

Kedar, S.C., Saini, R.K., Kmarnag, K.M. and Sharma, S.S. (2014). Records of natural enemies of whitefly, *Bemisia tabaci* Gennadius (Hemiptera;Areyrodidae) in some cultivated crops in Haryana. Journal of Biopest 7(1): 57–59.

Kempema, L.A., Cui, X., Holzer, F.M. and Walling, L.L. (2007). *Arabidopsis* transcriptome changes in response to phloem-feeding silver leaf whitefly nymphs: Similarities and distinctions in response to aphids. Plant Physiology 143: 849–865.

Khan, M.A., Akram, W., Khan, H.A.A., Asghar, J. and Khan, T.M. (2010). Impact of Bt cotton on whitefly, *Bemisia tabaci* (Genn.) population. Pakistan Journal of Agricultural Sciences 47: 327–332.

Khan, M.A., Khan, Z., Ahmad, W., Paul, B., Paul, S., Aggarwal, C. and Akhtar, M.S. (2015). Insect Pest resistance: An alternative approach for crop protection. Hakeem, K.R. (ed.). Crop Production and Global Environmental Issues, DOI 10.1007/978-3-319-23162-4_11.

Khan, M.A.U., Shahid, A.A., Rao, A.Q., Shahid, N., Latif, A., Din, S. and Husnain, T. (2015a). Defense strategies of cotton against whitefly transmitted CLCuV and Begomoviruses. Advancements in Life Sciences 2: 58–66.

Khan, S.M. and Ullah, Z. (1994). Population dynamics of insect pests of cotton in Dera Ismail Khan. Sarhad Journal of Agriculture 10: 285–90.

Khan, W.S., Ahmad, M., Waseem, S.M.I. and Bhatti, M.B. (1993). Inbuilt tolerance of cotton cultivars to sucking pests of cotton. Pakistan Cottons 37: 123–137.

Khatun, M.F., Jahan, S.M.H., Lee, S. and Lee, K.Y. (2018). Genetic diversity of geographic distribution of *Bemisia tabaci* species complex in Bangladesh. Acta Tropica 87: 28–36.

Kikuchi, Y. (2009). Endosymbiotic bacteria in insects: their diversity and cultivability. Microbes Environment 24: 195–204.

Kil, E.J., Kim, S., Lee, Y.J., Byon, H.S., Park, J., Seo, H., Kim, C.S., Lee, J.H., Kim, J.K. and Lee, K.Y. (2014). Sweet pepper confirmed as a reservoir host of Tomato Yellow Leaf Curl Virus both agro-inoculation and whitefly mediated–inoculation. Archives of Virology 159: 2387–2395.

Kil, E.J., Kim, S., Lee, Y.J., Byon, H.S., Park, J., Seo, H., Kim, C.S., Shim, J.K., Lee, J.H., Kim, J.K., Lee, K.Y., Choi, H.S. and Lee, S. (2016). Tomato Yellow Leaf Curl Virus (TYLCV-IL) a seed transmissible Geminivirus in tomato. Science Reporter 6: 19013. doi:10 1038/Srep.19013(2016).

Kiriticos, D., De Barro, P.J., Yonow, T., Ota, N. and Sutherst, R.W. (2020). The potential geographical distribution and phenology of *Bemisia tabaci* (Middle East Asia Mediterranean1) considering irrigation and greenhouse production. Bulletin Entomological Research, DOI:10.1017/S0007485320000061.

Kirk, A.A., Lacey, L.A., Brown, J.K., Ciomperlik, M.A., Goolsby, J.A., Vacek, D.C., Wendel, L.E. and Napompeth, B. (2000). Variation within the *Bemisia tabaci* species complex (Hemiptera Aleyrodidae) and its natural enemies leading to successful biological control of *Bemisia* biotype B in the USA. Bulletin of Entomological Research 90(4): 317–327.

Kliot, A., Cillia, M., Czosnek, H. and Ghanim, M. (2014). Implications of the bacterial symbiont, Rickettsia spp., and interacting whitefly, *Bemisia tabaci* with Tomato Yellow Leaf Curl Virus. Journal Virology 88: 5660–5662.

Koehler, P.G. and Patterson, R.S. (1991). Incorporation of Pyriproxyfen in a German cockroach (Dictyoptera: Blattellidae) management program. Journal of Economic Entomology 84: 917–921, https://doi.org/10.1093/jee/84.3.917.

Kogan, M. (1982). Plant resistance in pest management. pp. 93–134. *In*: Metcalf, R.L. and Luckmann, W. (eds.). Introduction to Insect Pest Management (second edition). John Wiley and Sons Inc, New York.

Kontsedalov, S., Zchori-Fein, E., Chicl, E., Gottlieb, Y., Inbar, M. and Ghanim, M. (2008). The presence of Rickettsia is associated with increased susceptibility of *Bemisia tabaci* (Homoptera: Aleyrodidae) to insecticides. Pest Management Science 64: 789–792.

Kontsedalov, S., Abu-Moch, F., Lebedov, G., Czosnek, H., Horowitz, R. and Ghanim, M. (2012). *Bemisia tabaci* Biotype dynamics and resistance to insecticides in Israel during the Years 2008–2010. Journal of Integrative Agriculture 11: 312–320.

Kranthi, K.R., Jadhav, D.R., Wanjari, R.R., Ali, S.S. and Russell, D.A. (2001). Carbamate and organophosphate resistance in cotton pests in India from 1995 to 1999. Bulletin Entomological Research 91: 37–42.

Kranthi, K.R., Jadhav, D.R.Y., Kranthi, S., Wanjari, R.R., Ali, S.S. and Russell, D. (2002a). Insecticide resistance in five major insect pests of cotton in India. Crop Protection 21: 449–460.

Kranthi, K.R., Russell, D., Wanjari, R., Kherde, M., Munje, S., Lavhe, N. and Armes, N. (2002). In-season changes in resistance to insecticides in *Helicoverpa armigera* (Lepidoptera: Noctuidae) in India. Journal of Economic Entomology 95: 134–142.

Kranthi, K.R. (2007). Insecticide resistance management in cotton to enhance productivity. Model training course on cultivation of long staple cotton (ELS). Central Institute for Cotton Research Regional Station, Coimbatore, pp. 214–231.

Kranthi, K.R. and Russell, D.A. (2009). Changing trends in cotton pest management in Integrated pest management: innovation-development process. Peshin, R. and Dhawan, A. (ed.). 499–541 (Springer, 2009).

Kranthi, K.R. (2012). Bt Cotton Questions and Answers. Indian society for cotton improvement. Mumbai. pp.71.

Krishna, V.V. and Qaim, M.B. (2012). Bt cotton and sustainability of pesticide reductions in India. Agriculture Systems 107: 47–55.

Kular, J.S. and Butter, N.S. (1991). Efficacy of different insecticides against whitefly, *Bemisia tabaci* Genn (Aleyrodidae; Hemiptera) on cotton. Pestology 15: 6–13.

Kular, J.S. and Butter, N.S. (1995). Screening of cotton germplasm against whitefly, *Bemisia tabaci* Genn. Journal of Entomology Research 19: 341–344.

Kular, J.S., Butter, N.S. and Chahal, G.S. (1995). A technique to measure the resistance in cotton against whitefly, *Bemisia tabaci* Genn. Plant Protection Bulletin 47: 1–4.

Kular, J.S. and Butter, N.S. (1996). An improved technique for screening cotton germplasm against whitefly, *Bemisia tabaci*. Indian Journal of Entomology 58: 210–214.

Kular, J.S. and Butter, N.S. (1999). Influence of some morphological traits of cotton genotypes on resistance to whitefly, *Bemisia tabaci* Genn. Journal of Insect Science 12: 81–83.

Kumar, P. and Poehling, H.M. (2006). UV blocking plastic films and nets influence vector and virus transmission in greenhouse tomatoes in the humid tropics. Environmental Entomology 35: 1069–1082.

Kumar, P. and Poehling, H.M. (2007). Efficacy of azadirachtin, abamectin and, spinosad on sweet potato whitefly (Homoptera: Aleyrodidae) on tomato plants under laboratory and greenhouse conditions in humid tropics. Journal of Entomology 100: 411–420.

Kumar, V., Palmer, C., McKenzie, C.L. and Osborne, L.S. (2017). Whitefly (*Bemisia tabaci*) management program for ornamentals. UF/IFAS Mid Florida Research and Education Center. *Bemisia* Biotypes A and Q. (Documents ANY 989) SOSBORN@UFLEDIS.

Kumashiro, B.R., Lai, P.Y., Funasaki, G.Y. and Teramoto, K.K. (1983). The efficacy of Nephaspis amnicola and *Encarsia haitiensis* is in controlling *Aleurodicus disperses* in Hawaii. Proceedings of Hawaii. Entomological Society 24: 261–69.

Kuwana, I. and Ishii, T. (1927). On *Prospaltella smithi* Silv., and *Cryptognatha* sp., the enemies of *Aleurocanthus spiniferus* Quaintance, imported from Canton, China. Journal of Okitsu Horticultural Society 22: 77–80.

Lahm, G.P., Cordova, D. and Barry, J.D. (2009). New and selective ryanodine receptor activators for insect control. Bioorg. Med. Chem. 17: 4127–4133.

Lambert, L., Chouffot, T., Turcotte, G., Lemieux, M. and Moreau, J. (2003). Biological control of greenhouse whitefly (*Trialeurodes vaporariorum*) with *Dicyphus hesperus* on the tomato for interplanting under supplemental lighting in Quebec (Canada). (Contrôle de l'aleurode (*Trialeurodes vaporariorum*) and Dicyphus hesperus pour la tomate de serre sous éclairage d'appoint et en contre-plantation au Québec (Canada).). pp. 203–207. *In*: Roche, L., Edin, M., Mathieu, V. and Laurens, F. (eds.). Colloque international tomate sous abri, protection intégrée - agriculture biologique, Avignon, France, 17–18 et 19 septembre 2003. Paris, France: Centre Technique Interprofessionnel des Fruits et Légumes.

Lambkin, T.A. and Zalucki, M.P. (2010). Long-term efficacy of *Encarsia* dispersa Polaszek (Hymenoptera: Aphelinidae) for the biological control of *Aleurodicus dispersus* Russell (Hemiptera: Aleyrodidae) in tropical monsoon Australia. Australian Journal of Entomology 49: 190–98.

Latreille. (1796). Precis des caracteres generigues des Insectes: 93.

Laurentin, H., Pereira, C. and Sanabria, M. 2003. Phytochemical characterization of six sesame (Sesamum indicum L.) genotypes and their relationships with resistance against the sweetpotato whitefly *Bemisia tabaci* Gennadius. Agronomy Journal 95: 1577–1582.

Lee, W., Park, J., Lee, G.S., Lee, S. and Akimoto, S. (2013). Taxonomic status of the *Bemisia tabaci* complex (Hemiptera: Aleyrodidae) and reassessment of the number of its constituent species. PloS One 8(5): e63817. https://doi.org/10.1371/journal.pone.0063817.

Legg, J.P., French, R., Rogan, D., Okao-Okiya, G. and Brown, J.K. (2002). A distinct *Bemisia tabaci* (Gennadius) (Homoptera: Aleyrodidae) genotype cluster is associated with the epidemic of severe cassava mosaic virus disease in Uganda. Molecular Ecology 11: 1219–1229.

Legg, J.P. and Fauquet, C.M. (2004). Cassava mosaic geminiviruses in Africa. Plant Molecular Biology 56: 585–599.

Legg, J.P. (2010). Epidemiology of a whitefly-transmitted cassava mosaic geminivirus pandemic in Africa. pp. 233–257. *In*: Stansly, P.A. and Naranjo, S.F. (eds.). Bemisia: Bionomics and Management of a Global Pest. Dordrecht Heidelberg-London-New York: Springer.

Li, S.J., Ahmed, M.Z., Lv, N., Shi, P.Q., Wang, X.M., Huang, J.L. and Qiu, B.L. (2017). Plant mediated horizontal transmission of Wolbachia between whiteflies. ISME Journal 11: 1019–1028.

Li, J.J. and Yan, F.M. (2013). Comparative morphometry of six biotypes of *Bemisia tabaci* Gennadius (Hemiptera :Aleyrodidae) from China. Journal of Integrative Agriculture 12: 846–852.

Li, X., Degain, B., Harpold, V., Marcon, P., Nichols, R., Fournier, A., Naranjo, S., Palumbo, J. and Ellsworth, P. (2012). Baseline susceptibilities of B- and Q-biotype *Bemisia tabaci* to anthranilic diamides in Arizona. Pest Management Science 68: 83–91.

Li, Y.H., Ahmad, Z.M., Li, S.J., Lv, N., Shi, P.Q., Chen, Y.s. and Qiu, B.l. (2019). Plant mediated horizontal endosymbiont transmission of *Rickettsia* endosymbiont between different whitefly species. FEMS Microbiology and Ecology 93(12): 1–11.

Liang, P., Tian, Y.A., Biondi, A., Desneus, N. and Gao, K.W. (2012). Short term transgenerational effects of the neonicotinoid nitenpyram on suisceptibility to insecticides a two of whitefly species. Ecotoxicology 21: 1889–1898.

Liedl, B.E., Lawson, D.M., White, U.K., Siiabiho, J.A., Cohen, D.E. and Carson, W.G. (1995). Acylsugars of wild tomato, *Lycopersicon pennellii* alters settling and reduces oviposition of *Bemisia tabaci* (Homoptera: Aleyrodidae). Journal of Economic Entomology 88: 742–748.

Lisha, V.S., Anatony, B., Palaniswami, M.S. and Henneberry, T.J. (2003). Ecology and behavior of *Bemisia tabaci* (Homoptera: Aleyrodidae) in biotypes in India. Journal of Economic Entomology 96: 322–327.

Liu, D.Q., Wang, S.M., Xin, S.R. and Zhang, S.G. (2003). Study on different pesticide application methods for the control of *Trialeurodes vaporariorum* on tomato. Journal of Henan Agricultural University 37(2): 158–160&164.

Liu, G.X., Ma, H.M., Xie, H.Y., Xuan, N., Guo, X. and Fan, Z.X. (2016). Biotype characterization, developmental profiling, insecticide response, and binding property of *Bemisia tabaci* chemosensory proteins: role of CSP in insect defense. PLoS One 11: e0154706.

Liu, S.S., Colvin, J. and De Barro, P.J. (2012). Species concepts as applied to the whitefly *Bemisia tabaci* systematics: how many species are there? Journal of Integrative Agriculture 11: 176–186.

Liu, T.X. and Stansly, P.A. (1999). Searching and feeding behavior of *Nephaspis oculatus* and *Delphastus catalinae* (Coleoptera: Coccinellidae), predators of *Bemisia argentifolii* (Homoptera: Aleyrodidae). Environmental Entomology 28: 901–906.

Liu, T.X., Stansly, P.A. and Gerling, D. (2015). Whitefly-parasitoids: Distribution, life-history, bionomics, and utilization. Annual Reviews of Entomology 60: 273–292.

Lloyd, L. (1922). The control of the greenhouse whitefly (*Asterochiton vaporariorum*) with notes on its biology. Annals of Applied Biology 9: 1–32.

Loconsole, G., Saldarelli, P., Doddapenen, H., Savino, V., Martelli, G.P. and Saponari, M. (2012). Identification of India sheen DNA virus associated with the citrus chlorotic dwarf virus a new member in family Geminiviridae. Virology 432: 162–172.

Lopez, S.N. and Botto, E. (1995). (Control biologico de *Trialeurodes vaporariorum* (Westwood) (Homoptera: Aleyrodidae) por entomofagos parasitoides (Hymenoptera: Aphelinidae)). *In*: Congreso Argentino de Entomologia. Resumenes del 3, 3 Congreso Argentino de Entomologia.

Lopez, S.N. and Botto, E. (2005). Effect of cold storage on some biological parameters of *Eretmocerus corni* and *Encarsia formosa* (Hymenoptera; Aphelinidae. Biological Control 33: 123–130.

Lourencao, A.L., Alves, A.C., Fugi, C.G.Q. and Matos, E.S. (2008). Outbreaks of *Trialeurodes vaporariorum* (West) (Hemiptera; Aleyrodidae) under field conditions in the state of Sao Paulo, Brazil. Neotropical Entomology 37: 89–911.

Lu, Y., Bei, Y. and Jhang, J. (2012). Are yellow sticky traps an effective method for control of sweet potato whitefly, *Bemisia tabaci* in the greenhouse or field? Journal of Insect Science 12: 113 (DOI 10.1673/031.012.11301).

Lucatti, A.F., Alvarez, A.E., Machado, C.R. and Gilardón, E. (2010). Resistance of tomato genotypes to the greenhouse whitefly *Trialeurodes vaporariorum* (West.) (Hemiptera: Aleyrodidae). Neotropical Entomology 39: 5. https://doi.org/10.1590/S1519-566X2010000500019.

Lucatti, A.F., Meijer-Dekens, F.R., Mumm, R., Visser, R.G., Vosman, B. and van Heusden, S. (2014). Normal adult survival but reduced *Bemisia tabaci* oviposition rate on tomato lines carrying an introgression from *S. habrochaites*. BMC Genetics 15: 142.

Lundgren, J.G., López-Lavalle, L.A.B., Parsa, S. and Wyckhuys, K.A.G. (2014). Molecular determination of the predator community of cassava whitefly in Colombia: pest-specific primer development and field validation. Journal Pest Science 87: 125–131.

Luo, Z.Y., Zhang, W.N. and Gan, G.P. (1989). Population dynamics of tobacco whitefly in cotton fields and the influence of insecticides application. Acta Entomol. Sinia, 32: 293–299.

Luo, C., Jones, C., Devine, G.J., Zhang, F., Denholm, I. and Gorman, K. (2010). Insecticide resistance in *Bemisia tabaci* biotype Q (Hemiptera: Aleyrodidae) from China. Crop Protection 29: 429–434.

Lykouressis, D.P., Perdikis, D.C. and Konstantinou, A.D. (2009). Predation rates of *Macrolophus pygmaeus* (Hemiptera: Miridae) on different densities of eggs and nymphal instars of the greenhouse whitefly *Trialeurodes vaporariorum* (Homoptera: Aleyrodidae). Entomologia Generalis 32: 105–112.

Lynch, J.A. and Johnson, M.W. (1989). Mass rearing of greenhouse whitefly parasitoid Encarsia Formosa for augumentation, releases in fresh market tomatoes in Hawaii. ADAP Crop Protection Conference Proceedings. Edited by Johnson, M.W., D.E. Ullman and A. Vargo!.

Ma, D., Gorman, K., Devine, G., Luo, W. and Denholm, I. (2007). The biotype and insecticide-resistance status of whiteflies, *Bemisia tabaci* (Hemiptera: Aleyrodidae), invading cropping systems in Xinjiang Uygur Autonomous Region, northwestern China. Crop Protection 26: 612–617.

Ma, W.I., Li, X., Demnehy, T.J., Lei, C., Wang, M., Dyain, B.A. and Nicholis. (2010). Pyriproxyfen resistance of *Bemisia tabaci* (Hemiptera; Aleyrodidae) biotype-3 metabolic mechanism. Journal of Economic Entomology 103(1): 153–165.

Mabbett, T., Nachapong, M. and Mekdaang, J. (1980). The within canopy distribution of adult cotton whitefly (*Bemisia tabaci* Genn) incorporating economic thresholds and meteorological conditions. Agricultural Science 13: 98–108.

Macel, M., Bruinsma, M., Dijkstra, S., Ooijendijk, T., Niemeyer, H. and Klinkhamer, P.L. (2005). Differences in effects of pyrrolizidine alkaloids on five generalist insect herbivore species. Journal of Chemical Ecology 31: 1493–1508.

Maharshi, N.k., Yadav, N.K., Swami, P., Singh, L. and Singh, J. (2017). Progression of cotton leaf curl disease and its vector whitefly under weather influences. International Journal of Agriculture and Microbiology Application Science 6: 2663–2670.

Makawana, D.K., Chudarama, k.A. and Belas, T.K. (2018). Estimation of losses due to sucking pests of Bt cotton. International Journal Current Microbiology and Applied Sciences 7(5): 956–959.

Malekan, N., Hatami, B., Ebodi, R., Akhavan, A., Aziz, A.B.A. and Radjabi, R. (2013). Effect of entomopathogenic fungi from *Beauveria bassiana* (Bals) and *Lecanicillium muscarium* (Petch) on *Trialeurodes vaporariorum* (Westwood). Indian Journal of Entomology 75(2): 95–98.

Manani, D.M., Ateka, E.M., Nyanjom, R.G., Boykin, L.M. and Cornelius, M.L. (2017). Phylogenetic relationships among whiteflies in the *Bemisia tabaci* Gennadius) species complex from major cassava growing areas in Kenya. Insects 8(1): 25.

Mani, M. (2010). Origin, introduction, distribution, and management of invasive spiral whitefly, *Aleurodicus disperses*. Karnataka Journal of Agricultural Sciences 23: 59–75.

Mann, G.S., Dhaliwal, G.S. and Dhawan, A.K. (2001). Field efficacy of neem-based insecticides against whitefly and their impact on insect pest complex on cotton. Pesticides Research Journal 13: 79–85.

Mansoor-ul-Hassan, Akbar, R. and Latif, A. (1998). Varietal response of mung and mash beans to insect attack. Pakistan Journal of Entomology 20: 43–46.

Mansoor-ul-Hasan, F., Ahmad, J. and Wakeel, W. (2000). Role of biochemical components in varietal resistance of cotton against sucking insect pests. Pakistan Entomologist 22(1/2): 69–71.

MAPA, Ministério da Agricultura, Pecuária e Abastecimento. (2017). AGROFIT: Sistema de Agrotóxicos Fitossanitários. Brasília, Brazil: MAPA/CGAF/DFIA/DAS. URL http://agrofit.agricultura.gov.br/agrofit_cons/principal_agrofit_cons [accessed on 1 April 2017].

Marimuthu, T., Subramanian, C.L. and Mohan, R. (1981). Assessment of yield losses due to yellow mosaic virus infestation in Mung bean. Pulse Crop Newsletter 1: 104.

Markham, P.G., Bedford, I.D., Liu, S., Frolich, D.F., Rosell, R. and Brown, J.K. (1996). The transmission of geminiviruses by biotypes of *Bemisia tabaci* (Gennadius). pp. 69–75. *In*: Gerling, D. and Mayrer, R.T. (eds.). *Bemisia*: Taxonomy, Biology, Damage, Control and Management, Intercept, Andover.

Martin, J.H. (1987). An Identification guide to common whitefly pest species of the world (Homoptera; Aleyrodidae). Tropical Pest Management 33: 298–322.

Martin, J.H. (2004). Whiteflies of Belize (Hemiptera: Aleyrodidae) part-1: Introduction and account of the sub-family Aleurodidae Quaintance and Baker. Zootaxa 681: 1–119.

Martin, J.H. (2005). Whiteflies of Belize (Hemiptera: Alcyrodidae): part 2—A review of the subfamily Aleyrodinae Westwood. Zootaxa 1098: 1–116.

Martin, J.H. (2007). Giant whiteflies (Sternorrhyncha; Aleyrodidae) A discussion of their taxonomic and evolutionary significance with the distribution of new species, *Udamoselis estrallamarinae* from Equador. Tijdschrift voor Entomologie 150: 13–29.

Martin, J.H. and Mound, L.A. (2019). An annotated checklist of the world's whiteflies. DOI: https // dx.doi..org/10.11646/Zoo taxa: 1492.1.1.

Martin, P.A., Gundersen-Rindal, D., Blackburn, M. and Buyer, J. (2007). Chromobacterium subtsugae sp. nov., a betaproteobacterium toxic to Colorado potato beetle and other insect pests. International Journal of Systematic and Evolutionary Microbiology 57(5): 993–999.

Martinez-Carrillo, J.L. (2006) Whitefly resurgence on cotton from the Yaqui Valley Sonora Mexico. Beltwide Cotton Conferences, January 3–6, San Antonio, Texas, USA.

Marubayashi, J.M., Yuki, V.A., Rocha, K.C.G., Mituti, T., Pelegrinotti, F.M., Ferreira, F.Z., Moura, M.F., Navas-Castillo, J.E., Moriones, E., Pavan, M.A. and Krause-Sakate, R. (2013). At least two indigenous species of *Bemisia tabaci* complex are present in Brazil. Journal of Applied Entomology 137: 113–121.

Marubayashi, J.M., Kilot, A., Yuki, V.A., Razende, J.A.M., Sakate, R.K., Pavan, M.A. and Ghanim, M. (2014). Diversity and localization of bacterial endosymbionts of whitefly species collected in Brazil. PLoS ONE (9): 9 e10.8363, HTTP/org/10837 Journal Pone.0108363.

Mascarin, G.M., Kobori, N.N., Quintela, E.D. and Delalibera, jr I. (2013). The virulence of Entopathogenic fungi against *Bemisia tabaci*, biotype–B (Hemiptera: Mascarin Aleyrodidae) and their conidia production solid substitute fermentation. Biological Control 66: 209–228.

Mascarin, G.M., Kobori, N.N., Quintela, E.D. and Delalibera, jr I. (2013). The virulence of Entopathogenic fungi against *Bemisia tabaci*, biotype–B (Hemiptera: Aleyrodidae) and their conidia production using solid substrate fermentation. Biological Control 66: 209–218.

Mascarin, G.M., Kobori, N.N., Quintela, E.D., Arthurs, S.P. and Junior, I.D. (2014). Toxicity of non-ionic surfactants and interactions with fungus entomopathogens toward *Bemisia tabaci* B-biotype. Biocontrol 59: 111–123.

Masuda, T. and Kikuchi, O. (1993). Control of whitefly, *Trialeurodes vaporariorum*, by a *Verticillium lecanii* preparation. Annual Report of the Society of Plant Protection of North Japan 191–193.

McAuslane, H.J. (2000). Sweetpotato Whitefly B Biotype of Silverleaf Whitefly, *Bemisia tabaci* (Gennadius) or *Bemisia argentifolii* Bellows and Perring (Insecta: Homoptera: Aleyrodidae). Entomology and Nematology Department, Florida +Cooperative Extension Service, Institute of Food and Agricultural Sciences, University of Florida.

McCutcheon, J.P., McDonald, B.R. and Moran, N.A. (2009). Convergent evolution of metabolic roles in bacterial co-symbionts of insects. Proceedings of National Academy of Sciences, USA 106: 15394–15399.

McHugh, J.B. (1991). Attack: Fungus gnats and shore flies. Greenhouse Grower 9: 67–69.

McKenzie, C.L., Anderson, P.K. and Villreal, N. (2004). An extensive survey of *Bemisia tabaci* (Homoptera: Aleyrodidae) in agricultural ecosystems in Florida. Florida Entomologist 87.3: 403–407.

McKenzie, C.L., Kumar, V., Palmer, C.L., Oetting, R.D. and Osborne, L.S. (2014). Chemical class rotations for control of *Bemisia tabaci* (Hemiptera: Aleyrodidae) on poinsettia and their effect on cryptic species population composition. Pest Management Science 70: 1573–1587.

Meena, R.S., Aneta, C.P. and Meena, B.L. (2013). Population dynamics of sucking pests and their correlations with weather parameters in chili (*Capsicum annum* L.) Bioscan 8: 17–18.

Melamed-Madjar, V., Cohan, S., Chen, M., Tam, S. and Rosilio, D. (1979). Observations on populations of *Bemisia tabaci* Genn. (Hemiptera: Aleyrodidae) on cotton adjacent to sunflower and potato in Israel. Israel Journal of Entomology 18: 71–78.

Melamed-Madjar, V., Cohan, S., Chen, M., Tam, S. and Rosilio, D. (1982). A method for monitoring *Bemisia tabaci* and timing spray applications against the pest in cotton fields in Israel. Phytoparasitica 10: 85–91.

Mellor, H.E. and Anderson, M. (1995). Antennal sensilla of whiteflies: *Trialeurodes vaporariorum* (Westwood), the glasshouse whitefly, and *Aleyrodes proletella* (Linnaeus), the cabbage whitefly, (Homoptera: Aleyrodidae). Part 2: Ultrastructure. International Journal of Insect Morphology Embryology 24: 145–160.

Meekes, E.T.M. (2001). Entomopathogenic fungi against whitefly: Tritrophic interactions between Aschersonia species, *Trialeurodes vaporariorum* and *Bemisia argentifolii*, and glasshouse crops. ISBN: 90-5808-443-4.

Menn, J.J. (1996). The Bemisia complex, an International crop protection problem waiting for a solution. pp. 381–383. *In*: Gerling, D. and R.T. Mayer (eds.). *Bemisia* 1995: Taxonomy, Biology, Damage, Control and Management. Intercept Ltd., Andover, Hants, UK.

Messelink, G., van Steenpaal, S.E.F. and Ramakers, P. (2006). Evaluation of phytoseiid predators for control of western flower thrips on greenhouse cucumber. BioControl 51: 753–768.

Messelink, G.J., van Maanen, R., van Steenpaal, S.E.F. and Janssen, A.R.M. (2008). Control of thrips and whiteflies by a shared predator. Two pests are better than one. Biological Control 44: 372–379.

Metcalf, R.L. and Metcalf, R.A. (1993). Destructive and Useful Insects: Their Habits and Control, fifth ed. McGraw-Hill, Inc., New York, USA.

Miles, F.W. (1999). Aphid saliva. Biological Reviews Cambridge, Philosophical Society 74: 41–85.

Miles, H.W. (1927). On the control of glasshouse insects with calcium cyanide. Annals of Applied Biology 14: 240–246.

Miles, P.W. (1972). The saliva of Hemiptera. Advances in Insect Physiology 9: 183–256.

Miller, M.A., Pfeiffer, W. and Schwartz, T. (2010). Creating the CIPRES Science Gateway for inference of large phylogenetic trees. In the Gateway Computing Environments Workshop (GCE) 14: 1–8.

Miller, M.A., Schwartz, T., Pickett, B.E., He, S., Klem, E.B., Scheuermann, R.H., Passarotti, M., Kaufman, S. and O'Leary, M.A. (2015). A restful API for access to phylogenetic tools via the CIPRES science gateway. Evolution and Biology Information Online 11: 43–4.

Miller, G.L., Jensen, A.S., Nakahara, S., Carlson, R.W., Miller, D.R. and Stoezel, M.B. (2000). Systematic entomology laboratory whitefly web page: http://www.sel.barc.usda.gov.

Milligan, S.B., Bodeau, J., Yaghoobi, J., Kaloshian, I., Zabel, P. and Williamson, V.M. (1998). The root-knot nematode resistance gene Mi from tomato is a member of the leucine zipper, nucleotide-binding, leucine-rich repeat family of plant genes. Plant and Cell 10: 1307–1319.

Mishra, S., Jagadeesh, K.S., Krishnaraj, P.U. and Prem, S. (2014). Biocontrol of Tomato Leaf Curl Virus (ToLCV) in tomato with chitosan supplemented formulations of *Pseudomonas* species under field conditions. Australian Journal of Crop Science 8: 347–355.

Miyazaki, J., Stiller, W.N. and Wilson, L.J. (2013). Identification of host plant resistance to silver leaf whitefly in cotton: Implications for breeding. Field Crop Resistance 154: 145–152.

Mohd Rasdi, Z., Fauziah, I., Fairuz, K., Saiful, M.S., Mohd Jamaludin, B., Che Salmah, M.R. and Kamaruzaman Jusoff. (2009). Population ecology of *Bemisia tabaci* (Homoptera: Aleyrodidae) on brinjal. Journal of Agricultural Science, DOI 10.5539/jas.v1n1p27.

Monci, F., Garccia-Andres, S., Sanchez-Campos, S., Franandez-Munoz, R., Diaz, J.A. and Moriones, E. (2019). Use of systematic acquired resistance and whitefly, with optical barriers to reduce Tomato yellow leaf curl virus disease damage to tomato crop. Plant Disease 107(6): 1181. DOI:10.1094/PDIS-06-18-1069-RE.

Morales, F.J. (2007). Tropical whitefly. IPM Project Advances in Virus Research 69: 249–311.

Moran, N.A. and Telang, A. (1998). Bacteriocyte-associated symbionts of insects. Bioscience 48: 295–304.

Moran, N.A., Russell, J.A., Koga, R. and Fukatsu, T. (2005). Evolutionary relationship of three new species of *Bacterio ckeriaceae* living as symbionts of aphids and other insects. Journal of Applied Microbiology 7: 3202–3210.

Mordue (Luntz), A.J. and Nisbet, A.J. (2000). Azadirachtin from the neem tree *Azadirachta indica*: its action against insects. Anais da Sociedade Entomológica do Brazil 29: 615–632.

Moreno, A., Garzo, E., Fernandez-Mata, G., Kassem, M., Aranda, M.A. and Fereres, A. (2011). Aphids secrete watery saliva into plant tissues from the onset of stylet penetration. Entomologia Experimentalis et Applcata 139: 145–153.

Morin, S., Ghanim, M., Zeidan, M., Czosnek, H., Verbeek, M. and Van den Heuvel, J.F. (1999). A GroEL homolog from endosymbiotic bacteria of whitefly, *Bemisia tabaci* is implicated in the circulative transmission of Tomato yellow leaf curl virus. Virology 256: 75–85.

Morin, S., Ghanim, M., Sobol, I. and Czosnek, H. (2000). The GroEL protein of the whitefly *Bemisia tabaci* interacts with the coat protein of transmissible and non-transmissible begomoviruses in the yeast two-hybrid system. Virology 276: 404–416.

Morita, M., Yoneda, T. and Akiyoshi, N. (2014). Research and development of novel insecticide, flonicamid. Journal of Pesticide Science 39: 179–180.

Mottern, J.L. and Heraty, J.M. (2014). Revision of the *Cales noacki* species complex (Hymenoptera, Chalcidoidea, Aphelinidae). Systematic Entomology 39: 354–79.

Mound, L.A. (1965). Effect of leaf hairs on cotton whitefly population in the Sudan Gezira. Empire Cotton Growing Review 42: 33–44.

Mound, L.A. and Halsey, S.H. (1978). Whitefly of the World, A Systematic Catalog of the Aleyrodidae (Homoptera) with Host Plant and Natural Enemy Data. New York: John Wiley & Sons.

Mound, L.A. (1984). Zoogeographical distribution of whiteflies. Current Topics in Vector Research 2: 185–197.

Mouton, L., Thierry, M., Henri, H., Baudin, R., Gnankine, O. and Reynaud, B. (2012). Evidence of diversity and recombination in *Arsenophonus* symbionts of the *Bemisia tabaci* species complex. BMC Microbiology, 12.

Mouton, L., Gnankiné, O., Henri, H., Terraz, G., Ketoh, G. and Martin, T. (2015). Detection of genetically isolated entities within the Mediterranean species of *Bemisia tabaci*: new insights into the systematics of this worldwide pest. Pest Management Science 71: 452–455.

Moya, A., Peretó, J., Gil, R. and Latorre, A. (2008). Learning how to live together: genomic insights into prokaryote-animal symbioses. National Reviews of Genetics 9: 218–229.

Muchhlebach, M., Buchholz, A., Zambach, W., Schaetzer, J., Daniels, M., Hoeter, D., Kloer, D.P., Lind, R., Maierfisch, P., Pierce, A., Pitterna, T., Smejkal, T., Stafford, D. and Wiildsmith, L. (2020). Spiro N methoxy piperdine ring containing aryediones for the control of sucking insects and mites. Discovery of Spiropidion Pest Management Science (in Press).

Mugerwa, H., Rey, E.C., Alicai, T., Tairo, F., Ateka, M.E., Atuncha, H., Nanguru, J. and Sseruwagi, P. (2012). Genetic diversity and geographic distribution of *Bemisia tabaci* (Gennadius) (Hemiptera: Aleyrodidae) genotypes associated with cassava in East Africa. Ecology and Evolution 2: 2749–2762.

Mugerwa, H., Seal, S., Wang, H.L., Patel, M.V., Kabaalu, R., Omongo, C.A., Alicai, F., Tairo, F., Naunguru, J., Sseruwagi, P. and Calvin, J. (2018). African ancestry of new world *Bemisia tabaci* whitefly species. Scientific Reports 8: 2734.

Mutisya, S., Saidi, M., Opiyo, A., Ngouajio, M. and Martin, T. (2016). Synergistic effects of Agronet covers and companion cropping on reducing whitefly infestation and improving the yield of open field-grown tomatoes. Agronomy 6: 42.

Nachapong, M. and Mabbett, T. (1979). A survey of some wild hosts of *Bemisia tabaci* Genn around cotton fields in Thailand. Thai Journal of Agricultural Sciences 12: 217–222.

Naik, L.K. and Lingappa, S. (1992). Distribution pattern of *Bemisia tabaci* (Gennadius) in cotton. Insect Science Applications 13: 277–279.

Nakahara, S. (1995). Taxonomic status of genera *Trialeurodes* (Homoptera: Aleyrodidae). Insecta Mundi 9: 105–1150.

Naranjo, S.E., Chu, C.C. and Henneberry, T.J. (1996). Economic injury level for *Bemisia tabaci* (Homoptera: Aleyrodidae) in cotton. Impact of crop price control costs and efficiency of control. Crop Protection 15(8): 779–788.

Naranjo, S.E. (2001). Conservation and evaluation of natural enemies in IPM systems for *Bemisia tabaci*. Crop Protection (Elsevier) 20: 835–852.

Nasruddin, A. and Mound, L.A. (2016). First record of *Trialeuodes vaporariorum* Westwood (Hemiptera: Aleyrodidae) severely damaging field-grown potato crops in South Sulawesi, Indonesia. Journal of Plant Protection Research 56(2). DOI:10.1515/jppr-2016-00230.

Natarajan, K. (1986). Influence of NPK fertilization on population density of cotton whitefly. *In*: Jayraj, S. (ed.). Resurgence of Sucking Pests, Tamil Nadu Agricultural University, Coimbatore.

Natarajan, K. and Surulivelu, T. (1986). Population dynamics and management of whitefly on cotton. Paper presented in a group discussion on whitefly in Cotton RARS, Iam, Guntur April 29–30, 1986.

Natarajan, K., Sundaramurthy, V.T. and Basu, A.K. (1986). Meet the menace of whitefly in cotton. Indian Farming 36: 37–39,44.

Natarajan, K. and Sundaramurthy, V.T. (1990). Effect of neem oil on cotton whitefly, *Bemisia tabaci*. Indian Journal of Agricultural Sciences 60: 290–291.

Natarajan, K., Sundaramurthy, V.T. and Chidambram, D. (1991). Usefulness of fish oil resin soap in the management of whitefly and other sap-feeding insects of cotton. Entomon 16: 229–232.

Nauen, R., Stumpf, N. and Elbert, A. (2002). Toxicological and mechanistic studies on neonicotinoid cross-resistance in Q-type *Bemisia tabaci* (Hemiptera: Aleyrodidae). Pest Management Science 58: 868–875.

Nauen, R. and Denholm, I. (2005). Resistance of insect pests to neonicotinoid insecticides: current status and prospects. Archives of Insect Biochemistry and Physiology 58: 200–215.

Nauen, R., Reckmann, U., Thomzik, J. and Thielert, W. (2008). Biological profile of spirotetramat (Movento®)—a new two-way systemic (ambimobile) insecticide against sucking pest species. Bayer CropScience Journal 61: 245–278.

Nauen, R., Ghanim, M. and Ishaaya, I. (2014). Whitefly special issue organized in two parts. Pest Management Science 70: 1438–1439.

Nauen, R., Wölfel, K., Lueke, B., Myridakis, A., Tsakireli, D., Roditakis, E., Tsagkarakou, A., Stephanou, E. and Vontas, J. (2015). Development of a lateral flow test to detect metabolic resistance in *Bemisia tabaci* mediated by CYP6CM1, a cytochrome P450 with broad-spectrum catalytic efficiency. Pesticide Biochemistry and Physiology 121: 3–11.

Navas-Castillo, J., Fiallo-Olive, E. and Sancheez-Campos, S. (2011). Emerging virus diseases transmitted by whiteflies. Annual Reviews of phytopathology 49: 219–248.

Navas-Castillo, J., Lopez-Moya, J.J. and Aranda, M. (2014). Whitefly-transmitted RNA viruses that affect intensive vegetable production. Annals of Applied Biology 165: 155–171.

Naveen, N.C., Kumar, N.C., Alam, D., Chaubey, W. and Rahul. (2012). A model study integrating time-dependent mortality in evaluating insecticides against *Bemisia tabaci* (Hemiptera: Aleyrodidae). Indian Journal Entomology 74: 384–387.

Naveen, N.C., Chaubey, R., Kumar, D., Rebiseith, K.B., Rajgopal, R. and Subramanian, S. (2017). Insecticide resistance status in the Whitefly, *Bemisia tabaci* genetic groups Asia1, Asia ii-1 and Asia ii-7 on the Indian continent. Scientific Reports 7: 40634, Doi: 10. 1038/ Srep, 40634.

Neiva, I.P., Silva, A.A., Resende, J.F., Carvalho, R., Oliveira, A.M. and Maluf, W.R. (2019). Tomato genotype resistance to whitefly mediated by allelochemicals and Mi gene. Chilean Journal Agricultural Research 79: 124–130.

Nemade, P.W., Budhvat, K.P. and Wadaskar, P.S. (2018). Population dynamics of sucking pests with relation to weather parameters in Bt cotton in Buldana district, Maharashtra, India. International Journal of Current Microbiology and Applied Sciences 7(01): 620–626.

Ng, J.C.K. and Falk, B.W. (2006). Virus vector interactions mediating non–persistent and semi-persistent transmission of plant viruses. Annual Reviews of Phytopathology 44: 219–248.

Ng, J.C.K. and Zhou, J.S. (2015). Insect vector plant interactions associated with the non-circulative, semi-persistent transmission; Current perspectives and future challenges. Current Opinion in Virology 15: 48–55.

Nguyen, R., Brazzel, J.R. and Poucher, C. (1983). Population density of the citrus blackfly, *Aleurocanthus woglumi* Ashby (Homoptera: Aleyrodidae), and its parasites in urban Florida in 1979–1981. Environmental Entomology 12: 878–84.

Nimbalkar, N., Yadav, D.B. and Prabhune, R.N. (1994). Evaluation of Bio efficacy, phytotoxicity and compatibility of Neemax on okra (*Abelmoschus esculentus* (L.)) Moench cultivar. PARBHANI KRANTHI, Pestology 18: 10–19.

Nirgianaki, A., Banks, G.K., Frohlich, D.R., Veneti, Z., Braig, H.R. and Miller, T.A. (2003). *Wolbachia* infections of the whitefly *Bemisia tabaci*. Current Microbiology 47: 0093 0101.

Nombela, G., Beitia, F. and Manij, M. (2001). A differential interaction study of *Bemisia tabaci* Q biotype on commercial tomato varieties with or without the Mi gene and the comparative response with B-biotype. Entomologia Experimentalis et Applicata 98: 339–344.

Nombela, G., Williamson, V.M. and Muñiz, M. (2003). The root-knot nematode resistance gene Mi-1.2 of tomato is responsible for resistance against the whitefly *Bemisia tabaci*. Molecular Plant Microbes Interaction 16. 645–649.

Nomikou, M. Arne Janssen and Sabelis, M.W. (2003). Phytoseiid predator of whitefly feeds on plant tissue. Experimental and Applied Acarology 31: 27–36.

Oberemok, V.V., Laikaova, K.V., Gninenko, Y.I., Zaitsev, A.S., Wyadar, P.M. and Adehyenii, T.A. (2015). A short story of insecticides. Journal of Plant Protection Research 55: 221–226.

Oerke, E. (2006). Crop losses to pests. The Journal of Agricultural Science 144(1): 31–43. doi:10.1017/ S0021859605005708.

Oliveira, M.R.V., Henneberry, T.J. and Anderson, P. (2001). History, current status, and collaborative research projects for *Bemisia tabaci*. Crop Protection 20: 709–723.

Oliver, K.M., Russell, J.A., Moran, N.A. and Hunter, M.S. (2003). Facultative bacterial symbionts in aphids confer resistance to parasitic wasps. Proceedings of National Academy of Sciences, USA 100: 1803–1807.

Orozco-Santos, M., Prez-Zamora and Lopez–Arriaga, o. (1999). Effect of transparent mulch on insect populations virus diseases, soil temperature, and yield of cantaloup in a tropical region. New Zealand Journal of Crop and Horticultural Sciences 23: 1999–204.

Osborne, L.S. and Landa, Z. (1992). Biological control of whiteflies with Entopathogenic fungi. Florida Entomologist 75: 456–471.

Ovalle, T.M., Parsa, S., Hernández, M.P. and Becerra Lopez-Lavalle, L.A. (2014). Reliable molecular identification of nine tropical whitefly species. Ecology and Evolution 4: 3778–3787.

Ozgur, A.F. and Sekeroglu, E. (1986). Population development of *Bemisia tabaci* (Homoptera Aleyrodidae) on various cotton cultivars in Cucurova, Turkey Agriculture Ecosystem, and Environment 7: 83–88.

Ozgur, A.F., Sekeroglu, E., Gencer, O., Goemen, H., Yelin, D. and Isler, N. (1988). Study of population development of important cotton pests in relation to various cotton varieties and plant phenology. Doga Turk Tarim ve Ormancilik Dergesi 12: 48–74.

Ozgur, A.F., Sekeroglu, E., Ohnesorg, B. and Golmen, H. (1990). Studies on host plant changes, migration and population dynamics of cotton whitefly (*Bemisia tabaci*) in Cukurova (Turkey). Allgemeine Angewandte Entomol. 7: 653–656.

Painter, R.H. (1951). Insect Resistance and Environment Stance in Crop Plants. The MacMillan Co, New York. MacMillan Co, New York.

Palaniswami, M.S., Antony, B., Vijayan, S.L. and Henneberry, T.J. (2001). Sweet potato whitefly *Bemisia tabaci*: ecobiology, host interaction, and natural enemies. Entomon 26: 256–262.

Palumbo, S.J., Horowitz, A.R. and Prabhaker, N. (2001). Insecticidal control and resistance management for *Bemisia tabaci* J.C. Crop Protection 20: 739–765.

Palumbo, J.C., Tenaska, Jr A. and Byrne, D.N. (1984). Sampling plans and action threshold for whiteflies on spring melon IPM Series 1 Publication No. 194021, University of Arizona, College of Agriculture and Life Sciences, Cooperative Extension, Tucson, Arizona the USA.

Pan, H.P., Chu, D., Yan, W.Q., Su, Q., Liu, B. and Wang, S. (2012a). Rapid spread of Tomato Yellow Leaf Curl Virus in China is aided differentially by two invasive whiteflies. PLoS One 7: e 0034817.

Pan, H.P., Li, X.C., Ge, D.Q., Wang, S., Wu, Q., Xie, W., Jiao, X., Chu, Liu, B., Xu, B. and Zhang, Y. (2012b). Factors affecting population dynamics of maternally transmitted endosymbionts in *Bemisia tabaci*. PLoS One 7: e0030760.

Pan JieRu, Zhang HongSheng, Zhu YanPing and Qiu JunZhi. (2007). The insecticidal activities of the metabolites from entomopathogenic fungus *Aschersonia* sp. Jos009. Journal of Fujian Agriculture and Forestry University (Natural Science Edition) 36(1): 25–27.

Pandey, N., Singh, A., Rana, V.S. and Rajagopal, R. (2013). Molecular characterization and analysis of bacterial diversity in *Aleurocanthus woglumi* (Hemiptera: Aleyrodidae). Environmental Entomology 42: 1257–1269.

Panickar, B.K. and Patel, J.R. 2001. Population dynamics of different species of thrips on chili cotton and pigeon pea. Indian Journal of Entomology 63: 170–75.

Panini, M., Tozzi, F., Zimmer, C.T., Bass, C., Field, L., Borzatta, V. and Moores, G. (2017). Biochemical evaluation of interactions between synergistic molecules and phase I enzymes involved in insecticide resistance in B- and Q-type *Bemisia tabaci* (Hemiptera: Aleyrodidae). Pest Management Science 73: 1873–1882. https://doi.org/10.1002/ps.4553.

Pappas, M.L., Migkou, F. and and Broufas, G.D. (2013). incidence of resistance to neonicotinoid insecticides in greenhouse populations of the whitefly, *Trialeurodes vaporariorum* (Hemiptera: Aleyrodidae) from Greece. Applied Entomology and Zoology 38: 373–378.

Park, M., Kim, J.G., Song, Y., Lee, J.H., Shin, K. and Cho, K. (2009). Effect of nitrogen levels of two cherry tomato cultivars on development, preference and honeydew production of *Trialeurodes vaporariorum* (Hemiptera: Aleyrodidae) Journal of Asia-Pacific Entomology 12: 227–232.

Parker, T. (1928). The use of tetrachloroethane for commercial glasshouse fumigation. Annals of Applied Biology 15: 251–257.

Parrella, M.P., Paine, T.D., Bethke, J.A. and Hall, J. (1991). Evaluation of *Encarsia formosa* (Hymenoptera; Aphelinidae) for biological control of sweet potato whitefly (Homoptera; Aleyrodidae) on poinsettia. Environment Entomology 20: 713–719.

Patel, S.M. (2010). Population dynamics and management of major sucking insect pests in Bt cotton. M. Sc. (Agri.) thesis (unpublished) submitted to Anand Agricultural University, Anand.

Patil, B.V., Nandihalli, B.S. and Hugar, P. (1990). Effect of synthetic pyrethroids on population buildup of cotton whitefly, *Bemisia tabaci* (Gennadius). Pestology 14: 17–21.

Patra, B.S.K., Alam, F. and Sannanta, A. (2016). Influence of weather factors on the incidence whitefly, *Bemisia tabaci* (Genn) on tomato in Darjeeling hills of West Bengal. Journal of Agroecology and Natural Resource Management 3: 243–146.

Paulson, G.S. and Beardsley, J.W. (1985). Whitefly (Hemiptera, Aleyrodidae) egg pedicel insertion into host plant stomata. Annals of the Entomological Society of America 78: 506–508.

Pedigo, L., Hutchins, S. and Higley, L. (1986). Economic injury levels in theory and practice. Annual Reviews of Entomology 1986(31): 341–368.

Pei-Xiang, Z., Chang, M.L., Chuan, Y.Q., Jun, J.X. and Hong, Y.L. (2011). Control effects of whitefly by intercropping celery in greenhouse. Chinese Journal of Applied Entomology 48: 375–378.

Perlman, S.J., Hunter, M.S. and Zchori-Fein, E. (2006). The emerging diversity of *Rickettsia.* Biological Sciences 273: 2097–2106.

Perring, T.M., Cooper, A.D., Rodrigues, R.J., Farrar, C.A. and Bellows, Jr. (1993). Identification of whiteflies species by genomic and behavioral studies. Science 259: 74–77.

Perring, T.M. (2001). The *Bemisia tabaci* species complex. Crop Protection, 20: 725–737.

Perring, T.M., Stansly, P.A., Liu, T.X., Smith, H.A. and Andreason, S.A. (2018). Whiteflies: Biology, ecology, and management. pp. 73–110. *In*: Wakil, W., Brust, G.E. and Perring, T.M. (eds.). Chapter 4 Sustainable Management of Arthropod Pests of Tomato. Academic Press, Elsevier, ISBN 9720128024416.

Perry, A.S. (1985). The relative susceptibility several insecticides of adult whiteflies (*Bemisia tabaci*) for various cotton-growing regions in Israel. Phytoparasitica 13: 77–78.

Phadke, A.D., Khandal, V.S. and Rahalkar, S.R. (1988). Effect of Neemark's formulations on incidence of whiteflies and yield of cotton. Pesticides 22: 36–37.

Pickett, C.H., Simmons, G.S. and Goolsby, J.A. (2008). Releases of exotic parasitoids of *Bemisia tabaci* in San Joaquin Valley, California. pp. 225–242. *In*: Gould, J., Hoelmer, K. and Goolsby, J. (eds.). Classical Biological Control of *Bemisia tabaci* in the United States. Springer.

Pickett, J.A. and Khan, Z.R. (2016). Plant volatile-mediated signaling and its application in agriculture: successes and challenges. New Phytopathology 212: 856–870.

Pico, B., Diez, J. and Nuez, K. (1997). Evaluation of whitefly mediated inoculation technique to screen *Lycopersicon esculentum* wild relative for resistance to tomato yellow leaf curl virus. Euphytica 101: 259–271.

Pollard, D.G. (1955). Feeding habits of cotton whitefly *Bemisia tabaci* Genn (Homoptera; Aleyrodidae). Annals of Applied Biology 43: 664–671.

Prabhaker, N., Coudriet, D.L. and Meyerdirk, D.E. (1985). Insecticide resistance in the sweet potato whitefly, *Bemisia tabaci* (Homoptera: Aleyrodidae). Journal of Economic Entomology 78: 748–752.

Prabhaker, N., Coudriet, D.L. and Toscano, N.C. (1988). Effect of synergists on organophosphate and permethrin resistance in sweet potato whitefly (Homoptera: Aleyrodidae). Journal of Economic Entomology 81: 34–39.

Prabhaker, N., Toscano, N.C. and Coudrie, D.L. (1989). Susceptibility of the immature and adult stages of the sweet potato whitefly (Homoptera: Aleyrodidae) to selected insecticides. Journal of Economic Entomology 78: 983–988.

Prabhaker, N., Toscana, N.C. and Henneberry, T.J. (1998). Evaluation of insecticide rotations and mixtures as resistance management strategies for *Bemisia argentifolii* (Homoptera: Aleyrodidae). Journal Economic Entomology 91: 820–6.

Prabhaker, N., Castle, S., Henneberry, T.J. and Toscano, N.C. (2005). Assessment of cross-resistance potential to neonicotinoid insecticides in *Bemisia tabaci* (Hemiptera: Aleyrodidae). Bulletin of Entomological Research 95: 535–543.

Prabhaker, N., Castle, S. and Perring, T.M. (2014). Baseline susceptibility of *Bemisia tabaci* B biotype (Hemiptera: Aleyrodidae) populations from California and Arizona to spirotetramat. Journal Economic Entomology 107: 773–780.

Prada, E.N., Innacone, J. and Gumez, H. (2008). Effect of the two Entomopathogenic fungus, *Clonostachys rosea* pathogenic fungi in controlling *Aleurodicus cocois* (Hemiptera: Aleyrodidae). Chilean Journal of Agricultural Research 68: 21–30.

Prasad, N.V.V.S.D., Rao, M. and Rao, N.H. (2009). Performance of Bt cotton and non Bt cotton hybrids against pest complex under unprotected conditions. J. Biopestic. 2: 107–110.

Prasad, V.D., Bharati, M. and Reddy, G.P.V. (1993). Relative resistance to conventional insecticides in three populations of cotton whitefly, *Bemisia tabaci* (Gennadius) in Andhra Pradesh. Indian Journal of Plant protection Protection 21: 102–103.

Price, R.P. Jr. (1999). Reflective mulches and yellow sticky tape control whiteflies in greenhouse poinsettia (*Euphorbia pulcherrima*). As reported in Williams, Greg, and Pat. From the 1999 ASHS Conference. HortIdeas. August. p. 85.

Prohit, M.S. and Deshpande, A.D. (1991). Effect of inorganic fertilizers on the population density of cotton whitefly (*Bemisia tabaci*). Indian Journal of Agricultural Sciences 61: 696–698.

Prokopy, R.J. and Owens, E.D. (1983). Visual detection of plants by herbivorous insects. Annual Reviews of Entomology 28: 337–364.

Pruthi, H.S. and Samuel, C.K. (1942). Entomological investigations on the leaf curl virus disease of tobacco North India v. Biology and population of whitefly vector (*Bemisia tabaci* Genn) in relation to incidence of disease. Indian Journal of Agricultural Sciences 54;59: 35–37.

Purcell, A.H., Suslow, K.G. and Klein, M. (1994). Transmission via plants of an insect pathogenic bacterium that does not multiply or move in plants. Microbiology Ecology 27: 19–26.

Puri, S.N., Butler, G.D. and Henneberry, T.J. (1991). Plant-derived oils and soap solution as control agents for whitefly on cotton. Journal Zoology Research 1: 1–5.

Puri, S.N., Bhosle, B.B., Fartade, M.K., Kolhal, R.N., Butler, G.D. and Henneberry, T.J. (1994). Wild brinjal (*Solanum khassianum*) as a potential crop for the management of *Bemisia tabaci* on cotton. Phytoparasitica 22: 358–359.

Puri, S.N., Murthy, K.S. and Sharma, O.P. (1998). Integrated management of cotton whitefly, *Bemisia tabaci* (Gennadius). pp. 286–296. *In*: Dhaliwal, G.S., Randhawa, S., Romesh Arora and Dhawan, A.K. (eds.). Ecological Agriculture and Sustainable Development Vol II, IES & CRID, Chandigarh.

Purohit, S.S. and Vyas, S.P. (2004). Medicinal Plant Cultivation A Scientific Approach. Agrobios, Jodhpur, India pp. 158179.

Puthoff, D.P., Holzer, F.M., Perring, T.M. and Walling, L.L. (2010). Tomato pathogenesis-related protein genes are expressed in response to *Trialeurodes vaporariorum* and *Bemisia tabaci* biotype–B feeding. Journal of Chemical Ecology 36: 1271–1285.

Qiao, Q., Jhand, Z.C., Qin, Y.H., Jhang, D.S., Tian, Y.T. and Wang, Y.J. (2011). First report of Sweet potato Chlorotic Stunt Virus infecting sweet potato in China. Plant Disease 95: 356.

Qiu, B.L., De Barro, P.J., He, Y.R. and Ren, S.X. (2007). Suitability of *Bemisia tabaci* (Hemiptera: Aleyrodidae) instars for the parasitization by *Encarsia bimaculata* and *Eretmocerus* sp. no. furuhashii (Hymenoptera: Aphelinidae) on glabrous and hirsute host plants. Biocontrol Science and Technology 17: 823–39.

Qiu, B.L., Chen, Y.P., Liu, L., Peng, W., Li, X., Ahmad, M.Z., Mathur, V., Du, Y. and Ren, S. (2009). Identification of three major *Bemisia tabaci* biotypes in China based on morphological and DNA polymorphisms. Progress National Sciences 19: 713–718.

Qiu, B.L., Dang, F., Li, S.J. and Ahmad, M.Z.L. (2011). Comparison of biological parameters between the invasive B biotype and a newly defined CV biotype of *Bemisia tabaci* (Hemiptera: Aleyrodidae) in China. Journal of Pest Sciences 84: 419–427.

Qu, C., Zhang, W., Li, F., Tetreau, G., Luo, C. and Wang, R. (2017). Lethal and sub-lethal effects of dinotefuran on two invasive whiteflies, *Bemisia tabaci* (Hemiptera: Aleyrodidae). Journal of Asia-Pacific Entomology 20: 325–330.

Qu, D., Ren, L.-M., Liu, Y., Ali, S., Wang, X.M., Ahmed, M.Z. and Qiu, B.L. (2019a). Compatibility and efficacy of the parasitoid *Eretmocerus hayati* and the entomopathogenic fungus *Cordyceps javanica* for biological control of whitefly, *Bemisia tabaci*. Insects 10(12): 425. Doi103390/insects 101204245.

Quaintance, A.L. (1900). Contributions toward a monograph of the American Aleurodidae. Bulletin of the United States Department of Agriculture Entomology Technician 8: 9–43.

Quaintance, A.L. and Baker, A.C. (1913). Classification of the Aleyrodidae. Part I. United States Department of Agriculture Technical Series 27: 1–93.

Queiroz, P.R., Martins, E.S., Klautau, N., Lima, L., Praça and Monnerat, R.G. (2016). Identification of the B, Q, and native Brazilian biotypes of the *Bemisia tabaci* species complex using Scar markers. Pesquisa Agropecuária Brasileira 51(5): 555–562. https://doi.org/10.1590/S0100-204X2016000500016.

Quesada-Moraga, E., Maranhao, E.A.A., Valverde-García, P. and Santiago-Álvarez, C. (2006). Selection of Beauveria bassiana isolates for control of the whiteflies *Bemisia tabaci* and *Trialeurodes vaporariorum* on the basis of their virulence, thermal requirements, and toxicogenic activity. Biological Control 36: 274–287. ISSN 1049-9644, https://doi.org/10.1016/j.biocontrol.2005.09.022.

Qureshi, M.S., Midmore, D.J., Syeda, S.S. and Playford, C.L. (2007). Floating row covers and pyriproxyfen help control Silverleaf whitefly Australian *Bemisia tabaci* (Gennadius) Biotype B (Homoptera: Aleyrodidae) in zucchini. Journal of Entomology 46: 313–319.

Rabello, A.R., Queiroz, P.R., Simoes, K.C.C., Hiragi, C.O., Lima, L.H.C., Oliveira, M.R.V. and Mehta, A. (2008). Diversity analysis of *Bemisia tabaci* biotypes, RAPD PCR-RELP and sequences ITSI, DNA region. Genetic Molecular Biology 31: 585–590.

Rafiq, M.W., Gaffar, A. and Arshad, M. (2008). Dynamics of whitefly (*Bemisia tabaci*) on cultivated chilli hosts and their role in regulating its carryover to cotton. Annual Journal of Agriculture and Biology 10(4).

Rakha, M., Zeheya, N., seven, S., Mubembi, M., Ramasamy, S. and Henson, P. (2017). Screening study identified whitefly/spider mite wild tomato resistant accession for resistance to *Tuta absoluta*. Plant Breeding 136(4). HTTP/doi;10. 1111 /pbr..12503/.

Rana, V.S., Singh, S.T., Priya, N.G., Kumar, J. and Rajagopal, R. (2012). *Arsenophonus* GroEL interacts with CLCuV and is localized in the midgut and salivary gland of whitefly *Bemisia tabaci*. PLoS One 7: e42168.

Rao, N.V. (1987). Seasonal occurrence and management of whitefly, *Bemisia tabaci* Genn. on cotton. Ph.D. Thesis, APAU, Hyderabad.

Rao, N.V., Reddy, A.S. and Rao, K.T. (1989). Natural enemies of cotton whitefly, *Bemisia tabaci* Genn in relation to pest population and weather factors. Journal of Biological Control 3: 10–12.

Rao, N.V., Reddy, A.S. and Reddy, P.S. (1990). Relative efficacy of some new insecticides to cotton whitefly Genn. Indian Journal of Plant Protection 18: 53–58.

Rao, N.V., Reddy, A.S. and Tirumala Rao, K. (1991). Influence of spray applications on cotton whitefly, *Bemisia tabaci*. Journal of Cotton Research and Development 5(1): 67–71.

Rao, Q., Rollat-Farnier, P.A., Zhu, D.T., Santos-Garcia, D., Silva, F.J., Moya, A., Latorre, A., Klein, C.C., Vavre, F. and Sagot, M.F. (2015). Genome reduction and potential metabolic complementation of the dual endosymbionts in the whitefly *Bemisia tabaci*. BMC Genomics 16: 226.

Rashmi, P., Sharma, K., Chaudhari, D. and Rai, M. (2008). Effect of weather parameters on the incidence of *Bemisia tabaci* and Myzus persicae on a potato. Annals of Plant Protection Sciences 16.1: 78–80.

Rauch, N. and Nauen, R. (2003). Identification of biochemical markers linked to neonicotinoid cross-resistance in *Bemisia tabaci* (Hemiptera: Aleyrodidae). Archives of Insect Biochemistry and Physiology 54: 165–176.

Raviv, M. and Antignus, Y. (2004). UV radiations effects on pathogens and insect pests of greenhouse-grown crops. Photochemistry Photobiology 79: 219–226.

Raza, A., Malik, H.J., Shafiq, M., Amin, I., Scheffler, J.A., Scheffler, B.E. and Mansoor, S. (2016). RNA interference based approach to down regulate osmoregulators of whitefly (*Bemisia tabaci*): Potential technology for the control of whitefly. PLoS ONE 11(4): e0153883. doi:10.1371/journal pone.0153883.

Reaumur. (1734). Memoires sur les insectes 2(25): 302–317.

Reddy, A.S., Azam, K.M., Rosaiah, B., Rao, T.B., Rao, B.R. and Rao, N.V. (1986). Biology and management of whitefly, *Bemisia tabaci* (Genn) on cotton. Proceedings of Group Discussion on Whitefly in Cotton RARS Lam Guntur, April 29–30, 1986.

Reddy, A.S., Rosaiah, B., Bhaskara Rao, T., Rama Rao, B. and Venugopal Rao, N. (1986b). The Problem of whitefly on cotton. Proceedings of Seminar on the Problem of Whitefly on Cotton, March 14, 1986, Pune.

Reddy, A.S. and Rao, N.V. (1989). Cotton whitefly (*Bemisia tabaci* Genn.)—A review. Indian Journal of Plant Protection 17: 171–179.

Reddy, G.P.V. and Krishnamurthy, R.M.M. (1989). Insect pest management in cotton. Pesticides 23: 18–19.

Regu, K., Tamilselvan, C., Sundaraja, R. and David, B.V. (1990). Influence of certain insecticides on the population build-up of whitefly, *Bemisia tabaci* (Genn.) on cotton. Pestology 14(4): 810.

Riley, D.G. and Palumbo, J.C. (1995). Action threshold for *Bemisia argentifolii* (Homoptera: Aleyrodidae) in cantaloupe. Journal Economic Entomology 88: 1733–1738.

Roditakis, E., Roditakis, N.E. and Tsagkarakou, A. (2005). Insecticide resistance in *Bemisia tabaci* (Homoptera: Aleyrodidae) populations from Crete. Pest Management Science 61: 577–582.

Roditakis, E., Tsagkarakou, A. and Vontas, J. (2006). Identification of mutations in the para sodium channel of *Bemisia tabaci* from Crete, associated with resistance to pyrethroids. Pesticide Biochemistry and Physiology 85: 161–166.

Roditakis, E., Grispou, M., Morou, E., Kristoffersen, J.B., Roditakis, N., Nauen, R., Vontas, J. and Tsagkarakou, A. (2008). Current status of insecticide resistance in Q biotype *Bemisia tabaci* populations from Crete. Pest Management Science 65(3): 313–322.

Roditakis, E., Morou, E., Tsagkarakou, A., Riga, M., Nauen, R., Paine, M., Morin, S. and Vontas, J. (2011). Assessment of the *Bemisia tabaci* CYP6CM1vQ transcript and protein levels in laboratory and field-derived imidacloprid-resistant insects and cross-metabolism potential of the recombinant enzyme. Insect Science 18: 23–29.

Rolania, K., Janu, A. and Jaglan, R.S. (2018). Role of abiotic factors on the population build-up of whitefly (*Bemisia tabaci*) on cotton. Journal of Agrometeorology 20(Special Issue): 292–296.

Roopa, H.K., Asokan, R., Rebijith, K.B., Hande, R.H., Mahmood, R. and Kumar, N.K.K. (2015). Prevalence of a new genetic group, MEAM-K, of the whitefly *Bemisia tabaci* (Hemiptera: Aleyrodidae) in Karnataka, India, as evident from mt COI sequences. Florida Entomologist 98: 1062–1071.

Rose, M. and DeBach, P. (1992). Biological control of Para*bemisia myricae* (Kuwana) (Homoptera: Aleyrodidae) in California. Israel Journal of Entomology 26: 73–95.

Rosen, R., Kanakala, S., Kliot, A., Pakkianathan, B.C., Farich, B.A., Santana-Magal, N., Elimelech, M., Svetlana Kontsedalov, S., Lebedev, G., Cilia, M. and Ghanim, M. (2015). Persistent, the circulative transmission of begomoviruses by whitefly vectors. Current Opinion in Virology 15: 1–8.

Rote, N.B. and Puri, S.N. (1992). Effect of fertilizer application on the incidence of whitefly on different cotton cultivars. Journal of Maharashtra agricultural university 17: 45–48.

Rowland, M.B., Ackett, B. and Stribley, M. (1991). Evaluation of insecticide in field-control simulators and standard laboratory bioassays against resistant and susceptible *Bemisia tabaci* (Homoptera: Aleyrodidae) from Sudan. Bulletin of Entomological Research 81: 189–199.

Roychoudhury, R. and Jain, R.K. (1993). Neem for the control of aphids and whitefly vector and virus diseases of plants. World Neem Conference (Bangalore, India), Abstract p-31.

Ruan, Y.M., Xu, J. and Liu, S.S. (2006). Effects of antibiotics on the fitness of the B biotype and a non-B biotype of the whitefly *Bemisia tabaci*. Entomologia Experimentalis eT Applicata 121: 159–166.

Ruder, F.J., Guyer, W., Benson, J.A. and Kayser, H. (1991). The thiourea insecticide/acaricide diafenthiuron has a novel mode of action: Inhibition of mitochondrial respiration by its carbodiimide product. Pesticide and Biochemical Physiology 41: 207–219.

Russell, L.M. (1957). Synonyms of *Bemisia tabaci* (Gennadius) (Homoptera; Aleyrodidae). Bulletin Brooklyn Entomological Society 52: 122–123.

Sacchetti, P., Rossi, E., Bellini, L., Vernieri, P., Cioni, P.L. and Flamini, G. (2015). Volatile organic compounds emitted by bottlebrush species affect the behavior of the sweet potato whitefly. Arthropod-Plant Interactions 9: 393–403.

Saeed, F., Haider, S. and Shafiq, M. (2012). Biotypes of Whitefly (*Bemisia tabaci*) in Pakistan: Molecular Study. LAP LAMBERT Academic Publishing.

Safdar, M.Z., Naeem, Mamoon, M. and Niaz, U. (2019). Effect of abiotic factors on population dynamics of whitefly and jassid on Bt cotton. Current Investigations in Agriculture, Current Research (6): 2019. doi. 10.32474/CIACR, 2019.06.000 226.

Salati, R., Nahkla, M.K., Rojas, M.R., Guzman, P., Jaquez, J., Maxwell, D.P. and Gilbertson, R.L. (2002). Tomato Yellow Leaf Curl Virus in the Dominican Republic: characterization of an infectious clone, virus monitoring in whiteflies, and identification of reservoir hosts. Phytopathology 92: 487–496.

Salman, M., Masood, A., Arif, M.J., Saeed, S. and Hamed, M. (2011). The resistance levels of different cotton varieties against sucking pests complex in Pakistan. Pakistan of Journal of Agriculture, Agricultural Engineering and Veterinary Science 27: 168–175.

Sanahuja, E., Banakar, R., Twyman, R.M., Capell, T. and Christou, P. (2011). The *Bacillus thringiensis:* A century of research, development and commercial applications. Plant Biotechnology Journal 9: 287–300.

Sandhu, G.S., Sharma, B.R., Singh, B. and Bhalla, G.S. (1974). Sources of resistance to jassid and whitefly in okra germplasm. Crop Improvement 1: 77–81.

Sangha, K.S., Shera, P.S., Sharma, S. and Kaur, R. (2018). Natural enemies of whitefly, *Bemisia tabaci* (Gennadius) on cotton in Punjab, India. Journal of Biological Control 32(4): 270–274.

Santillan-Ortega, C., Rodríguez-Maciel, J., Lopex-Collados, J., Díaz-Gomez, O., Lagunes-Tejeda, A., Carrillo-Martínez, J., Bernal, J., Robles-Bermúdez, A., Aguilar-Medel, S. and Silva-Aguao,

G. (2011). Susceptibility of females and males of *Bemisia tabaci* (Gennadius) B-biotype and *Trialeurodes vaporariorum* (Westwood) to thiamethoxam. Southwestern Entomologist 36: 167–176.

Santos-Garcia, D., Farnier, P.A., Beitia, F., Zchori-Fein, E., Vavre, F., Mouton, L., Moya, A., Latorre, A. and Silva, F.J. (2012). Complete genome sequence of *"Candidatus Portiera aleyrodidarum"* BT-QVLC, an obligate symbiont that supplies amino acids and carotenoids to *Bemisia tabaci*. Journal of Bacteriology 194: 6654–6655.

Santos-Garcia, D., Latorre, A., Moya, A., Gibbs, G., Hartung, V., Dettner, K., Kuechler, S.M. and Silva, F.J. (2014). Small but powerful, the primary endosymbiont of moss bugs, *Candidatus Evansia muelleri*, holds a reduced genome with large biosynthetic capabilities. Genome Biological Evolution 6: 1875–1893.

Santos-Garcia, D., Vargas-Chavez, C., Moya, A., Latorre, A. and Silva, F.J. (2015). Genome evolution in the primary endosymbiont of whiteflies sheds light Uluon their divergence. Genome Biological Evolution 7: 873–888.

Satar, G., soy, M.R., Nauen, R. and Dong, K. (2018). Neonicotinoid insecticide resistance among populations of *Bemisia tabaci* in the Mediterranean region of Turkey. Bulletin of Insectology 71(2): 171–177.

Sattar, M.S. and Abro, G.H. (2011). Mass rearing of *Chrysoperla carnea* (Stephens) (Neuroptera; Chrysopidae) adults for integrated pest management prorammes. Pakistan Journal of Zoology 43: 483–487.

Sattelle, D.B., Cordova, D. and Cheek, T.R. (2008). Insect ryanodine receptors: molecular targets for novel pest control chemicals. Invertebrate Neuroscience 8: 107–119.

Sayed, T.S., Ahro, G.H., Khuhro, A.D. and Dhaceroo, M.H. (2003). Relative resistance of cotton varieties against sucking pests. Pakistan Journal of Biological Sciences 6: 1232–1233.

Schilmiller, A.L., Charbonneau, A.L. and Last, R.L. (2012). Identification of a BAHD acetyltransferase that produces protective acyl sugars in tomato trichomes. Proc. Natl. Acad. Sci. USA 109: 16377–16382.

Schlaeger, S., Pickett, J.A. and Birkett, M. (2018). Prospects for management of whitefly using plant semiochemicals compared with related pests. Pest Management Science 74: 2405–2411.

Schoeller, E.N. and Redak, R.A. (2020). Climate and seasonal effects on phenology and biological control of giant whitefly *Aleurodicus dugesii* (Hemiptera: Aleyrodidae) with parasitoids in southern California, USA. Biocontrol 65: 559–570.

Schuster, D.J., Stansly, P.A. and Polston, J.E. (1996). Expressions of plant damage by *Bemisia*. pp. 153–165. *In*: Gerling, D. and Mayer, R.T. (eds.). *Bemisia* 1995: Taxonomy, Biology, Damage Control and Management Intercept Ltd. Andover, MA.

Schuster, D.J. (2004). Squash as a trap crop to protect tomato from whitefly-vectored Tomato Yellow Leaf Curl. International Journal of Pest Management 50: 281–284.

Schuster, D.J., Mann, R.S., Toapanta, M., Cordero, R., Thompson, S. and Cyman, S. (2010). Monitoring neonicotinoid resistance in biotype B of *Bemisia tabaci* in Florida. Pest Management Science 66: 186–195.

Schuster, D.J. and Smith, H.A. (2015). Scouting for Insects, Use of Thresholds, and Conservation of Beneficial Insects on Tomatoes This document is ENY685, one of a series of the Entomology and Nematology Department, UF/IFAS Extension. December 2015. http://edis.ifas.ufl.edu.

Seal, S.E., Van Den Boosh, F.C. and Jeger, M. (2006). Factors influencing Begomoviruses evolution and their increasing global significance implications for sustainable control. Critical Reviews in Plant Science 25: 23–46.

Seiter, N. (2018). Integrated Pest Management: What are the economic threshold and how are they developed. Farmdoc Daily, (8): 197, Department of agricultural agronomic economies and University of Illinois-at Urbana campaign. Oct 24, 2018.

Sengonca, C. and Liu, B. (1998). Biological studies on *Eretmocerus longipes* Compere (Hymenoptera; Aphelinidae), a parasitoid of *Alurotuberculatus takahashi* David et Subramaniam (Homoptera: Aleyrodidae) in the laboratory. Journal of Applied Entomology 122. DOI. org/10.1111/j.1439-0418.1998.tb01485.x.

Sethi, A. and Dilawari, V.K. (2008). Spectrum of insecticide resistance in whitefly from upland cotton in Indian Subcontinent. Journal of Entomology 5: 138–147.

Sexena, H.P. (1983). Losses in black gram due to insect pests.- crop losses due to insect pests. Indian Journal Entomology (Special issue): 294–297.

Shadmany, M., Omar, D. and Muhamad, R. (2015). Biotype and insecticide resistance status of *Bemisia tabaci* populations from Peninsular Malaysia. Journal Applied Entomology 139: 67–75.

Shan, H.W., Lu, Y.H., Bing, X.L., Liu, S.S. and Liu, Y.Q. (2014). Differential responses of the whitefly *Bemisia tabaci* symbionts to unfavorable low and high temperatures. Microbiology Ecology 68: 472–82.

Shan, H.W., Zhang, C.R., Yan, T.T., Tang, H.Q., Wang, X.W. and Liu, S.S. (2016). Temporal changes of symbiont density and host fitness after rifampicin treatment in whitefly of the *Bemisia tabaci* species complex. Insect Science 23: 200–214.

Sharma, M. and Budha, P.B. (2015). Host preference vegetables of Tobacco Whitefly *Bemisia tabaci* (Gennadius, 1889) in Nepal. Journal of Institute of Science and Technology 20(1): 133–137.

Sharma, R.K., Singh, S., Kumar, R. and Pandher, S. (2020). Status of insecticide resistance in whitefly, *Bemisia tabaci*. Annals of Entomology 30(2): 113–127.

Sharma, S., Kooner, R. and Arora, R. (2017). Insect pests and crop losses. pp. 45–66. *In*: Arora, R. and Sandhu, S. (eds.). Breeding Insect Resistance Crop for Sustainable Agriculture. Springer, Singapore Pvt Ltd.

Sharma, S.S. and Batra, G.R. (1995). Whitefly outbreak and failure of insecticides in its control in Haryana state: a note. Haryana Journal of Horticultural Sciences 24.2: 160–161.

Sharma, S.S. and Kumar, R. (2014). Influence of abiotic weather parameters in population dynamics of whitefly *Bemisia tabaci* (Genn) in cotton. Journal of Cotton Research and Development 28: 286–288.

Shelka, S.S., Mali, A.R. and Ajri, D.S. (1987). Effect of different schedules of insecticidal sprays on pest incidence, the yield of seed cotton and quality of seed in Laxmi cotton. Current. Research. Report Mahatma Phule Agricultural University 3(2): 3945.

Shera, P.S., Kumar, V. and Aneja, A. (2013). Seasonal abundance of sucking insect pests on transgenic Bt cotton vis-à-vis weather parameters in Punjab, India. Acta Phytopathologica et Entomologica Hungarica 48(1): 63–74.

Shi, P.Q., Wang, L., Liu, Y., An, X., Chen, X.S., Ahmed, M.Z., Qiu, B.L. and Sang, W. (2018). Infection dynamics of endosymbionts reveal three novel localization patterns of *Rickettsia* during the development of whitefly *Bemisia tabaci*. FEMS Microbiology Ecology 94(11): 165. https://doi.org/10.1093/femsec/fiy165.

Shi, X.B., Chen, G., Tian, L.X., Peng, Z.K., Xie, W., Wu, Q.J., Shaoli Wang, S., Zhou, X. and Zhang, Y. (2016). The salicylic acid-mediated release of plant volatiles affects the host choice of *Bemisia tabaci*. International Journal of Molecular Science 17: 1048.

Shukla, A.K., Upadhyay, S.K., Mishra, M., Saurabh, S., Singh, R., Singh, H., Thakur, N., Rai, P., Pandey, P., Hans, A.L., Srivastava, S., Yadav, S.K., Singh, M.K., Kumar, J., Chandrashekar, K., Verma, P.C., Singh, A.P., Nair, K.N., Bhadauria, S., Wahajuddin, M., Singh, S., Sharma, S., Omkar, Upadhyay, R.S., Ranade, S.A., Tuli, R. and Singh, P.K. (2016). Expression of an insecticidal fern protein in cotton protects against whiteflies. Nature Biotechnology 34: 1046–1051.

Sidhu, A.S. and Dhawn, A.K. (1980). Incidence of sucking pests on different varieties of cotton Journal of Research Punjab Agricultural University 17: 152–156.

Silva, L.D., Omoto, C., Bleicher, E. and Dourado, P.M. (2009). Monitoring the susceptibility to insecticides in *Bemisia tabaci* (Gennadius) (Hemiptera: Aleyrodidae) population from Brazil. Neotropical Entomology 38: 116–125.

Silva-Aguao, A., Muscarin, G.M., Castro, R.P.V., Castilho, L.R. and Feifre, D.M.G. (2019). Novel combination of biosurfactant with Entomopathogenic IFAS enhances efficacy against *Bemisia* whitefly. Pest Management Science. https /DOI: Org/10 1002 es 5458.

Simmons, A.M. and Levi, A. (2002). Sources of whitefly (Homoptera: Aleyrodidae) resistance in Citrullus for improvement of cultivated watermelon. Hortscience 37: 581–584.

Simmons, A.M., Kousik, C.S. and Levi, A. (2010). Combining reflective mulch and host plant resistance to sweet potato whitefly (Hemiptera: Aleyrodidae) management in watermelon. Crop Protection 29: 898–902.

Simon, B., Cenis, J.L., Beitia, F., Khalid, S., Moreno, I.M., Fraile, A. and García-Arenal, F. (2003). Genetic structure of field populations of begomoviruses and their vector *Bemisia tabaci* in Pakistan. Phytopathology 93: 1422–1429.

Simon, E.B., Briddon, R.W., Sserubombwe, W.S., Ngugi, K., Markham, P.G. and Stanley, J. (2006). Genetic diversity and phylogeography of Cassava Mosaic Viruses in Kenya. Journal of General Virology 87: 3053–3065.

Singh, H. (2017). Status of insecticide resistance in *Bemisia tabaci* (Gennadius) on Bt kinds of cotton. MSc thesis submitted to Punjab Agricultural University Ludhiana (Unpublished).

Singh, D., Denholm, I., Russell, D., Sharma, N. and Sarao, P.S. (1998). Insecticide resistance pattern in whitefly, *Bemisia tabaci* (Gennadius) from Punjab. *In*: National Level Discussion on Critical Issues of IPM in the Changing Scenario in India. November 28–29, 1998, Department of Entomology, PAU Ludhiana.

Singh, D., Singh, O., Denholm, I., Russell, D., Sohal, B.S., Sarao, P.S. and Sharma, N. (2001a). Monitoring of insecticide resistance and esterase variations in cotton whitefly *Bemisia tabaci* (Gennadius) from Punjab. pp. 22–23. Proceedings National Conference: Plant Protection New Horizons in the Millennium, Maharana Pratap University of Agriculture and Technology, Udaipur.

Singh, J., Butter, N.S. and Madan, V.K. (1983). Assessment of losses due to whitefly in hirsutum cotton. Indian Journal Entomology (Special Issue) 2: 362–366.

Singh, J. and Butter, N.S. (1985). Influence of climatic factors on th build-up of whitefly, *Bemisia tabaci* Genn on cotton. Indian Journal Entomology 47: 359–360.

Singh, P., Beattie, G.A.C., Clift, A.D., Watson, D.M., Furness, G.O., Tesoriero, L., Rajakulendran, V., Parkes, R.A. and Scanes, M. (2000). Petroleum spray oils and tomato integrated pest and disease management in southern Australia. General and Applied Entomology 29: 69–93.

Singh, S.T., Priya, N.G., Kumar, J., Rana, V.S., Ellango, R., Joshi, A., Priyadarshini, G., Asokan, R. and Rajagopal, R. (2012). Diversity and phylogenetic analysis of endosymbiotic bacteria from field-caught *Bemisia tabaci* from different locations of North India based on 16S rDNA library screening. Infections, Genetics and Evolution 12: 411–419.

Singh, T.H. and Butter, N.S. (1997). Present status and future strategies for the management of whitefly, *Bemisia tabaci* Genn in cotton. Journal of Indian Society for Cotton Improvement 22: 80–89.

Sintupachee, S., Milne, J.R., Poonchaisri, S., Baimai, V. and Kittayapong, P. (2006). Closely related Wolbachia strains within the pumpkin arthropod community and the potential for horizontal transmission via the plant. Microbiology Ecology 51: 294–301.

Skaljac, M., Žanić, K., Ban, S.G., Konstedalov, S. and Ghanim, M. (2010). Co-infection and localization of secondary symbionts in two whitefly species. BMC Microbiology 10: 142, 2010/PMID 20482452.

Skaljac, M., Žanić, K., Hrnčić, S., Radonjić, S., Perović, T. and Ghanim, M. (2013). Diversity and localization of bacterial symbionts in three whitefly species (Hemiptera: Aleyrodidae) from the east coast of the Adriatic Sea. B Entomology Research 103: 48–59.

Skaljac, M., Kanakala, S., Zanic, K., Puizina, J., Pieik, I.L. and Ghanim, M. (2017). Diversity and phylogenetic analyses of bacterial symbionts in three insect species from southeast Europe. Insects, 8: 113. Doi 10. 3390 insects 8040.113.

Sloan, D.B. and Moran, N.A. (2012). Endosymbiotic bacteria as a source of carotenoids in whiteflies. Biology Letters 8: 985–989.

Smith, H.A., Koenig, R.L., McAuslane, H.J. and McSouley. (2000). Effect of silver reflective mulch and summer squash trap crop on the severity of immature *Bemisia argentifolii* (Homoptera: Aleyrodidae) on the organic bean. Journal Economic Entomology 93: 726–731.

Smith, H.A., Seijo, T.E., Vallad, G.E., Peres, N.A. and Druffel, K.L. (2015). Evaluating weeds as hosts of Tomato yellow leaf curl virus. Environmental Entomology 44(4): 1101–1107.

Smith, H.A., Nagle, C.A., MacVean, C.M., Vallad, G.E., VanSanten, E. and Hutton, S.F. (2019). Comparing host plant resistance repellent mulches and plant insecticide for the management of *Bemisia tabaci* EAM1 (Hemiptera: Aleyrodidae) and Tomato yellow leaf curl virus. Journal of Economic Entomology 112(1): 236–243.

Smith, H.D., Maltby, H.L. and Jimenez, E.J. (1964). Biological control of the citrus blackfly in Mexico. Technical Bulletin 1311, US Dept. Agric., Washington.

Srinivas, R., Udikeri, S.S., Jayalakshmi, S.K. and Sreeramulu, K. (2004). Identification of factors responsible for insecticide resistance in *Helicoverpa armigera*. Comparative Biochemistry and Physiology Part C Toxicology and Pharmacology 137: 261–269.

Srivastava, A., Mangal, M., Saritha, R.K. and Kalia, P. (2017). Screening of chilli pepper (*Capsicum* spp.) lines for resistance to the begomoviruses chilli leaf curl disease in India. Crop Prot. 100: 177–185.

Sseruwagi, P., Maruthi, M.N., Colvin, J., Rey, M., Brown, J.K. and Legg, J.P. (2006). Colonization of non-cassava plant species by cassava whiteflies (*Bemisia tabaci*) in Uganda. Entomologia Experimentalis et Applicata 119(2): 145–153.

Stacey, D.L. (1977). Banker plant production of *Encarsia formosa* Gahan and its use in control of glasshouse whitefly on tomatoes. Plant Pathology 26: 136–139.

Stadler, E. (1986). Oviposition and feeding stimuli in leaf surface waxes. pp. 105–121. *In*: Jupiner, B.E. and Southwood, T.R.E. (eds.). Insects and the Plant Surface. London: Chapman & Hall.

Stansly, P.A. and McKenzie, C.L. (2007). Fourth International *Bemisia* Workshop International Whitefly Genomics Workshop December 3–8, 2006, Duck Key, Florida, USA.

Stansly, P.A. and Natwick, E. (2010). Integrated Systems for Managing *Bemisia tabaci* in Protected and Open Field Agriculture Biology and Epidemiology of *Bemisia*-vectored Viruses (pp. 467–497).

Stansly, P.A. and Naranjo, S.E. (2010). *Bemisia*: Bionomics and Management of a Global Pest. Springer, New York, USA.

Steiner, M.Y. (1993). IPM practices in greenhouse poinsettia crops in Alberta, Canada. Bulletin OILB/SROP Route de Marseilles - BP 91, Netherlands; International Organization for Biological Control of Noxious Animals and Plants (IOBC/OILB), West Palaearctic Regional Section (WPRS/SROP), 16(8): 133–134.

Sterk, G., Bolckmans, K. and Eyal, J. 1996. A new microbial insecticide, *Paecilomyces fumosoroseus* strain Apopka 97, for the control of the greenhouse whitefly. Brighton Crop Protection Conference: Pests & Diseases – 1996: Volume 2: Proceedings of an International Conference, Brighton, UK, 18–21 November 1996, 461–466.

Streibert, H.T., Drabek, J. and Rindlisbacher, A. (1988). CGA106630-A new type of acaricide/insecticide for the control of the sucking pest complex in cotton and other crops. pp. 25–33. Proceedings Brighton Crop Protection Conference-Pests and Diseases.

Su, Q., Xie, Wang, Wu, Liu, Fang, Xu, B. and Zhang, Y. (2014). The endosymbiont *Hamiltonella* increases the growth rate of its host *Bemisia tabaci* during periods of nutritional stress. PLoS One 9: e89002.

Sukhija, H.S., Butter, N.S. and Singh, J. (1986). Determination of economic threshold of whitefly, *Bemisia tabaci* Genn on American cotton in the Punjab, India. Tropical Pest Management 32: 134–136.

Sukhija, H.S., Butter, N.S., Kular, J.S. and Singh, T.H. (1989). Efficacy of triazophos (Hostathion 40EC) against sucking pests of upland cotton. Journal of Indian Society for Cotton Improvement 14: 1–4.

Summers, C.G. and Stapleton, J.J. (2002). Use of UV reflective mulch to delay the colonization and reduce the severity of *Bemisia argentifolii* (Homoptera: Aleyrodidae) infestations in cucurbits. Crop Protection 21: 921–928.

Summers, C.G., Mitchell, J.P. and Stapleton, J.J. (2004). Management of aphid borne viruses and *Bemisia argentifolii* (Homoptera: Aleyrodidae) in Zucchini squash by using UV reflective plastic and wheat straw mulches. Environmental Entomology 33: 1447–1457.

Summers, C.G., Mitchell, J.P. and Stapleton, J.J. (2004a). Management of aphid borne viruses and *Bemisia argentifolii* (Homoptera: Aleyrodidae) in zucchini squash using UV reflective plastic and wheat straw mulches. Environmental Entomology 33: 1644–51.

Summers, C.G., Mitchell, J.P. and Stapleton, J.J. (2004b). Non-chemical insect and disease management in cucurbit production systems. Acta Horticulturae, 632004-b8: 119–25.

Summers, C.G., Mitchell, J.P. and Stapleton, J.J. (2005). Mulches reduce aphid-borne viruses and whiteflies in cantaloupe. California Agriculture 59: 90–04.

Sundaramurthy, V.T. and Chitra, K. (1992). Integrated pest management in cotton. Indian Journal of Plant Protection 20: 1–17.

Sundaramurthy, V.T. (1992). Upsurge of whitefly, *Bemisia tabaci* Genn, and its management in the cotton ecosystem in India. Paper Presented in Cotton Development in India Jointly Organized by PKV and Vasant Rao Naik, Smurti Pratishthan, Pusad, Nagpur, December 5–6, 1992.

Sundararaj, R. (2014). Species diversity of whiteflies (Aleyrodidae: Homoptera) in India. htt// wwwresearchgate.net/publication 1237116768: 1–12.

Suwwan, M.A., Akkawi, M., Al-Musa. A.M. and Mansour, A. (1988). Tomato performance and incidence of Tomato yellow leaf curl (TYLC) virus as affected by type of mulch. Scientia Horticulturae 37: 39–45.

Taggar, G.K. and Gill, R.S. (2011). A novel technique for screening of black gram against whitefly, *Bemisia tabaci* (Gennadius). Crop Improvement 39: 119–124.

Taggar, G.K., Gill, R.S. and Sandhu, J.S. (2013). Evaluation of black gram (*Vigna mungo* (L.) Hepper) genotypes against whitefly, *Bemisia tabaci* (Gennadius) under screen-house conditions. Acta Phytopathologica et Entomologica Hungarian Journal 48(1): 53–62.

Taggar, G.K. and Gill, R.S. (2016). Host plant resistance in *Vigna* species toward whitefly, *Bemisia tabaci* (Gennadius): a review. Entomologia Generalls 36(1): 1–24.

Takikawa, Y., Matunda, Y., Kakutani, K., Nonomura, T., Kusakari, S., Okuda, K., Kimbara, J., Osamura, K. and Toyoda, H. (2015). Electrostatic insect sweeper for eliminating whiteflies colonizing host plants a complimentary pest control device in an electric field screen greenhouse guarded glasshouse. Insects 6: 442–4445.

Takikawa, Y., Matsuda, Y., Nonomura, T., Kakutani, K., Okada, K., Morikawa, S., Shibao, M., Kusakari, S. and Toyoda, H. (2016). An electrostatic nursery shelter for raising pest and pathogen-free tomato seedlings in an open-window greenhouse environment. Journal of Agricultural Science (Toronto) 8: 13–25.

Tan, X., Hu, N., Zhang, F., Ramirez-Romero, R., Desneux, N., Wang, S. and Ge, F. (2016). Mixed release of two parasitoids and a polyphagous ladybird as a potential strategy to control the tobacco whitefly, *Bemisia tabaci*. Scientific Reports 6(1): 28245.

Tanaka, N., Matsuda, Y., Kato, E., Kokabe, K., Furukawa, T., Nonomura, T., Ken-ichiro, H., Kusakari, S., Imura, I., Kimbara, J. and Toyoda, H. (2008). An electric dipolar screen with oppositely polarized insulators for excluding whiteflies from greenhouses. Crop Protection 27: 215–221.

Tang, X.T., Cai, Li, Shen, Y. and Du, Y.Z. (2018). Diversity and evolution of endosymbionts of *Bemisia tabaci* in China. Paleontology and Evolutionary Science PubMed 30186690.

Tay, W.T., Evans, G.A., Boykin, L.M. and De Barro, P.J. (2012). Will the real *Bemisia tabaci* please stand up? PLoS ONE 7: e50550. 10.1371/journal.pone.0050550.

Tcach, M.A., Spoljaric, M.V., Bela, D.A. and Acuna, C.A. (2019). Joint segregation of high glanding with nectariless and frego bract in cotton. The Journal of Cotton Science 23: 177–181.

Thao, M.L., Baumann, L., Hess, J.M., Falk, B.W., Ng, J.C., Gullan, P.J. and Baumann, P. (2003). Phylogenetic evidence for two new insects associated Chlamydia of the family, Simkaniaceae. Current Microbiology 47: 46–50.

Thao, M.L. and Baumann, P. (2004). Evidence for multiple acquisitions of *Arsenophonus* by whitefly species (Sternorrhyncha: Aleyrodidae). Current Microbiology 48: 140–144.

Thomas, A., Kar, A. and Ramamurthy, V.V. (2014). An analysis of leaf trichome density and its influence on the morphology of dorsal setae in the puparium of *Bemisia tabaci* (Hemiptera: Aleyrodidae) on a single cotton leaf. Indian Journal Entomology 76: 128–131.

Thomas, P., Freman, J.S., buckner, C.-C., Scot apayne, J. and Moore, A. (2008). Ultrastrutural characteristics of *Bemisia tabaci* adult and nymph feeding. Journal of Insect Science 8(1).

Thompson, C.R., Cornell, J.A. and Sailer, R.I. (1987). Interactions of parasites and a hyperparasite in biological control of citrus blackfly, *Aleurocanthus woglumi* (Homoptera: Aleyrodidae), in Florida. Environmental Entomology 16: 140–44.

Thompson, W.M. (2011). The Whitefly, *Bemisia tabaci* (Homoptera: Aleyrodidae) Interaction with Geminivirus-Infected Host Plants: *Bemisia tabaci*, Host Plants and Geminiviruses: Springer Science & Business Media.

Thompson, S.N. (1999). Nutrition and culture of entomophagous insects. Annual Reviews of Entomology 44: 565–592.

Thriveni, K.P. (2019). Correlation of whitefly population with weather parameters and management of Leaf Curl of Chili. Journal of Pharmacognosy and Phytochemistry 8: 4624–4628.

Togni, P.H.B., Laumann, R.A., Medeiros, M.A. and Sujii, E.R. (2010). Odour masking of tomato volatiles by coriander volatiles in host plant selection of *Bemisia tabaci* biotype B. Entomologia Experimentalis et Applicata 136: 164–173.

Tomizawa, M. and Casida, J.E. (2005). Neonicotinoids insecticide toxicology mechanism of selective action. Annual Review of Pharmacology and Toxicology 45: 247–268.

Trehan, K.N. (1944). Further notes on bionomics of *Bemisia gossypiperda* M and L, the whitefly on cotton in Punjab. Indian Journal of Agricultural Sciences 14: 53–63.

Tsagkarakou, A., Mouton, L., Kristoffersen, J.B., Dokianakis, E., Grispou, M. and Bourtzis, K. (2012). Population genetic structure and secondary endosymbionts of q *Bemisia tabaci* (Hemiptera: Aleyrodidae) from Greece. Bulletin Entomological Research 102: 353–365.

Tsueda, H., Tsuduki, T. and Tsuchida, K. (2014). Factors that affect the selection of tomato leaflets by two whiteflies, *Trialeurodes vaporariorum* and *Bemisia tabaci* (Homoptera: Aleyrodidae). Applied Entomology and Zoology 49: 561–570.

Tu, H.T. and Qin, Y.C. (2017). Repellent effects of different celery varieties in *Bemisia tabaci* (Hemiptera: Aleyrodidae) biotype Q. Journal of Economic Entomology 110: 1307–1316.

Ucko, O., Cohen, S. and Ben-Joseph, R. (1998). Prevention of virus epidemics by a crop free period in the Arava region of Israel. Phytoparasitica 26: 313–321.

Ueda, S. and Brown, J.K. (2006). First report of the Q biotype of *Bemisia tabaci* in Japan by mitochondrial cytochrome oxidase I sequence analysis. Phytoparasitica 34: 405. https://doi.org/10.1007/BF02981027.

Ueda, S., Kitamura, T., Kijima, K., Honda, K.-I. and Kanmiya, K. (2008). Distribution and molecular characterization of distinct Asian populations of *Bemisia tabaci* (Hemiptera: Aleyrodidae) in Japan. Journal Applied Entomology 133: 355–366.

Ullah, F., Baloch, A.F. and Badshah, H. (2006). Studies on varietal resistance and chemical control of whitefly (*Bemisia tabaci* Genn.) in cotton. Journal of Biological Sciences 6: 261–264.

Umar, M.S., Arif, M.J., Murtaza, M.A., Gogi, M.D. and Salman, M. (2003). Effect of abiotic factors on the population of whiteflies *Bemisia tabaci* (Gen.) in nectaries and nectariless genotypes of cotton. International Journal of Agriculture and Biology 5: 559–563.

Usta, C. (2013). Microorganisms in biological pest control—A review (Bacterial toxin application and effect of environmental factors) DOI: 10.5772/55786.

Van Giessen, W.A., Mollema, C. and Elsey, K.D. (1995). Design and use of a simulation model to evaluate germplasm for antibiotic resistance to the greenhouse whitefly (*Trialeurodes vaporariorum*) and the sweet potato whitefly (*Bemisia tabaci*). Entomologia Experimentalis et Applicata 75: 271–286.

Van Lenteren, J.C., van Vianen, A., Gast, H.F. and Kortenhoff, A. (1987). The parasite-host relationship between *Encarsia formosa* Gahan (Hymenoptera: Aphelinidae) and *Trialeurodes vaporariorum* Westwood (Homoptera: Aleyrodidae). XVI. Food effects on oogenesis, oviposition, life-span, and fecundity of *Encarsia formosa* and other hymenopterous parasites. Journal of Applied Entomology 103: 69–84.

Varma, S. and Bhattacharya, A. (2015). Whitefly destroys 2/3rd of Punjab cotton crop, 15 farmers commit suicide, Times of India, October 8, 2015.

Vassiliou, V., Emmanouilidou, M., Perrakis, A., Morou, E., Vontas, J., Tsagkarakou, A. and Roditakis, E. (2011). Insecticide resistance in *Bemisia tabaci* from Cyprus. Insect Science 18: 30–39.

Vautrin, E. and Vavre, F. (2009). Interactions between vertically transmitted symbionts: cooperation or conflict? Trends in Microbiology 17: 95–99.

Vega, F.E. (2018). The use of fungal entomopathogens as endophytes in biological control. Annual Reviews of Mycologia 110: 4–30.

Venugopal, R.N. (1987). Seasonal occurrence and management of whitefly, *Bemisia tabaci* Genn in cotton D Thesis, Andhara Pradesh Agricultural University Hyderabad.

Verbeek, M., Petra, J., Van Beckkum, V., Dullemans, A., Rene, A.A. and Vlugt, V.D. (2014). Torradoviruses transmitted in a semi-persistent and Stylet borne manner by thee whitefly vectors. Virus Research 186: 55–60.

Verma, A.K., Mitra, P., Saha, A.K., Ghatak, S.S. and Bajpai, A.K. (2013). Effect of trap crops on the population of the whitefly *Bemisia tabaci* (Genn.) and the diseases transmitted by it. Bulletin of Indian Academy of Sericulture, Unit of Association for Development of Sericulture 17: 37–44.

Vieira, S.S., Bueno, R.C.D.D., Bueno, A.D.F., Boff, M.I.C. and Gobbi, A.L. (2013). Different timings of whitefly control and soybean yield. CiencIa Rural 43(2) Feb. 2013. http//dx DOI./10.159 0i50103-84782013000200009.

Vincent, C., Hallman, G., Panneton, B. and Lessard, F.F. (2003). Management of agricultural insect pests with physical control methods. Annual Reviews of Entomology 48: 261–281.

Viscarret, M.M., Botto, E.N. and Polaszek, A. (2000). Whiteflies (Hemiptera: Aleyrodidae) of economic importance and their natural enemies (Hymenoptera: Aphelinidae, Signiphoridae) in Argentina. Revista Chilena de Entomología 26: 5–11.

Viscarret, M.M., Torres-Jerez, I., Maneo, E.A., Lopez, S.N., Botto, E.E. and Brown, J.K. (2003). Mitochondrial DNA evidence for a distinct New World group of *Bemisia tabaci* (Gennadius)

(Hemiptera: Aleyrodidae) indigenous to Argentina and Bolivia and the presence of the old world B biotype in Argentina. Annals of Entomological Society of America 96: 65–72.

Visser, J.H. (1988). Host-plant finding by insects: orientation, sensory input, and search patterns. Journal of Insect Physiology 34: 259–268.

Visser, J.H. and de Jong, R. (1988). Olfactory coding in the perception of semiochemicals. Journal of Chemical Ecology 14(11): 2005–2018. doi:10.1007/BF01014246.

Von Arx, R., Baurngaertner, J. and Delucchi, V. (1983). Developmental biology of *Bemisia tabaci* (Genn.) (Sternorrhyncha, Aleyrodidae) on cotton at constant temperatures. Bulletin Society of Entomology. Suisse 56: 389–399.

Vyskocilova, S., Seal, S. and Colvin, J. (2019). Relative polyphagy of "Mediterranean" cryptic *Bemisia tabaci* whitefly species and global pest status implications. Journal of Pest Science 92: 1071–1088. https://doi.org/10.1007/s10340-019-01113-9.

Wadhero, H.B., Hussain, T., Talpur, M.A., Rustamani, M.A. and Qureshi, K.H. (1998). Relative resistance of sesame varieties to whitefly and red pumpkin beetle. Pakistan Entomol. 20: 98–100.

Wafaa, A. and Al-Kherb. (2011). Efficacy of some neonicotinoid insecticides on whitefly, *Bemisia tabaci* (Homoptera; Aleyrodidae) and its natural enemies in cucumber and tomato plants in Al-Qassim regions in KSA. Journal of Entomology 8: 429–439.

Walker, G.P. (1988). The role of leaf cuticle in leafage preference by bayberry whitefly (Homoptera: Aleyrodidae) on lemon. Annals of Entomological Society of America 81: 365–369.

Walker, G.P. and Natwick, E. (2006). Resistance to Silverleaf whitefly, *Bemisia argentifolii* (Hem., Aleyrodidae), in *Gossypium thurberi*, a wild cotton species. Journal of Applied Entomology 130(8): 429–436.

Walker, G.P., Perring, T.M. and Freeman, T.P. (2010). Life history, functional anatomy, feeding, and mating behavior. pp. 109–160. *In*: Stansly, P.A. and Naranjo, S.E. (eds.). *Bemisia*: Bionomics and Management of a Global Pest. Springer, New York, USA.

Walling, L.L. (2000). The myriad plant responses to herbivores. Journal of Plant Growth Regulation 19: 195–216.

Walling, L.L. (2008). Avoiding effective defenses: strategies employed by phloem-feeding insects. Plant Physiology 146(3): 859–866. doi: 10.1104/pp.107.113142Wan F.

Wang, Z., Liu, Y., Shi, M., Huang, J.H. and Chen, X. (2019). Parasitoid wasps as effective biological control agents. Journal of Integrative Agriculture 18: 705–715.

Wang, J.R., Song, Z.Q. and Du, Y.Z. (2014). Six new records of whiteflies (Hemiptera; Aleyrodidae) infesting *Morus alba* in China. Journal of Insect Science 14(274). DOI:10.1093/jisesa/leu136.

Wang, K.Q., Li, X.M., Liu, C.L. and Xu, G.Q. (2000). Preliminary study on controlling *Trialeurodes vaporariorum* with toxin extracted from *Verticillium lecanii*. Plant Protection 26(4): 44–45.

Wang, R., Fang, Y., Mu, C., Qu, C., Li, F., Wang, Z. and Luo, C. (2017a). Baseline susceptibility and cross-resistance of cycloxaprid, a novel cis-nitro methylene neonicotinoid insecticide, in *Bemisia tabaci* MED from China. Crop Protection. https://doi.org/10.1016/j.cropro.2017.02.012.

Wang, S., Zhang, Y., Yang, X., Xie, W. and Wu, Q. (2017b). Resistance monitoring for eight insecticides on the sweet potato whitefly (Hemiptera: Aleyrodidae) in China. Journal of Economic Entomology 110: 660–666.

Wang, Y., Gao, N., Shi, L., Qin, Y., He, P., Hu, D.Y., Tan, X.F. and Chen, Z. (2015). Evaluation of the attractive effect of coloured sticky traps for *Aleurocanthus spiniferus* (Quaintance) and its monitoring method in tea garden in China. DOI: 10.4081/jear.2015.4603.

Wang, L.D., Huang, J., You, M.S., Guan, X. and Liu, B. (2007). Toxicity and feeding deterrence of crude toxin extracts of *Lecanicillium* (*Verticillium*) *lecanii* (Hyphomycetes) against sweet potato whitefly, *Bemisia tabaci* (Homoptera: Aleyrodidae). Pest Management Science 63: 381–387.

Wang, R., Li, F., Zhang, W., Zhang, X., Qu, C., Tetreau, G., Sun, L., Luo, C. and Zhou, J. (2017). Identification and expression profile analysis of odorant-binding protein and chemosensory protein genes in *Bemisia tabaci* MED by head transcriptome. PloS One 12(2): e0171739. https://doi.org/10.1371/journal.pone.0171739.

Wang, Z., Yao, M. and Wu, Y. (2009). Cross-resistance, inheritance, and biochemical mechanisms of imidacloprid resistance in B-biotype *Bemisia tabaci*. Pest Management Science 65: 1189–1194.

Wang, Z., Yan, H., Yang, Y. and Wu, Y. (2010). Biotype and insecticide resistance status of the whitefly *Bemisia tabaci* from China. Pest Management Science 66: 1360–1366.

Wanghoonkong, S. (1981). Chemical control of cotton pests in Thailand. Tropical Pest Management 27: 495–500.

Wardlow, L.R., Ludlam, A.B. and French, N. (1972). Insecticide resistance in glasshouse whitefly. Nature 239: 164–65.

Wardlow, L.R., Ludlam, A.B. and Bradley, L.F. (1976). Pesticide resistance in glasshouse whitefly, *Trialeurodes vaporariorum*. Pesticide Science Banner 7(3): 320–324.

Watanabe, L.K.M., Bele, V.H., De Marchi, B.B., De Silva, F.B., Fusco, L.M., Sartori, M.M.P., Pavan, M.A., Sartori, M.M.P. and Sakate, R.K. (2019). Performance and competitive displacement of *Bemisia tabaci* MEAM1 and MED cryptic species on different hosts. Crop Protection 124: 104860.

Watve, C.M. and Lienk, S.E. (1976). Resistance to carbaryl and six organophosphorus insecticides of *Amblyseius fallacis* and *Typhlodromus* pyri from New York apple orchards. Environmental Entomology 5: 368.

Weeks, A.R., Velten, R. and Stouthamer, R. (2003). Incidence of a new sex–ratio–distorting endosymbiotic bacterium among arthropods. Proceedings of Royal Society of London B Biological Sciences 270: 1857–1865.

Weintraub, P.G. and Horowitz, A.R. (2001). Vacuuming Insect Pests: The Israeli Experience, Physical Control Measures in Plant Protection, pp. 294–302, Springer ISBN 978-3-662-0586-2.

Weintraub, P.G. (2009). Physical control: An important tool in pest management programs. pp. 317–324. *In*: Ishaaya, I. and Horowitz, A. (eds.). Biorational Control of Arthropods Pests. Springer, Dordrecht.

Weintraub, P.G. and Berlinger, C. (2009). Physical control in greenhouses and field crops. Insect Pest Management, pp. 301–318.

Weintraub, P.G., Racht, E., Mondaca, L.L., Harari, A.R., Diaz, B.M. and Bennison, J. (2017). Arthropod pest management in organic vegetables in greenhouses. Journal of Integrated Pest Management 8(29): 1–14.

Were, H.K., Winter, S. and Maiss, E. (2007). Characterisation and distribution of cassava viruses in Kenya. African Crop Sciences Conference Proceedings 8: 909–912.

Williams, G. and Pat. (1995). Oil, soap, surfactant, and garlic vs. whiteflies on tomatoes. HortIdeas. May. pp. 55–56.

Wisler, G.C., Li, R.H., Liu, H.Y., Lowry, D.S. and Duffus, J.E. (1998). Tomato chlorosis virus: a new whitefly-transmitted, Phloem-limited, bipartite closterovirus of tomato. Phytopathology 88(5): 402–9.

Wraight, S.P., Carruthers, R.L., Jaronski, S.T., Bradley, C.A., Garza, C.J. and Wraight, S.G. (2000). Evaluation of Entomopathogenic fungi *Beauveria bassiana* and *Paecilomyces fumosoroseus* for microbial control of Silver whitefly, *Bemisia argentifolii*. Biological Control 17: 203–217.

Xie, M., Xie, W. and Wu, G. (2011). Effects of temperature on the growth and reproduction characteristics of *Bemisia tabaci* B-biotype and *Trialeurodes vaporariorum*. Journal of Applied Entomology 135: 252–257.

Xie, W., Liu, Y., Wang, S., Wu, Q., Pan, H., Yang, X., Guo, L. and Zhang, Y. (2014). The sensitivity of *Bemisia tabaci* (Hemiptera: Aleyrodidae) to several new insecticides in China: effects of insecticide type and whitefly species, strain, and stage. Journal of Insect Science 14: 261.

Xu, W.H., Guo, R.Z., You, J.Z., Quing, J.W., Bao, Y.X. and Gui, L.L. (2003). Analysis of the life table parameters of *Bemisia tabaci* feeding on seven species of host plants. Entomology Knowledge 40: 453–455.

Yadav, S.K., Archana, Singh, R., Singh, P.K. and Vasudev, P.G. (2019). Insecticidal fern protein Tma12 is possibly a lytic polysaccharide monooxygenase. Planta 249(6): 1987–1996. doi: 10.1007/s00425-019-03135-0. Epub 2019 Mar 22.

Yang, C. and Everitt, J.H. (2011). Remote sensing for detecting and mapping whitefly *Bemisia tabaci* (Homoptera: Aleyrodidae) infestations. pp. 137–381. *In*: Thompson, W.H.O. (ed.). The Whitefly, *Bemisia tabaci* (Homoptera Aleyrodidae) Interactions with Geminivirus Infested Plants *Bemisia tabaci*, Host Plants and Geminiviruses. Springer Doredrecht Heidelberg, London, New York.

Yang, T.C. and Chi, H. (2006). Life tables and development of *Bemisia argentifolii* (Homoptera: Aleyrodidae) at different temperatures. Journal of Economic Entomology 99: 691–698.

Yang, N.W., Li, A.L., Wan, F.H., Liu, W.X. and Johnson, D. (2010). Effects of plant essential oils on immature and adult sweet potato whitefly, *Bemisia tabaci* biotype B. Crop Protection 29: 1200–1207.

Yasarakinci, N. and Hincal, P. (1996). Izmir'de örtüaltinda yetistirilen domateslerde bulunan ana zararlilarin (*Trialeurodes vaporariorum* (Westw.)), *Bemisia tabaci* Gern ve Liriomyza spp.

populasyon gelismesi üzerinde arastirmalar. Türkiye III. Entomoloji Kongresi Bildirileri 24–28 Eylül 1996, Ankara: 150–157.

Yasui, M., Fukada, M. and Maekawa, S. (1985). Effect of buprofezin on different developmental stages of greenhouse whitefly, *Trialeurodes vaporariorum* (Westwood) (Homoptera; Aleyrodidae). Agricultural Entomology and Zoology 20: 340–347.

Yokomi, R.K., Hoelmer, K.A. and Osborne, L.S. (1990). Relationship between the sweetpotato whitefly and the squash Silverleaf disorder. Phytopathology 80: 895–900.

Yousaf, M.J., Nadeem, I., Niaz, T., Ahmed, R., Akhtar, M.F. and Raza, A. (2015). The response of some new cotton genotypes against insect pests complex and cotton leaf curl virus. Journal of Entomology and Zoology Studies 3(2): 211–214.

Youssef, H.I., Hady, S.A. and Ismail, H. (2001). Intercropping pattern against *Bemisia tabaci* (Gennadius) infestation (Homoptera: Aleyrodidae). Annals of Agricultural Science, Moshtohor 39: 651–654.

Yu, G., Nguyen, T.T., Guo, Y., Schauvinhold, I., Auldridge, M.E., Bhuiyan, N., Ben-Israel, I., Iijima, Y., Fridman, E., Noel, J.P. and Pichersky, E. (2010). Enzymatic functions of wild tomato methyl ketone synthases 1 and 2. Plant Physiology 154: 67–77.

Yuan, L., Wang, S., Zhou, J., Du, Y., Zhang, Y. and Wang, J. (2012). Status of insecticide resistance and associated mutations in Q-biotype of whitefly, *Bemisia tabaci*, from eastern China. Crop Protection 31: 67–71.

Zakhidov, F.M. (2001). Whitefly in Uzbekistan. Zashchita I Karantin Rastenii, No. 11: 21.

Zang, L.S., Chen, W.Q. and Liu, S.S. (2006). Comparison of performance in different host plants between B biotype and a nonbiotype of *Bemisia tabaci* from Zhejiang, China. Entomologica Experimentalis et Applicata 121: 221–227.

Zanik, K., Dumicic, D., Mandusic, M., Selak, G.W., Bocina, I., Popovic, V.B. and Ban, S.G. (2018). *Bemisia tabaci* MED population density and affected by rootstock in modified leaf anatomy and amino acid profiles and hydroponically grown tomato. Frontiers of Plant Science 9: 88. DOI:10. 3889 /fpls, 201. 00086.

Zarate, S.I., Kempema, L.A. and Walling, L.L. (2007). Silverleaf whitefly Induces salicylic acid defenses and suppresses effectual jasmonic acid defenses. Plant Physiology 143: 866–875.

Zchori-Fein, E., Gottlieb, Y., Kelly, S.E., Brown, J.K., Wilson, J.M., Karr, T.L. and Hunter, M.S. (2001). A newly discovered bacterium associated with parthenogenesis and a change in host selection behavior in parasitoid wasps. Proceedings of National Academy of Sciences, USA 98: 12555–12560.

Zchori-Fein, E. and Brown, J.K. (2002). Diversity of prokaryotes associated with *Bemisia tabaci* (Gennadius) (Hemiptera: Aieyrodidae). Annals of the Entomological Society of America 95: 711–718.

Zchori-Fein, E. and Perhnam, S.J. (2004). Distribution of the bacterial symbiont Cardinium in arthropods. Mol. Ecol. 13: 2009–2016. 10.1111/j.1365-294X.2004.02203.x.

Zchori-Fein, E., Lahav, T. and Freilich, S. (2014). Variations in the identity and complexity of endosymbiont combinations in whitefly hosts. Frontiers of Microbiology 5: 310. 10.3389/fmicb.2014.00310.

Zeshan, M.A., Khan, M.A., Ali, S. and Arshad, M. (2015). Correlation of conducive environmental conditions for the development of whitefly, *Bemisia tabaci* population in different tomato genotypes. Pakistan Journal of Zoology 47: 1511–1515.

Zhang, C.-Y. and Liu, T.-X. (2016). Leaf morphological characters can be a factor for intra-varietal preference of whitefly *Bemisia tabaci* (Hemiptera: Aleyrodidae) among eggplant varieties. PLoS ONE 11(4): e0153880. doi:10.1371/journal.

Zhang, G., Liu, S., Luo, C., Chu, D., Zhang, Y.J., Zang, L.S., Jiu, M., Lü, Z.C., Cui, X.H., Zhang, L., Zhang, F., Zhang, Q.W., Liu, W.X., Liang, P., Lei, Z.R. and Zhang, Y.J. (2009). Invasive mechanism and management strategy of *Bemisia tabaci* (Gennadius) biotype B: Progress report of 973 Program on invasive alien species in China. Science China Series C-Life Sciences 52: 88–95.

Zhang, G., Li, D., Liu, T., Wan, F. and Wang, J. (2011). Interspecific interactions between *Bemisia tabaci* biotype B and *Trialeurodes vaporariorum* (Hemiptera: Aleyrodidae). Environmental Entomology 40: 140–150.

Zhang, P.J., Ku, C.T., Jhang, J.M., lu, V.B., Wei, J.N., Liu, Y.Q., David, A., Boland, W. and Turlings, T.C.J. (2013). Phloem feeding whiteflies can fool their host plants, but not their parasitoids. Functional Ecology 27: 1304–1312.

Zhang, P.J., Wei, J.N., Zhao, C., Zhang, Y.F., Li, C.U.Y., Liu, S.S., Dicke, M., Yu, X.P. and Turlings, T.C.J. (2019). Air-borne host–plant manipulates by whiteflies via an inducible blend of plant volatiles. PNAS 116(115): 7387–7396.

Zhang, X., Lv, J., Hu, Y., Wang, B., Chen, X., Xu, X. and Wang, E. (2015). Prey preference and life table of Amblyseius orientalis on *Bemisia tabaci* and *Tetranychus cinnabarinus*. PLoS One 10(10): e0138820.

Zhang, X., Ferrante, M., Wan, F., Yang, N., Gabor, L. and Lovei, G.L. (2020). The parasitoid *Eretmocerus hayati* is compatible with barrier cropping to decrease whitefly (*Bemisia tabaci* MED) densities on cotton in China. Insects 11(57).

Zhang, X.M., Wang, S., Li, S., Luo, C., Li, Y.X. and Zhang, F. (2015). Comparison of the antennal sensilla ultrastructure of two cryptic species in *Bemisia tabaci*. PLoS One 10: e0121820.

Zhang, Y., Kanakala, S. and Ghanim, M. (2019a). Global genetic diversity and geographical distribution of *Bemisia tabaci* and its bacterial endosymbionts. PLoS One 14(3): e0213946. March 2019, DOI: 10.1371/journal.pone.0213946.

Zhao, Q., Zhu, J.J., Qin, Y., Pan, P., Tu, H., Du, W., Zhou, W. and Baxendale, F.P. (2014). Reducing whiteflies on cucumber using intercropping with less preferred vegetables. Entomologia Experimentalis et Applicata 150: 19–27.

Zheng, H., Xie, W., Wang, S., Wu, Q., Zhou, X. and Zhang, Y. (2017). Dynamic monitoring (B versus Q) and further resistance status of Q-type *Bemisia tabaci* in China. Crop Protection 94: 115–122.

Zhu, D.T., Wang, X.R., Ban, F.X., Zhu, C., Liu, S.S. and Wang, X.W. (2017). Methods for the extraction of endosymbionts from the whitefly, *Bemisia tabaci*. Journal of Visualised Experiments 19(124) Doi: 10.379.

Zhu-Salzman, K., Salzman, R.A., Ahn, J.E. and Koiwa, H. (2004). Transcriptional regulation of sorghum defense determinants against a phloem-feeding aphid. Plant Physiology 134: 420–431.

Zia, K., Hafeez, F., Khan, R.R., Arshad, D.M. and Naeem-Ullah, U. (2008). Effectiveness of *Chrysoperla carnea* (Stephens) (Neuroptera: Chrysopidae) on the population of *Bemisia tabaci* (Homoptera: Aleyrodidae) in different cotton genotypes. Journal of Agriculture & Social Science ISSN Print: 1813–2235; ISSN Online: 1814–960X 07–423/SAE/2008/04–3–112–116. http://www.fspublishers.org.

Ziam K,, Ashfaqm M,, Arifm M.J. and Sahi, S.T. (2011). Effect of physical-morphic characters on the population of whitefly *Bemisia tabaci* in transgenic cotton. Pakistan Journal of Agricultural Sciences 48(1): 63–69.

Zolnerowich, G. and Rose, M. (1996). A new species of *Entedonon ecremnus* (Hymenoptera: Chalcidoidea: Eulophidae) parasitic on the giant whitefly, *Aleurodicus dugesii* Cockerell (Homoptera: Aleyrodidae). Proceedings of Entomological Society Washington 98: 369–713.

Subject Index

4-leaf stage 78

A

Acephate 56, 58, 63, 66, 68, 71, 153
Acyl sugars 84, 113
Aleurodicinae 4, 5, 42
Aleurotrachelus socialis 45, 85, 88, 126
Aleyrodes 1, 2, 5, 10, 12, 17, 112
Aleyrodidae 1, 4, 6, 7, 12, 17, 19, 25, 38, 40,
    42–45, 62, 84, 88, 89, 103, 114, 136
Aleyrodoidea 4, 6, 7
Allelochemicals 34, 111, 130
Allomones 111, 112, 130
Allozyme marker 37
Alteration of sowing/harvesting time 122
Aluminum mulches 109, 110, 118, 156
Amblyseius swirskii 95, 101, 105, 129, 155
Antibiosis 75, 80, 85–90, 113
Antixenosis 75, 80, 85, 86, 88
Apanteles ruficrus 98
Arabidopsis thaliana 113, 115
Arsenophonus 31–33, 38, 40–46, 140
Aschersonia 102, 103, 107, 151, 156

B

Barrier crops 125–127, 148
Basil 43, 87, 111, 114, 130, 131, 156
B-biotype 5, 6, 11, 12, 26–30, 34–39, 41, 45, 46,
    52, 65, 66, 69, 70, 88, 90, 99, 102, 104, 106,
    113, 124, 136, 140, 152
Beauveria 101–103, 105–107, 129, 131, 151, 156
Begomovirus 1, 23, 62, 138, 140, 141, 143
Bemisia argentifolii 1, 5, 25, 27, 29, 34, 37, 39,
    41, 46, 51, 80, 88, 90, 96, 99, 105, 111,
    113, 152
Bemisia tabaci 1, 2, 4–6, 8–14, 16–18, 20, 23,
    25–39, 41–46, 51, 52, 55, 57–59, 62, 64–67,
    69–73, 80, 84–86, 88–90, 93–106, 110–114,
    119, 120, 123–125, 132–136, 138–143, 149,
    151, 152

Biotechnology 90
Biotype A 1, 38, 45
Biotype Q 1, 5, 6, 29, 34, 36–39, 43, 45, 46, 65,
    66, 69–71, 84, 90, 114, 140
Biotypes 1, 6, 25–31, 34–39, 41, 43–46, 64, 65,
    70, 90, 114, 140, 141
Blackgram 86
Blasting 120, 121, 157

C

Candidatus 31, 42, 44
Carbamates 55, 58, 67–69, 71, 73, 117, 145, 151
Cardinium 30–33, 38, 40–46, 140
Cassava 13, 16, 23, 28, 29, 36–38, 43, 50, 52,
    76–78, 85, 86, 92, 95, 103, 126, 140, 143,
    154
Chaperone 140, 141
Chlorinated hydrocarbons 55, 67–69, 73
Chrysoperla carnea 94, 96, 106, 107, 119, 153,
    155
Citronella grass 114
Closterovirus 23, 138, 140–142, 144
Coat protein 140
Collateral glands 38, 42
Clonostachys 102, 105, 106, 151
Compound eyes 7, 35
Coriander 114, 126
Correlation coefficient 81, 83, 135
Cotton 1–5, 12, 14, 18, 20–24, 27, 29, 37, 41,
    43, 45, 48, 50–56, 58–64, 67, 69–74,
    76–78, 80–83, 85–92, 95, 98, 104, 107, 118,
    120–125, 131–136, 142, 143, 147–149, 151,
    152, 154, 155, 157
Crop free period 123, 126, 127, 146
Crop Geometry 122, 155
Crop sanitation 123, 127
Cryptic species 25, 29–31, 33, 35, 64
Cucumber vein Yellowing Virus 138
Cucurbit Yellow Stunting Disorder Virus 138,
    141, 142, 144
Cultural measures 122, 127, 130, 143, 146, 148,
    150, 153, 154, 158, 159

D

Damage 2, 3, 10, 18, 20–25, 48, 50, 52, 60, 62, 75–77, 79, 90, 91, 102, 118, 124, 126, 131, 143
Deltamethrin 55, 58, 59, 63, 65, 67, 68, 71, 145
Deterrent 55, 74, 94
Diafenthiuron 66, 68, 70, 73, 157
Dihydroxyphenols 83
Dill seed 130, 131

E

Economic damage 50
Economic injury level 50, 120
Edible fern 83, 91, 147, 154
Electrostatic Insect Sweeper 118, 119, 156
Empodium 7
Encarsia 93, 94, 96–101, 112, 113, 115, 119, 129, 130, 153, 155–157
Endosymbionts 14, 15, 25, 26, 30–35, 37–46, 140, 141
Entomopathogenic organism 93, 101, 103
Economic threshold 47–53, 98, 149, 152, 153, 158
Eradication of alternate hosts 126, 154
Eretmocerus 93, 94, 96–101, 103, 119, 129, 130, 153, 155, 156
Ethylene 115

F

Fagopyrum esculentum 110
Farnesene 111
Fenvalerate 55, 58, 59, 62, 68, 73, 120, 145
Flavonols 82, 83
Fluvalinate 56, 58, 59
Fritschea 31–33, 38, 40, 41, 43–46, 140

G

Geminivirus 23, 24, 63, 143, 144, 154
Glandular trichomes 80, 83, 84, 89, 115
Greengram 76

H

Hair density 80–82, 85, 89
Hair length 81, 82, 85
Hamiltonella 30–33, 37, 38, 40–46, 140
Heat shock protein 70 140
Hirsutella 101, 106, 151
Horizontal transmission 43
Host plant Resistance 75, 86, 91, 158

I

Imidacloprid 37, 42, 57, 58, 63–72, 110, 146, 151, 153
Insect growth regulators 50, 55–57, 60, 64, 72–74, 90, 145–147, 150, 151, 153, 155, 157
Inter crop 61, 126, 154
Invasive species 32, 33, 35, 65, 158
Irregular ripening 23, 90

J

Jasmonic acid 115

K

Kairomones 100, 111–114, 130

L

Leaf lamina 47, 78, 80–85, 88–90, 141, 147
Leaf thickness 81, 85, 89
Lecanicillium 94, 101–103, 105, 106, 151
Lingula 6, 7, 35
Live mulches 110
Losses 4, 20, 22, 23, 26, 54, 56, 90, 125
Lipophilic molecules 113
Lytic Polysaccharide 83, 91

M

Marigold 87, 126, 130, 131
Mediterranean 1, 6, 20, 26–28, 33, 36–38, 69, 70, 90
Metarhizium 101–103, 105–107, 131, 151, 156
Mitochondrial 16s 27, 34
Mixed crop 126, 127, 143, 155
Modification of habitat 126, 127
monooxygenases 65, 66, 71
Mulches 109–111, 116, 118, 121, 130, 131, 147
Myrcene 113–116

N

Neem oil 104, 105, 121, 130, 131, 147, 152, 153
Nepovirus 23
Nitrogen 81, 83, 84, 115, 124
Non-preference 47, 75, 80, 83, 86, 87, 124
Nuvaluron 72, 146

O

Oils 49, 58, 73, 104, 105, 108, 114, 115, 118, 120, 121, 128–131, 147, 150, 152, 153, 156
Olfactory 7, 9, 112

Orentia-like organism 38, 140
Organophosphates 58, 63, 70
Outbreak 1, 14, 22, 23, 55, 59, 62, 63, 68, 77, 91, 123, 133, 136–138

P

Paecilomyces 101–103, 105–107, 129, 151, 156
Parabemisia myricae 4, 5, 16, 100, 138, 139, 141
Parasitoids 19, 34, 39, 43, 73, 93–101, 106, 107, 112, 113, 115, 119, 121, 125, 127, 129–131, 147, 149, 155–158
Parsley 130, 131
Parthenogenetic 2, 16, 59, 60, 73
Pesticide resistance 15, 34, 35, 37, 41, 59, 60, 64, 68, 109, 110, 116, 152
Pesticide resurgence 54, 58–60
Pesticides 1, 15, 20, 31, 4, 35, 37, 38, 40, 41, 46, 47, 49, 50, 52–61, 63–65, 67–74, 77, 91, 93, 98, 102–107, 109, 110, 112, 116, 118, 120–123, 126–131, 133, 145–147, 149–159
Phenols 47, 81–83, 91, 113, 147
Phosphorus 47, 81–83, 91, 124, 147, 148, 155
Physical measures 117, 119, 121, 131, 149, 150
Poikilothermic 10, 16, 17, 23, 133, 145
Polyacrylamide Gel Electrophoresis (PAGE) 35
Polymerase chain reaction 6, 34
Polyphagous 1, 9, 14, 18, 20, 36, 61, 62, 64, 73, 125, 154
Portiera 30–33, 38, 40–46, 140
Portiera aleyrodidarum 31, 32, 38, 40–42, 44
Potyvirus 23
Predators 93–96, 100, 105–107, 111, 119, 129–131, 136, 147, 149, 153, 155, 156, 158
Pretarsus 7
Primary Endosymbionts 31, 40–42, 44–46
Primary symbionts 140
Pseudo-pupae 2, 17, 98, 99, 107
Putative species 27, 28, 35, 70
Pyriproxyfen 42, 57, 64–68, 71, 72, 146, 153, 155, 157

R

Rainfall 9, 82, 125, 133–136
Ratio 14, 17, 59, 63, 68, 71, 85, 88, 99, 115, 126, 130
Relative humidity 16, 23, 132–136
Resistance 15, 17, 34, 35, 37, 39–42, 54–60, 63–86, 88–91, 98, 104, 109, 110, 112, 113, 116, 121, 122, 124, 125, 129, 145, 147, 148, 150–153, 155–158
Resurgence 54–56, 58–60, 62, 63, 73, 91, 121, 127
Rickettsia 30–32, 37–46, 140

S

Salicylic acid 115
Sampling 47–50, 52, 53, 152
Secondary Endosymbionts 31, 38–41, 43–46, 140
Secondary symbiont 31, 44, 46
Semiochemicals 83, 100, 109, 111–114, 116, 130, 149
Sensilla 7, 9, 112
Setae 6, 15
Soap 58, 105, 108, 118, 128, 129, 131, 147, 150, 153, 156, 158
Soybean 13, 29, 50, 52, 60, 76, 80, 87, 91, 92, 105, 129, 153, 154
Stylet-borne virus 110, 126, 139, 156
SubSahara Africa 37
Sucrose Gene Alpha Glucosidase 116
Synthetic mulches 110
Synthetic pyrethroids 55, 56, 58, 59, 62, 63, 65, 67–69, 71, 73, 77, 145, 151

T

Tannin 47, 81–83, 91, 147, 148
Tectaria macrodonta 91, 147
Temperature 9, 11, 12, 16, 17, 19, 20, 23, 38, 40, 42, 60, 84, 95, 106, 107, 121, 132–136, 145
Thiamethoxam 37, 42, 57, 65–68, 70–72, 146, 151, 153
Tma 12 147
Tissue culture 116
Tolerance 38–40, 42, 75, 80, 85, 111, 136
Tomato Torrado Virus 138–142, 144
Transgenic 20, 67, 83, 91, 111, 116, 147, 148, 150, 151, 153, 154
Trap crop 111, 122, 126, 127, 146, 148, 155
Trialeurodes abutilonea 1, 4–6, 18, 23, 51, 96, 99, 132, 138, 139, 141, 142
Trialeurodes ricini 5, 19, 20, 136, 138, 139, 141
Trialeurodes vaporariorum 1, 2, 5, 6, 10, 11, 15–18, 20, 23, 27, 34, 35, 38, 39, 41, 43–45, 51, 52, 57, 58, 65, 66, 71, 73, 88, 90, 93–99, 102, 105, 106, 112–116, 124, 132, 136, 138, 139, 141, 142
Triazophos 55, 56, 58, 60, 63, 66, 68, 73, 145, 153
Trichogramma 98, 119
Trichomes 9, 80, 81, 83, 84, 88, 89, 115, 147

U

Udamoselinae 4
UV Wavelength 110

V

Vacuuming 118, 130, 158
Vascular bundles 88–90
Vasiform orifice 3, 4, 6, 7, 9, 35
Vertical transmission 43
Verticillium 101–103, 105, 151, 156

W

Wax glands 4, 38, 42
Weather parameters 17, 23, 107, 132–136

Whitefly resistant Index 79
Wind Speed 133, 134, 136
Wolbachia 30–33, 38, 40, 41, 43–46, 140

Y

Yellow Sticky Traps 48, 49, 99, 115, 120, 121,
          129–131, 145–147, 150, 153, 156, 158

About the Authors

Dr. N.S. Butter

Professor Nachhattar Singh Butter was born to a rural family in a small village, Hakam Singh Wala, District Bathinda, Punjab, India on 13 October 1948. After attaining his school education in 1967, he joined the Punjab Agricultural University, Ludhiana, and obtained his BSc (Agriculture) in 1971, MSc (Entomology) in 1973, and Ph.D. (Entomology) degrees in 1976. Being a meritorious student, he won a merit scholarship for his bachelor's and master's degrees and was awarded the Senior Fellowship, Council of Scientific and Industrial Research during his Ph.D., which was converted into a post-doctorate fellowship after the completion of his doctorate in 26 November 1976. Dr. N.S. Butter served in this position up till 23 February 1977. The fellowship was relinquished on taking on the post of Assistant Entomologist where he continued to work till 30 June 1987. Dr. Butter got promoted as Entomologist (Cotton) and continued to work in that post till 30 March 1994. Dr. Butter was selected as Professor of Plant Protection and joined that post on 31 March 1994, after which he was elevated to the post of Head, Department Entomology on 20 June 2006, where he discharged the administrative duties till his superannuation on 31 October 2008. During this period, Dr. Butter worked on cotton insect pests and made several recommendations for their management, guided nine postgraduate students in entomology, and taught both under-graduate and post-graduate courses in entomology, delivered special lectures in seminars/ symposia, delivered forty-two TV/radio talks on pest management, established a plant clinic at the university level (this model was emulated in the country). Dr. Butter has published 250 papers, including 85 research papers extension articles, books, book chapters, review articles in journals of repute to his credit. Dr. Butter is a Fellow of the Entomological Society of India and the Indian society for the Advancement of Insect Science. As an administrator, he has organized seminars on important topics, and conducted workshops and review meetings of research teams/projects. Besides this, he has helped in upgrading the infrastructure facilities in the Department of Entomology. Considering his outstanding research contributions in agricultural research in general and entomology in particular, Prof. Butter was honored with the 'Hexamar Award' given by the Indian Society for Cotton Improvement, in Mumbai, 1992, the 'Punjab Sarkar Parman Patra, 2002' by the state government of Punjab, and 'The Lifetime Achievement Award' by the Cotton Research and Development Association of India in 2018. Dr. Butter has also visited the erstwhile USSR as a member of a two man delegation to study integrated pest and disease management in cotton from 4th–24th July 1987.

Dr. A.K. Dhawan

Dr. A.K. Dhawan was the former Additional Director of Research and Head Department of Entomology at Punjab Agricultural University Ludhiana. After obtaining his postgraduate and doctorate degrees from PAU, joined the same Institute as an Assistant Entomologist in 1973 and continued to work on the management of cotton pests and made more than seventy-one recommendations to benefit farmers. He won the ICAR Team Award 2004, National Award for Transfer of Technology for Street Play and PAU Best Extension Worker Award 2008, and Award of Honor (2008). Dr. Dhawan also served as the State Coordinator for IPM (cotton) from 2004 to 2011 in Punjab. He has authored 15 books, 402 research papers, and 80 extension publications. He has been instrumental in developing a methodology to disseminate technology to farmers, which was adopted at the national level. He is a member of various scientific panels for different levels of the government and the private sector. Dr. Dhawan is Fellow, Indian Society for the Advancement of Insect Science (2007), Entomological Society of India (2005), Indian Ecological Society (1995), Society for Sustainable Cotton Production (2006). He is a Life member of eleven professional societies, and has held positions of president, vice president, general secretary, joint secretary, managing editor, editor, and associate editor and served as a Member, Executive Board with headquarter at Ludhiana. He has been instrumental in organizing ten international/national seminars/workshops/conferences to demonstrate the concept of IPM and transfer of technology.

9780367559076